Drinking Water Treatment 3

Drinking Water Treatment 3

Organic and Mineral Micropollutants

Kader Gaid

iSTE

WILEY

First published 2023 in Great Britain and the United States by ISTE Ltd and John Wiley & Sons, Inc.

ISTE Ltd
27-37 St George's Road
London SW19 4EU
UK

www.iste.co.uk

John Wiley & Sons, Inc.
111 River Street
Hoboken, NJ 07030
USA

www.wiley.com

Library of Congress Control Number: 2022945176

British Library Cataloguing-in-Publication Data
A CIP record for this book is available from the British Library
ISBN 978-1-78630-785-9

Contents

10

Removal of Micropollutants

10.1. Introduction

Since aquatic environments have always been intimately related to economic growth (which, in turn, involves population growth, industrial development, intensive agriculture and mass tourism), they have become the ultimate dump for the waste generated by our lifestyles. Nowadays, micropollutants (MPs) in the aquatic environment are a major problem, not only for the human population who uses water resources, but also for aquatic ecosystems.

Estimates for the European Community have shown that among the more than 110,000 chemical substances placed on the market, more than 100,000 substances have little information concerning the hazards they may pose. These substances are involved in the composition of many formulations and take part in a number of processes (industrial, agricultural practices, etc.). They also appear in the daily activities of households.

MPs present in water comprise a multitude of mineral and organic compounds likely to have a toxic action on humans and/or aquatic organisms, even at very low concentrations in water (ranging from 1 ng to 100 μg/L).

These MPs may have various origins (Table 10.1):

– craft or industrial economic activities (plasticizers, flame retardants, electrical insulators, etc.);

– agricultural (e.g. use of phytosanitary products, veterinary drugs);

– domestic use: cleaning products, detergents, cosmetics, drugs, etc.;

– contamination provoked by atmospheric fallout and transfer made by water runoff on urban surfaces;

– natural origin (metals relating to environmental geology).

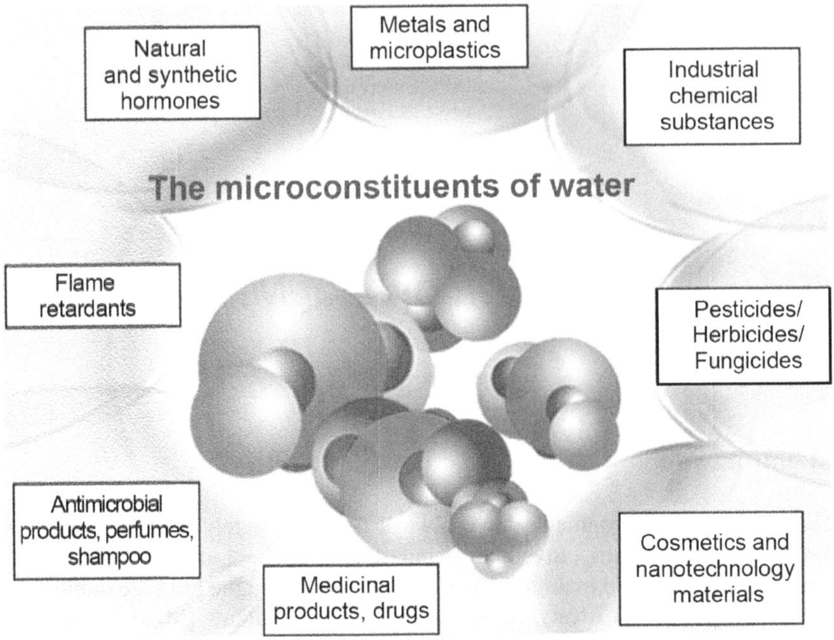

Figure 10.1. *Constituents present in water. For a color version of this figure, see www.iste.co.uk/gaid/watertreatment3.zip*

10.2. Pesticides

Agriculture is largely the primary pesticide consumer. The list of biocides chiefly reveals three main families: insecticides, fungicides and herbicides. Insecticides can have an action on the nervous system (neurotransmission), the respiratory system or molting. Fungicides act selectively on the respiratory route, biosynthesis (sterols, RNA, melanin), cell growth or cell permeability. Finally, herbicides can have a deleterious impact on photosynthesis, on the synthesis of lipids and amino acids, on cell division or growth. Among the most widely represented families of pesticides, we can find:

– organophosphates (chlorpyrifos, malathion), which have an insecticidal effect by inhibiting the acetylcholinesterase enzyme (disruption in the transmission of nervous impulses);

– organochlorines (DDT, chlordecone, lindane), insecticides affecting the nervous system by disrupting nervous impulses at the synapse level. Although at present these compounds are all banned in France, they are highly persistent in the environment (e.g. the problem of chlordecone exposure in the West Indies);

– carbamates (aldicarb, carbofuran), which have an effect on the nervous system in the same way as organophosphates;

– triazines (atrazine, simazine), which have a herbicidal effect by blocking the transport of electrons and light energy;

– phenylureas (diuron, isoproturon), which have a comparable mode of action to that of triazines;

– chloroacetanilides (metolachlor, acetochlor), which act on plants by inhibiting elongases, the enzymes involved in lipid synthesis (long chain fatty acids);

– aryloxyacids (2,4 D, mecoprop, dichlorprop), disrupting the regulation of the AIA auxin (indole-3-acetic acid), an essential phytohormone responsible for plant growth;

– sulfonylureas (chlorsulfuron, nicosulfuron), which affect the synthesis of branched amino acids (valine, leucine, isoleucine) via the inhibition of acetolactate synthetase (ALS).

Main families	Compounds
Drug molecules	Anxiolytics, analgesics, anti-inflammatories, antibiotics, hypolipidemics, antihypertensives, antiepileptics, anticancers, iodinated contrast agents, etc.
Cosmetics and personal hygiene products	Musks (galaxolide, tonalide), triclosan, etc.
Endocrine-disrupting compounds	Hormones, alkyl phenols, bisphenol A (BPA), phthalates, flame retardants, etc.
Illegal drugs	Cocaine and metabolites, THC, amphetamines, etc.
Pesticides	Insecticides, herbicides
Priority substances (Annex X of the European Directive) and dangerous substances (European circular on the Reduction of Dangerous Substances Discharged into Water 050109)	Polycyclic aromatic hydrocarbons (PAHs), dichloromethane
Disinfection by-products	Trihalomethanes, haloacetic acids

Table 10.1. *Main families and organic compounds*

Agricultural activity is not the only source of pesticide emissions. Among the secondary sources, we can mention the maintenance of public open spaces, golf courses, individual use, annual mosquito management in wetlands, the veterinary sector, road maintenance, etc.

10.3. Pharmaceuticals and industrial waste

France is one of the world's largest consumers of medicinal products. More than 3,000 pharmaceutical molecules for human use are commonly used in medicines and 300 veterinary drugs are currently available on the French market. The biologically active substances contained in each pharmaceutical product feature a great diversity of chemical structures. The parent compounds of pharmaceutical residues are essentially excreted via the stool and urine, or in the form of one or more active/inactive metabolites, and then emitted to the sewage system and the soil.

The main sources of water pollution due to pharmaceutical products are:

– diffuse sources, which represent the majority of drugs discharged into the environment; this includes the improper storage and disposal of expired or unused medicines; the metabolic excretion of drugs consumed by humans and animals, via the urinary or digestive tract;

– point sources, which are the cause of far more concentrated, but geographically limited, emissions. The point sources include the direct discharge of medicinal products (and chemical products used during drug manufacture) into the wastewater from pharmaceutical and chemical factories, the direct or indirect disposal of the pharmaceutical agents used in healthcare establishments, the direct dispersion of veterinary medicinal products released to the environment from use in aquaculture and treatment of pasture animals, or indirectly during the land application of manure and slurry from livestock facilities, the direct dispersion of therapeutic molecules in the form of feed additives poured directly into fish ponds.

Drug substances are grouped into several classes, reuniting a wide range of chemical families as well as degradation metabolites produced by humans or in the environment:

– analgesics and anti-inflammatories are the most widely consumed molecules in France for human use, such as diclofenac and ibuprofen, acetylsalicylic acid and paracetamol;

– psychotropics include antidepressants, anxiolytics and sleeping pills;

– anticonvulsants;

– lipid-lowering agents from the fibrate family, whose role is to lower the plasma concentration of cholesterol and triglycerides;

– due to their action, beta-blockers reduce blood pressure and the fixing of adrenaline on receptors. They are cardioselective, in that they act specifically by reducing the nervous impulses toward the heart;

– antibiotics destroy, inactivate or block the action of bacteria by disrupting, for example, the synthesis of proteins associated with bacterial growth;

– the action of bronchodilators stimulates the beta-2 receptors of bronchi, provoking their short-term dilation.

Other classes can be mentioned, such as diuretics, antifungals, antivirals and laxatives.

The use of different antibiotics for animal production and for human medicine may explain the observation of distinct correlations between the composition of the bacterial community and antibiotic-resistant populations, due to the concentration of tetracyclines, sulfonamides and penicillins or fluoroquinolones in wastewater. In fact, the antibiotic resistance of bacteria has been detected in several raw and treated wastewater sources.

These substances are found in the various environmental compartments (water, air, soil), by direct or indirect emission (surface runoff, drainage, atmospheric fallout, etc.), with potential direct or indirect effects on human health and ecosystems, namely due to contaminants in the food chain (Figure 10.2). Concentration levels in natural waterways vary depending on chemical stability, biodegradability, the physicochemical characteristics of molecules and treatment station performances.

In general, the concentration of pesticides and pharmaceutical products in the environment vary not only from one compound to another, but also from site to site. Regional differences in the use of medicinal products, as well as the biodegradability of pharmaceuticals and their metabolites, could explain the differences in concentrations.

The analysis campaigns performed (Veolia) have detected the presence of some of these substances at the inlet and outlet of wastewater treatment stations, as well as in the resources used for the production of drinking water.

Due to the diversity and complexity of the substances considered, data are still limited on the effects of pharmaceutical residues and endocrine-disrupting compounds (EDCs) on the environment, as well as their occurrence patterns in water resources. The great diversity of the molecules considered and the difficulties related to their analysis – which have sometimes been associated with their low concentrations – as well as the studies carried out until now to assess the presence of these compounds in the environment, have been limited to a few groups of molecules. These molecules were chosen on the basis of production amounts and their estimated persistence in the environment (for pharmaceutical residues [PRs]), as well as the potential endocrine activity for EDCs. In point of fact, some of these molecules have an effect on the hormonal balance of many living species, in which case they are qualified as EDCs. Recent toxicological studies have shown that in males, EDCs influence the number of spermatozoa, genital malformations, impaired sexual functions, neuronal development, obesity and even cancer.

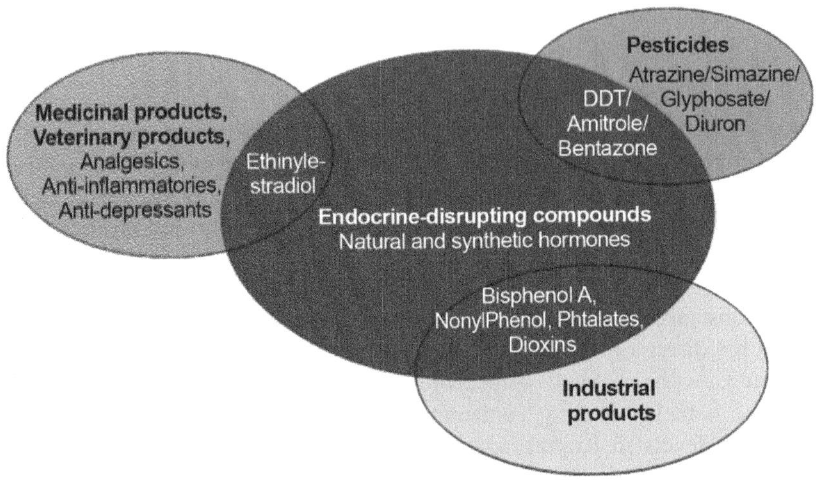

Figure 10.2. *Different categories of micropollutants. For a color version of this figure, see www.iste.co.uk/gaid/watertreatment3.zip*

The harmful effects of EDC and PR discharges into the environment of aquatic and terrestrial organism populations (from simple gastropods to mammals) are now definitely beyond doubt. Numerous examples demonstrate the impact of these substances on fish (e.g. incomplete development of genital organs, hermaphroditism, modification of the male/female ratio), birds (e.g. thinning of egg shells) and mammals (e.g. decline in seal populations feeding on PCB-contaminated fish). Many questions remain among the scientific community regarding this cause-and-effect relationship, which is often difficult to determine (still fragmented knowledge about the toxic and ecotoxic effects of metabolites, substance cocktails, disturbance mechanisms, etc.).

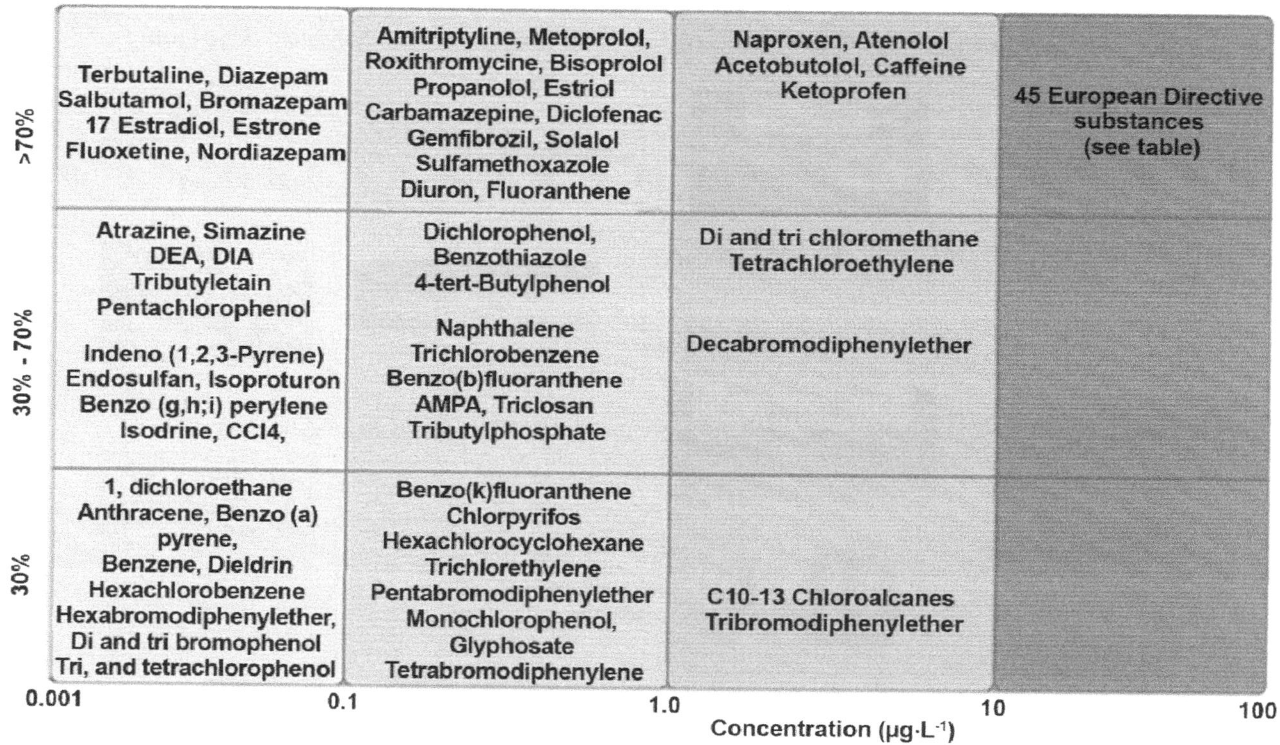

Figure 10.3. *Example of the evaluation of micropollutant concentrations in waterways. For a color version of this figure, see www.iste.co.uk/gaid/watertreatment3.zip*

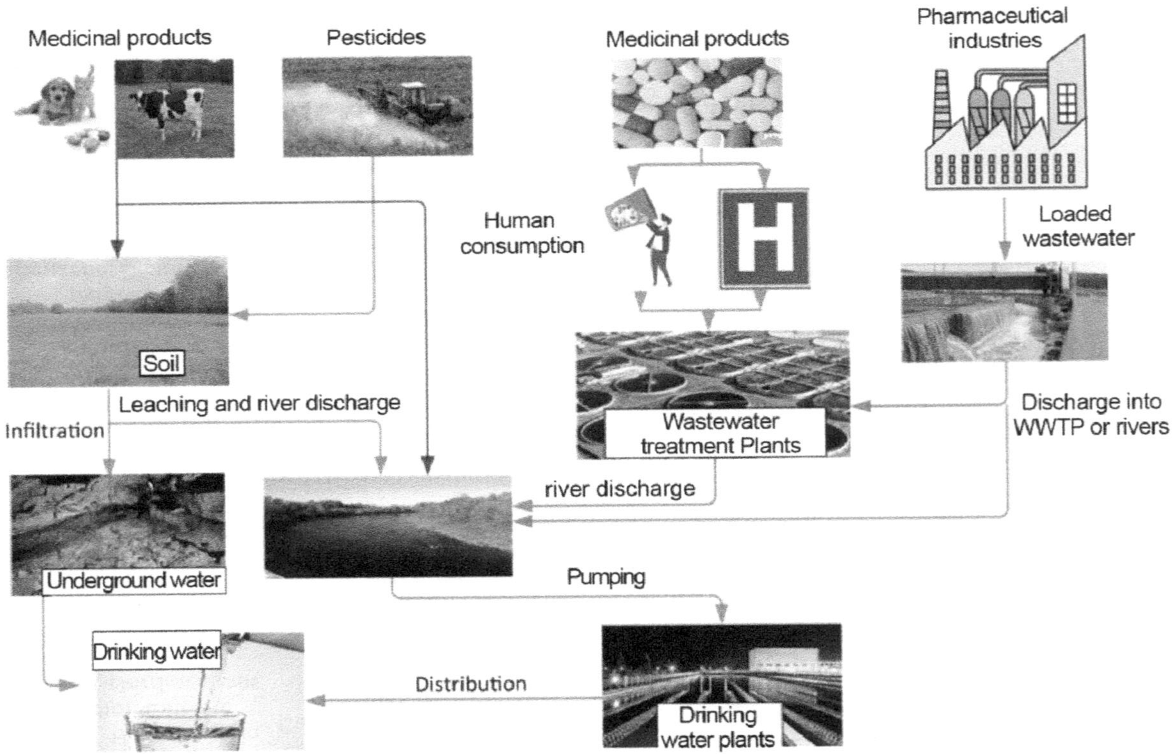

Figure 10.4. *Simplified diagram of the cycle of water containing pesticides, pharmaceutical residues, endocrine-disrupting compounds. For a color version of this figure, see www.iste.co.uk/gaid/watertreatment3.zip*

According to the European Commission, an endocrine disruptor is "an exogenous substance or mixture that alters function(s) of the endocrine system and consequently causes adverse health effects in an intact organism, or its progeny, or (sub)populations".

In humans, the connection between the presence of pharmaceutical residues and EDCs in the environment is a subject of major concern for populations. European authorities have admitted the link between EDCs and increasingly frequent pathologies, such as testicular, breast and prostate cancers, a drop in the number of spermatozoa, malformations of the reproductive organs, thyroid disorders. Their massive use, as well as their discharge during the manufacturing process, makes them present in concentration levels ranging from a few tens of nanograms per liter to a few micrograms per liter.

The presence of pharmaceutical residues in drinking water is a major concern for the scientific community, the public authorities and the public. Everybody wonders about the presence of pharmaceutical residues in aquatic environments (surface water, groundwater, transfer media, soil) and drinking water, in trace amounts, as well as their effects on the environment and human health. This question is set against a broader context involving the preservation of the environment and water resources, particularly in relation to MPs, substances likely to have a toxic action, even at low doses, in a given environment (ranging from nanogram to microgram per liter of water).

The French public authorities have prioritized source reduction to the point of completely banning the emission of certain substances. Source reduction is essential to prevent the dissemination of substances in the environment, but its effectiveness will take many years to be demonstrated, and will not suffice to control all the risks involved by the use of MPs. This is not only due to the large number of molecules, uses and practices, but also to the fact that many MPs mostly result from the diffuse pollution emitted by individuals (pharmaceutical residues, hormones, biocides, etc.), and will thus become difficult to fully remove.

In Europe, the establishment of a regulatory framework aims to reduce the discharge of hazardous substances into natural environments and to monitor their course within the biosphere. The numerous efforts carried out, particularly on a European scale, focus on the scrutiny of the environmental (chemical) quality of the various water resources available, as well as their evolution.

In France, the regulatory requirements aimed at reducing the emissions of chemical substances and their presence in the natural environment arise from

European regulations and directives such as the REACH regulation and the regulations on biocidal and phytopharmaceutical products, the EU Water Framework Directive (WFD), the Industrial Emissions Directive (IED), etc. In essence, these regulations aim to regulate the launch on the market of potentially hazardous substances, to reduce their emissions in the various environmental compartments and to achieve environmental quality goals. These increasingly restrictive requirements will tend to be reinforced in the upcoming years, with the further development of scientific knowledge and with the impact on health and the environment becoming more and more evident.

Apart from transcribing these fundamental European texts into French law, the government has implemented an MP plan aiming to preserve water quality and biodiversity. This plan aims to reduce MP emissions whose toxicity has been proven, to consolidate knowledge and to compile a list of pollutants that require monitoring.

Despite having significantly improved from a technical point of view, wastewater treatment stations have not been designed to remove the entirety of pharmaceutical molecules. For example, even if paracetamol is degraded to more than 90% in treatment stations, traces of this substance can be found in treated wastewater, which is then discharged into surface water. Some compounds, such as diclofenac (anti-inflammatory) or carbamazepine (an anti-epileptic), only poorly degrade. These pharmaceutical product residues, discharged into surface waters, can pose a significant environmental risk due to the fact that their metabolites or degradation by-products are sometimes even more hazardous than the original product.

For this reason, it is necessary to implement so-called "high-performance" sectors, including the use of activated carbon (AC) in its different forms, chemical oxidation with ozone or advanced oxidation (O_3/H_2O_2, UV/H_2O_2) or high-pressure membranes, such as nanofiltration or low-pressure reverse osmosis (LPRO).

These treatments (AC and high-pressure membranes) are well known in this era because they have already been tried and tested for the treatment of drinking water. As described in Chapter 7, they have proven their effectiveness against organic matter (humic substances) after being tested in the field. In France, the use of ozone for pesticide removal is discouraged due to the by-products generated.

The French Government Resolution of July 8, 2010, enlisting priority substances and establishing the procedures and deadlines for the gradual reduction and removal of discharges, spills, direct or indirect discharges, respectively, of priority substances and hazardous substances referred to in article R. 212-9 of the Environmental Code, consolidated version as of May 8, 2018.		
Alachlor	Hexachlorobenzene	Naphthalene
Anthracene	Dicofol	Nickel and its compounds
Atrazine	Perfluorooctanesulfonic acid and its derivatives (perfluorooctane sulfonate PFOS)	Nonylphenols
Benzene	Quinoxyfen	Octylphenols (6)
Brominated diphenyl ethers	Dioxins and dioxin-like compounds	Lead and its compounds
Cadmium and its compounds	Aclonifen	Pentachlorobenzene
Chloroalkanes, C10-13	Cypermethrin (10)	Pentachlorophenol
Chlorfenvinphos	Dichlorvos	Polycyclic aromatic hydrocarbons (PAHs) (7)
Chlorpyrifos (ethylchlorpyrifos)	Bifenox	Simazine
1,2-Dichloroethane	Cybutryne	Tributyltin compounds
Dichloromethane	Hexachlorobutadiene	Trichlorobenzene
Di(2-ethylhexyl)phthalate (DEHP)	Hexachlorocyclohexane	Trichloromethane (chloroform)
Diuron	Isoproturon	Trifluralin
Endosulfan	Mercury and its compounds	Hexabromocyclododecanes (HBCDD)
Fluoranthene	Heptachlor and heptachlor epoxide	Terbutryne

Table 10.2. *List of hazardous priority substances*

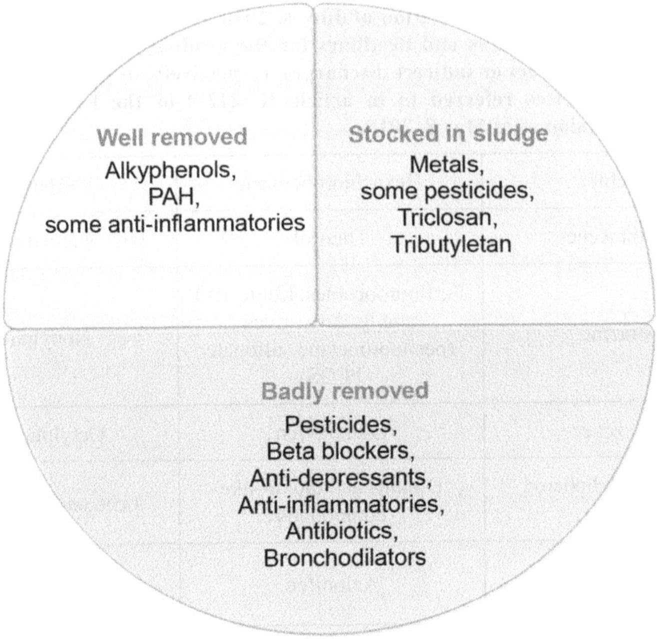

Figure 10.5. *The evolution of some substances in wastewater treatment stations. For a color version of this figure, see www.iste.co.uk/gaid/watertreatment3.zip*

10.4. Pesticide removal technologies and emerging MPs

Within their structure, MPs have functional groups which determine their reactivity toward adsorption or chemical oxidation. The most important groups are as follows:

– -OH functional groups (hydroxyl, phenol);

– C-O functional groups (carboxyl, carbonyl);

– -N- (amino) functional groups.

Reactivity is strongly influenced by the polarity of the functional groups and the positive or negative charge acquisition when dissociation or a proton association takes place within these functional groups. The hydroxyl group is present in alcohols and phenols. Alcohols are considered hydroxyl-alkene compounds. In most cases, alcohols are neutral, because the hydroxyl group does not ionize easily. Phenols include hydroxyl components that are attached to the aromatic ring.

The phenol group is made of a hydroxyl group directly attached to the carbon of an aromatic ring. In general, phenols are highly reactive.

The carbonyl group includes aldehydes, ketones and carboxylic acids. Due to the double bond between carbon and oxygen, and because the electrons from this double bond are not uniformly distributed, the carbonyl group is polarized. Carbonyls acquire protons. In general, compared to ketones, aldehydes are more easily oxidized. The carboxyl group constitutes the weak acid group, incorporating the reactive properties of the carbonyl and hydroxyl group. The amine group is found in amines. Most often, these are organic bases, which can be aliphatic and aromatic.

In point of fact, the presence of these groups influences the treatment's efficacy and has to be acknowledged during the selection of the process to be implemented.

The treatment techniques presented in this chapter concern:

– adsorption onto AC;

– simple or advanced chemical oxidations;

– combinations of chemical oxidation and adsorption onto AC;

– the membrane barrier (nanofiltration or low-pressure [LP] reverse osmosis);

– specially adapted biological treatments.

10.4.1. *Adsorption onto activated carbon (AC)*

AC has long been known for its ability to adsorb organic substances in water, as well as the molecules responsible for tastes and odors. Because of this, it is naturally used for the removal of organic MPs (EDCs, pharmaceutical residues, industrial waste), following the same provisions used for eliminating pesticides.

The adsorption curves onto AC depend on the characteristics of the following MPs, on whether they are adsorbable or not, and/or biodegradable or not (Figure 10.6).

At present, adsorption onto AC is the most widely used technique because the pollution transfer onto the porous medium is currently the most reliable process, provided that it is properly designed and well exploited.

Veolia technologies involving the use of AC are based either on powdered or micrograin (μgrain) AC in specific reactors operating with recirculation, or by its occasional addition into a settler, or with granular AC placed in a filter.

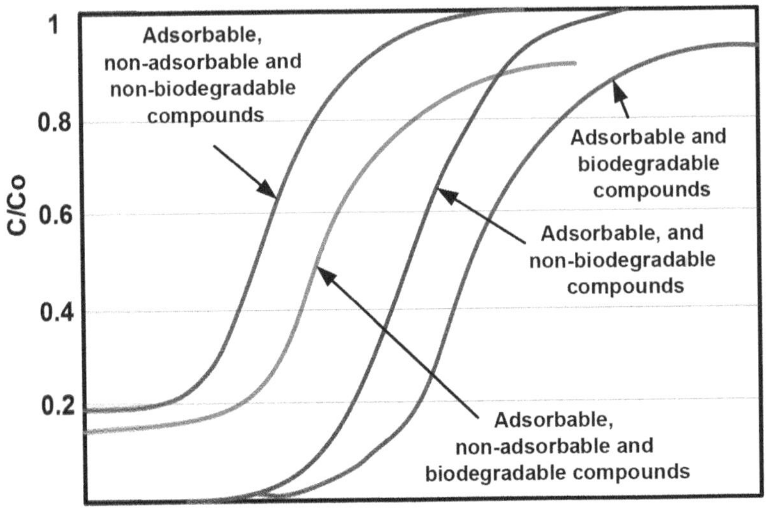

Figure 10.6. *Types of adsorption curves depending on the nature of organic compounds. For a color version of this figure, see www.iste.co.uk/gaid/watertreatment3.zip*

AC is available in different grain sizes: powdered activated carbon (PAC) with diameters between 5 and 35 µm, µgrain activated carbon with diameters between 0.2 and 0.6 mm and granular activated carbon (GAC) with mean diameters between 0.85 and 1.2 mm.

10.4.1.1. *Powdered activated carbon and µgrain activated carbon*

AC is extensively used in the form of powder (PAC) or µgrain because it makes it possible to associate and to control the dosages in accordance with inlet pollution. Consequently, it is the tradeoff between produced water quality and operating cost that is often chosen by environmental consulting firms and local authorities, as it enables them to optimize the operating cost of AC throughout the year in function of pollution's seasonal variations.

Water treaters have marketed various processes based on PAC reactors and environmental consulting firms now tend to formalize this type of process in their specifications, not only for tertiary or quaternary wastewater treatment processes, but also for refining in drinking water plants.

In France, the market is now dominated by PAC or µgrain activated carbon reactors having an action on natural organic matter (NOM) as well as on pesticides and MPs. This is especially relevant since all these substances are found at the same time in surface waters. These reactors are also being developed in Europe, the United States and China. They are applied to both drinking water and wastewater. They are used for wastewater in tertiary treatments, after the clarification stage.

Veolia has several processes including PAC (Actiflo® Carb, Multiflo® Carb, Opacarb®MF) and µgrain activated carbon (Filtraflo® Carb, Opacarb® FL and Opacarb®MG). They are detailed in Chapter 9, describing adsorption onto AC.

These processes pursue the same treatment goals, but their integration into the water treatment station depends on the quality of inlet water and the flow to be treated. These processes are equally applicable to drinking water, municipal wastewater and industrial effluents.

The removal efficiency of MPs is based on the adsorption capacity of AC.

Adsorption is a surface phenomenon whereby molecules become attached to the surface of the AC due to different forces (electric charges, dipole–dipole interaction, van der Waals forces, covalent bonds, hydrogen bonds, etc.). Hydrophobic interaction processes on the carbon's surface are the dominant mechanisms behind MP removal.

This treatment system aims to create a reaction volume in which the AC is highly concentrated and can therefore adsorb micro-pollutants more effectively. It is necessary to let AC react with the water to be treated with enough contact time and to favor a high concentration of adsorbent in the contact tank (1.2–5 g of activated carbon·L^{-1}). It is necessary to couple the retention times of water and AC through the separation and recirculation of AC.

Once the MP adsorption reaction has been completed, the AC is separated by settling or by membrane filtration (see Chapters 7 and 9). The type of AC influences MP removal and its choice is based on technical characteristics (pore size, specific surface, iodine number, etc.), as well as on the results of laboratory tests or on-site pilot tests.

Adsorption with PAC or µgrain activated carbon is governed by three parameters:

– PAC dosage;

– the hydraulic retention time (HRT) in the AC reactor ensures the adsorption equilibrium between the AC and organic MPs;

– the age of AC in the AC reactor favors the optimal use of the adsorption capacity of carbon (this is separated after passing through the reactor and is then recirculated upstream of the reactor).

The general mechanisms for the adsorption of pesticides and other MPs are similar to those described for NOM and differ in that the MPs have far smaller sizes and it is the mesopores and micropores which are essential.

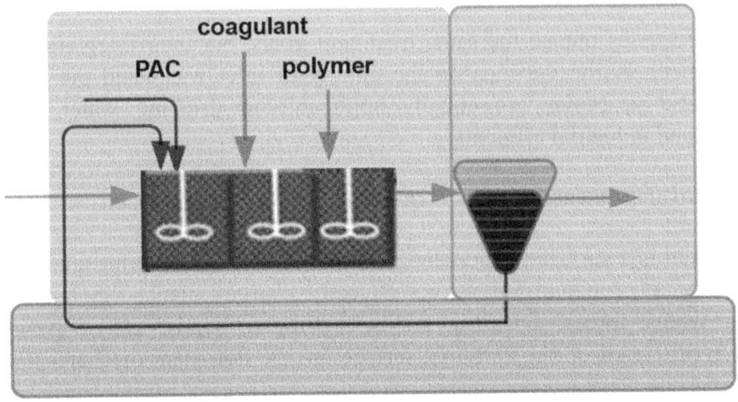

Figure 10.7. *PAC reactor with recirculation. For a color version of this figure, see www.iste.co.uk/gaid/watertreatment3.zip*

The parameters which influence the adsorption of MPs are as follows:

– *molecule structure*: the presence of functional groups such as NO_2, NOH, COOR, halogens, as well as double bonds, coplanarity, symmetry, electron delocalization;

– *molecular weight* of MPs (the lightest ones are adsorbed, but also desorbed, before others) and *molecular size* (which must not exceed the maximum pore size);

– *MP solubility*: the higher the solubility, the lesser the importance of adsorption. Hydrophobic molecules (high log Kow) are better adsorbed, whereas highly polar compounds are difficult to remove;

– *MP concentration*: equilibrium depends on the concentration in the dissolved phase. The higher the concentration, the greater the proportion adsorbed (the reduction percentage remains relatively constant, though);

– *intermolecular competition*: in the event of mixture, the substances the more difficult to adsorb are retained even less than when tested separately. The presence of natural particulate organic matter (comparable to MPs) in water decreases the efficiency of MP adsorption, because it competes for binding sites or can block the structure's pores;

– *temperature*: an increase in temperature shifts the sorption equilibrium and reduces MP adsorption.

10.4.1.2. *Design parameters: choice of carbon*

The choice of AC is important because water has high molecular weight molecules (>1,000 Da) and MPs (50–200 Da). It is crucial to choose the carbon with the suitable specificities matching the characteristics of the water to be treated, and to carefully calculate the concentration to be injected. The adsorption phenomenon is relevant only if the adsorbent medium has a developed porous structure and a sufficient adsorption capacity for these molecules. The adsorption capacity depends on temperature, pressure, the concentration of organic molecules, the functional groups on the surface, the presence of other constituents likely to be adsorbed, such as dissolved organic carbon (DOC), and intermolecular competition.

10.4.1.2.1. Design parameters evaluated out of isotherms

The parameters obtained out of adsorption isotherms indicate the affinity between the adsorbate and the adsorbent. Freundlich's empirical model, frequently accepted for solid adsorbents, takes into account the possible formation of several successive molecule layers on the solid's surface, implying heterogeneity in the energy of the adsorbent's adsorption sites. In this case, the isotherm is described in terms of the following equation:

$$\frac{(Cin - Cout)V}{Mac} = qe = kCout^1/n$$

where:

– qe: adsorption capacity (mg adsorbate/g adsorbent);

– Cin: substance concentration in raw water (ng or $\mu g \cdot L^{-1}$);

– Cout: substance concentration in treated water (ng or $\mu g \cdot L^{-1}$);

– Mac: mass of PAC or µgrain activated carbon in the reactor (g);

– V: volume of the AC reactor (L);

– k: Freundlich adsorption constant related to the adsorption capacity (µg micropollutant$\cdot g^{-1}$ AC) ($L^{1/n} \cdot \mu g^{-1/n}$ micropollutant);

– 1/n: second constant related to the active carbon surface's heterogeneity and describing the affinity between adsorbent and pollutant.

Freundlich's equation leads to the following linear form:

$$\log qe = 1/n \log Cout + \log k$$

By plotting the straight line of log qe as a function of log (Cout) for each PAC concentration studied, it is possible to determine 1/n from the slope of the line, log k being the ordinate of origin.

The value of 1/n is between 0 and 1. The more heterogeneous the carbon's surface, the closer the value of 1/n to 0.

According to Langmuir, the saturation of adsorption sites takes place in a single layer. These sites can only contain one molecule per site. There are no interactions between the adsorbed molecules. In this case, we assume that adsorption sites are equivalent and that adsorption is homogeneous. The Langmuir model is modeled by the equation:

$$qe = \frac{qmax\ bCout}{1 + bCout}$$

where:

– qe: quantity adsorbed per mass of adsorbent onto ($mg \cdot kg^{-1}$) AC;

– Cout: outlet concentration ($ng \cdot L^{-1}$ or $\mu g \cdot L^{-1}$), b is a Langmuir constant describing the adsorption energy ($L \cdot \mu g^{-1}$ or ng^{-1}); it depends on temperature and offers an indication of the adsorbate's affinity with the adsorbent: the higher this is, the stronger the affinity;

– qmax: maximum quantity which can be adsorbed on a given mass of solid, in terms of monolayer adsorption. It also refers to the number of active sites per unit mass in the solid phase and translates the complete occupation of adsorption sites ($\mu g \cdot g^{-1}$).

By plotting 1/qe as a function of 1/Cout, we obtain a straight line whose slope leads to 1/qmax·b and the ordinate of origin results in 1/qmax.

The prediction of the simultaneous adsorption of several constituents is one of the most complex problems in the field of adsorption. This is particularly the case of surface waters where the matrices of the water to be treated are complex and involve a mixture of pesticides, MPs and humic substances.

In general, it appears that the Freundlich equation is more representative of adsorption mechanisms than the Langmuir equation. This is because the principle of a heap of adsorbed organic substances seems more realistic. And this is also why it is generally accepted that there are few restrictions to the use of the Freundlich model.

10.4.1.2.2. Implementation parameters

Table 10.3 presents the AC implementation parameters in reactors.

Parameters	**Values**
AC concentration ($mg \cdot L^{-1}$)	10–25
Hydraulic retention time (min)	15–60
Inlet SS concentration ($mg \cdot L^{-1}$)	5–25
Coagulant dosage ($mg\ metal \cdot L^{-1}$)	1–3
AC concentration in the reactor ($g \cdot L^{-1}$)	1–20
AC age (d)	6–15

Table 10.3. *Design magnitudes of AC reactors*

The total amount of sludge (kg) present in the reactor is the sum of AC, metal hydroxides and SS.

That is:

total amount of sludge (kg) = Volume (m^3)·[AC ($kg \cdot m^{-3}$) + SS ($kg \cdot m^{-3}$) + $Me(OH)_3$ ($kg \cdot m^{-3}$)]

The applied load ($kg \cdot d^{-1}$) is given by:

applied load ($kg \cdot d^{-1}$) = Flow ($m^3 \cdot d^{-1}$) [AC ($kg \cdot m^{-3}$) + SS ($kg \cdot m^{-3}$) + $Me(OH)_3$ ($kg \cdot m^{-3}$)]

AC age (d):

AC age (d) = Volume (m^3)Conc·AC ($kg \cdot m^{-3}$)/extracted Q AC ($kg \cdot d^{-1}$)

Q AC ($kg \cdot d^{-1}$): quantity of AC extracted

AC Hydraulic retention time (min) = Volume (m^3)/Inlet flow ($m^3 \cdot min^{-1}$)

10.4.1.2.3. Operating parameters and performances

The parameters to be optimized are:

– carbon retention time in the contact reactor (recirculation rate, reactor volume);

– AC dosage;

– coagulant and polymer dosage;

– carbon age and frequency of spent carbon extractions.

10.4.1.2.4. Performances of AC reactors in relation to pesticides and other micropollutants

Polarity can be defined as the distribution of positive and negative charges within a molecule. Electronegativity differences in space increase the polar character of a molecule. In the case of organic molecules, polar molecules mainly contain alcohol, acid or amine functions. To determine the polarity of a molecule, the main index is log Kow (octanol-water partition coefficient, also called log P). The log Kow (logarithm of the octanol-water partition coefficient) can be a first indicator of the adsorbability of organic compounds such as pesticides and other MPs. It indicates whether a molecule is rather hydrophilic (low log Kow) or hydrophobic (high log Kow). The higher the log Kow, the more hydrophobic the molecule. Hydrophobic compounds are better adsorbed than hydrophilic compounds. It is also used with AC as an adsorption indicator.

10.4.1.2.5. The case of pesticides

The first design parameters are the isotherm constants, which can predict the efficiency of AC in relation to the pesticide involved. The tests must be carried out on resource water because the AC doses obtained on synthetic waters do not take into account the competition brought in by the DOC when it is present in this water. On the other hand, the tests are more representative for groundwater because the DOC there is present in small concentrations (DOC < 1 $mg \cdot L^{-1}$). In general, pesticides with a Freundlich coefficient greater than 150 ($mg \cdot g^{-1}$) ($L^{1/n}$ /$mg^{-1/n}$) are removed.

Table 10.5 presents the values of k and 1/n for a number of pesticides which can predict the PAC doses to be injected into reactors in order to reduce them to a residual concentration smaller than or equal to 0.1 $\mu g \cdot L^{-1}$ for an inlet concentration <10 $\mu g \cdot L^{-1}$. The more soluble the compound, the less adsorbable it is and this is reflected in the values of the Kow partition coefficient (Figure 10.8).

Glyphosate and AMPA are very lowly or not adsorbed onto AC. But they are adequately removed by clarification (40%), ozonation (>90%) and chlorination (>90%). They are also fairly well removed (40–50%) in PAC reactors with coagulant injection. It is the coagulant's effect that enables this reduction.

Well adsorbed pesticides		Moderately adsorbed pesticides	Pesticides with low or no adsorption
Acetochlor	Ethylbenzene	Hydroxyalachlor	Aminotriazole
Acetochlor oxanilic acid (OXA)	Fenoprop	Aldicarb	Glyphosate
Alachlor	Flusilazole	Bentazone	AMPA
Aldrin	Heptachlor	Bromacil	Mesotrione
Ametryne	Heptachlor epoxide	Carbofuran	Nicosulfuron
Atrazine	Isoproturon	Desethylhydroxy-atrazine	Sulcotrione
Chlordane	Ioxynil	Imidachloprid	
Cyanazine	Linuron	Iodosulfuron	
Carbendazim	Lindane	Metaldehyde	
Chlorfenvinphos	Malathion	Metolachlor ESA	
Chloridazone	Metholachlor	Metolachlor OXA	
Chlortoluron	Hydroxy metolachlor	Metribuzin	
Chlorpyrifos ethyl	Metazachlor	Metamitrone	
Deisopropyltatrazine	Parathion	Metsulfuron	
Desethylatrazine	Parathion methyl		
Carbofuran	Pendimethalin		
2.4 D	Pentachlorobenzene		
DDT	Pentachlorophenol		
Dianizon	Phenmedipham		
Dicamba	Pichloram		
Dichloprop	Propachlor		
Dichlorvos	Prosulfocarb		
Dimethachlor	Propazine		
Dimethenamid	Simazine		
Dinoseb	2.4.5T		
Dieldrin	Terbutylazine		
Diuron	Terbutryne		
Diflufenicacil	Thiodicarb		
Endrin	Triadimenol		
Alpha endosulfan	Triclopyre		
Ethofumesate	Trifluralin		

Table 10.4. *Adsorption of pesticides onto activated carbon*

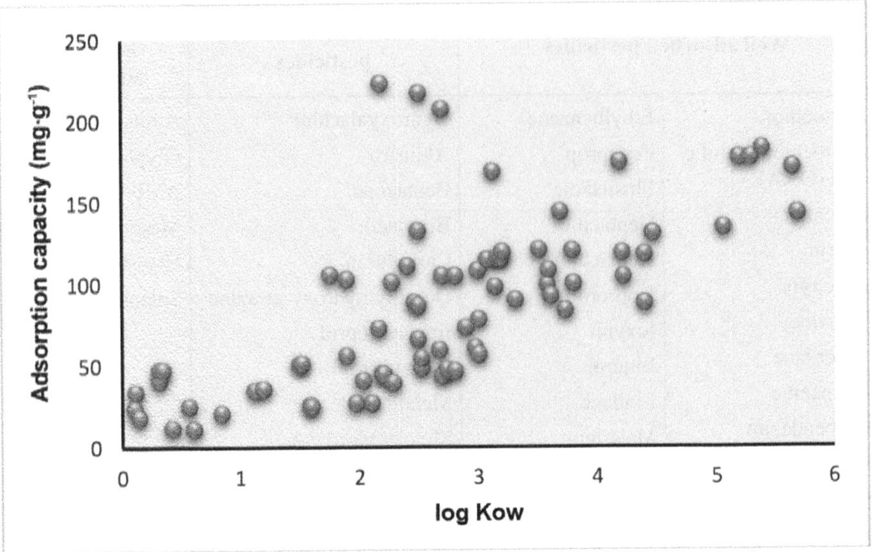

Figure 10.8. *Adsorption capacity (mg·g^{-1}) of pesticides based on log Kow*

Pesticides	MW	log Kow	log Koc	Solubility (ppm)	Freundlich parameters k 1/n (mg·g^{-1})(L·mg^{-1})1/n	
Acetochlor	269.11	3.17	2.49	282	350	0.48
Acetochlor ethanesulfonic acid (ESA)	315.11	0.32	2.49	282	132	0.51
Acetochlor oxanilic acid (OXA)	265.13	2.28	1.38	–	298	0.47
Alachlor	269.77	3.52	2.52	242	400	0.52
Alachlor OXA	265	–	–	–	–	–
Hydroxyalachlor	285.77	2.74	–	–	127	0.43
Deschloroalachlor	–	3.08	–	–	331	0.46
Aldicarb	190.26	1.13	1.47	6,000	102	0.47
Aldrin	364.91	3.01	–	0.027	651	0.92
Ametryn	227	2.98	–	209	201	0.52
Aminotriazole	84.08	–0.97	–	–	100	0.65

Pesticides	MW	log Kow	log Koc	Solubility (ppm)	Freundlich parameters k 1/n $(mg \cdot g^{-1})(L \cdot mg^{-1})1/n$	
AMPA	186.17	–2.17	–	–	–	–
Atrazine	215.69	2.68	1.94	33	179	0.48
Bentazone	240.3	–0.46	1.32	500	102	0.62
Bromacil	261.12	2.11	1.86	815	44	0.47
Carbofuran	221.6	1.98	1.46	700	51.1	0.28
Chlordane	409.78	3.32	5.14	600	320	0.55
Cyanazine	240.7	2.04	1.74	171	208	0.71
Carbendazim	191.19	1.52	3.44	29	190	0.57
Chlorfenvinphos	359.57	4.22	–	145	452	0.58
Chloridazone	221.65	1.50	2.08	–	152	0.48
Chlorsulfuron	357.8	–0.99	1.40	12,500	–	–
Chlortoluron	212.68	2.41	2.29	74	285	0.41
Chlorpyrifos ethyl	350.58	4.23	3.86	–	425	0.61
Desethylatrazine	187.60	1.5	1.38–3.47	3,200	131	0.41
Deisopropylatrazine	17.60	1.2	–	670	123	0.54
Desethylhydroxy-atrazine	169	0.43	–	–	59	0.71
2.4 D	221	2.81	1.17	680	125	0.47
DDT	354.5	5.7–6.90	5.17	0.003	322	0.35
Dianizon	304.35	3.81	3.17	60	191	0.28
Dicamba	221.04	2.21	0.30	4,500	131	0.47
Dichloprop	235.07	2.29	3	350	139	0.55
Dichlorvos	221	1.90	1.47	18,000	290	0.45
Dimetachlor	255.18	2.17	1.83	2,300	229	0.50
Dimethenamid	275	2.47	–	1,450	238	0.43
Dinoseb	240	3.15	1.47	52	205	0.44
Diquat	184.24	–4.6	4.5–6.9	712,000	–	–
Diazinon	304.34	3.01	2.72	40	220	0.59

Pesticides	MW	log Kow	log Koc	Solubility (ppm)	Freundlich parameters k 1/n $(mg \cdot g^{-1})(L \cdot mg^{-1})1/n$	
Dieldrin	380.91	5.40	–	0.25	606	0.52
Diflufenican	394.3	4.20	–	0.05	400	0.36
Diuron	233.1	2.81	2.68	42	128	0.44
Endrin	380.91	3.20	4	0.23	304	0.42
Ethylene dibromide	187.85	1.76	1.35	4,300	666	0.80
Alpha endosulfan	406.92	–	2.47–3.30	0.06	194	0.51
Ethofumesate	286.34	2.7	2.50	50	185	0.63
Ethylbenzene	106.17	2.53	–	–	178	0.57
Fenoprop	269.5	3.80	2.47	140	295	0.39
Flusilazole	315.40	3.74	3.22	41.9	265	0.50
Glyphosate	169.07	–3.20	–	–	–	–
Heptachlor	373.32	4.41	4.71	0.06	530	0.65
Heptachlor epoxide	389.3	2.7	2.34	0.35	1,038	0.70
Imidacloprid	255.05	0.57	2.32	610	143	0.76
Iodosulfuron	529.24	1.59	–	25,000	112	0.64
Ioxynil	370.92	2.20	–	3,034	135	0.48
Isoproturon	206.29	2.5	2.24	67.6	233	0.43
Lindane	290.83	–	3.10	7.3	256	0.49
Linuron	249.1	2.19	2.93	75	537	0.38
Mecoprop	214	–0.19	1.30	250,000	–	–
Metaldehyde	176	0.12	2.38	188	95	0.45
Methoxychlor	345.65	3.62	5.90	0.1	212	0.36
Metolachlor	283.8	3.13	2.35	530	559	0.52
Metolachlor ESA	329.11	0.32	1.20	212,461	142	0.48
Metolachlor OXA	279.14	<1	0.9	8,500	142	0.48
Hydroxymetolachlor	300.8	2.81	–	–	289	0.44
Metribuzin	214.29	1.6	1.27	1,100	98	0.60

Pesticides	MW	log Kow	log Koc	Solubility (ppm)	Freundlich parameters k 1/n $(mg \cdot g^{-1})(L \cdot mg^{-1})1/n$	
Malathion	330.35	2.70	–	148	296	0.45
Mecroprop	215	–0.19	–	620	–	–
Mesotrione	283.8	<1	–	160	86	0.68
Metamitrone	202.22	0.85	1.95–2.60	1,700	32	0.19
Metazachlor	277.76	2.5	2.34	450	400	0.48
Metsulfuron	381	<1	1.54	2,800	78	0.52
Nicosulfuron	410.1	0.61	1.47	7,500	35	0.50
Pendimethalin	281.31	5.20	4.55	0.38	500	0.45
Parathion methyl	263.20	2.9	3.7	37.7	265	0.56
Pentachlorobenzene	249.5	5.30	–	–	500	0.45
Pentachlorophenol	270.37	5.66	4.62	–	451	0.42
Phenmedipham	300.31	3.59	2.94	1.80	405	0.61
Pichloram	241.46	1.9–2.30	1.20	4.30	185	0.52
Propachlor	211.69	4.48	1.90	–	352	0.43
Propachlor	211.69	4.48	1.90	–	352	0.43
Prosulfocarb	251.39	4.41	3.22	13.2	285	0.51
Quinmerac	221.5	–1.41	1.93	107,000	–	–
Simazine	201.66	2.51	2.09	3.5	387	0.25
Sulcotrione	–	–1.70	0.03–0.95	1,670	12	0.12
2,4,5-T	255.48	3	4.90	238	358	0.52
Terbuthylazine	229.72	3.21	0.53	6.6	314	0.42
Terbutryne	241.36	3.6	3.30	22	412	0.58
Thiodicarb	354.46	2.5	2.54	25–30	268	0.61
Triadimenol	295.77	3.7	2.47	72	445	0.49
Triclopyr	256.47	2.53	2.90	210–440	183	0.53
Trifluralin	335.28	5.07	3.90	0.3	425	0.50

Table 10.5. *Freundlich model parameters k and 1/n*

10.4.1.2.6. The special case of metolachlor and metabolites

For potable water, French authorities estimated that five metabolites (alachlor ESA, acetochlor ESA, acetochlor OXA, metazachlor ESA and metazachlor OXA) can be classified as not relevant, whereas three metabolites (alachlor OXA, metolachlor ESA and metolachlor OXA) can be classified as "relevant". For relevant metabolites, a regulatory value of 0.1 $\mu g \cdot L^{-1}$ is settled for potable water

10.4.1.2.7. The physicochemical characteristics of metabolites

Alachlor and metolachlor are chloroacetamide herbicides. Metolachlor is highly soluble in water. This makes it a potentially significant pollutant for waterways and groundwater. Its low adsorption coefficient in soils indicates that it should be present in lower concentrations in groundwater than in surface waters.

The sulfonic (ESA) and oxanilic (OXA) acids of alachlor and metolachlor are degradation products of parent molecules present in our environment. The general formula for these molecules is illustrated by Figure 10.9 where R1, R2, R3 and R4 refer to groups including various substituents.

Groups	Alachlor	Alachlor OXA	Metolachlor	Metolachlor ESA	Metolachlor OXA
R1	CH_3CH_2	CH_3CH_2	CH_3	CH_3	CH_3
R2	CH_3CH_2	CH_3CH_2	CH_3CH_2	CH_3CH_2	CH_3CH_2
R3	CH_2OCH_3	CH_2OCH_3	$CH(CH3)OC_2H_5$	$CH(CH_3)OC_2H_5$	$CH(CH_3)OC_2H_5$
R4	CH_2Cl	CO_2H	CH_2Cl	CH_2SO_3H	CO_2H

Table 10.6. *Semi-structural formula of chloroacetamides and their metabolites considered to be relevant*

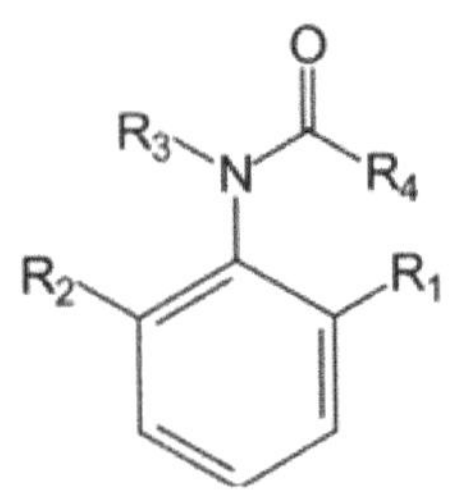

Figure 10.9. *Chloroacetamides and various possible groups*

The degradation of alachlor in soil leads to the major metabolites shown in Figure 10.10.

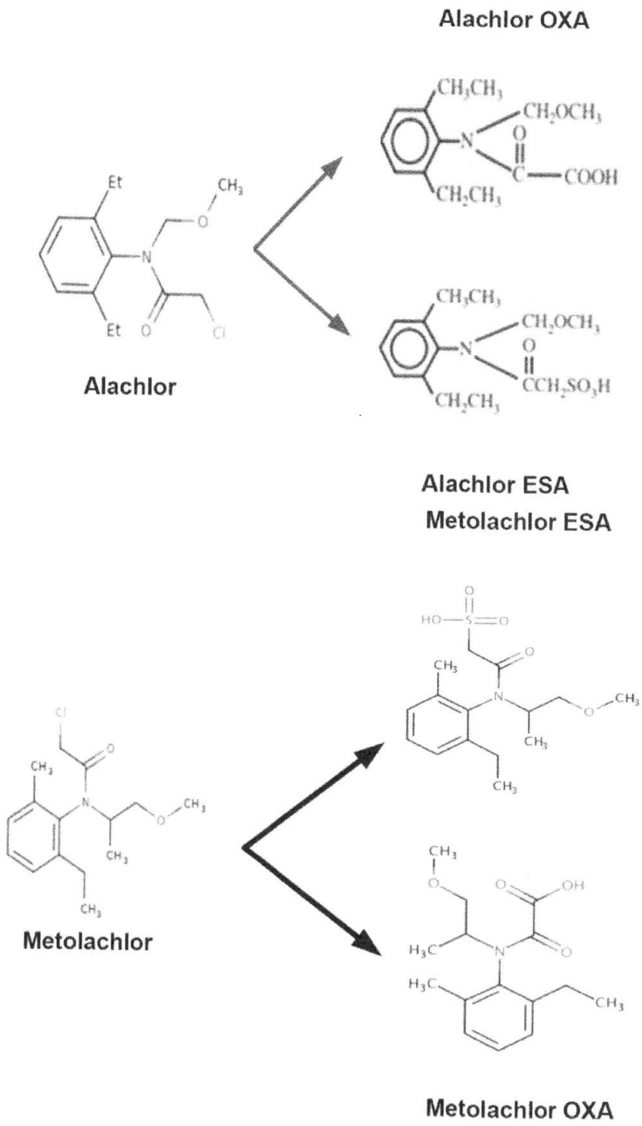

Figure 10.10. *The main metabolites of alachlor and metolachlor. For a color version of this figure, see www.iste.co.uk/gaid/watertreatment3.zip*

Metolachlor biodegrades in the soil (Figure 10.9) with at least five metabolites having been identified, among which the two principal ones are metolachlor ethane sulfonic acid (ESA) and metolachlor oxanilic acid (OXA). It has been suggested that the transformation of metolachlor into metabolites by the microorganisms present in the soil results from the displacement of the chlorine atom by glutathione, followed by the formation of ESA and OXA through different enzymatic pathways. ESA and OXA are highly soluble in water and more resistant to microbial degradation than the parent herbicide. This explains why ESA and OXA are detected more frequently and in higher concentrations than metolachlor itself. The physicochemical characteristics of alachlor, metolachlor and metabolites recognized as relevant are summarized in Table 10.7.

	Mother molecules and metabolites				
	Metolachlor	**Alachlor**	**Alachlor OXA (alachlor oxanilic acid)**	**Metolachlor ESA (metolachlor ethane sulfonic acid)**	**Metolachlor OXA (metolachlor oxanilic acid)**
Formula	$C_{15}H_{22}ClNO_2$	$C_{14}H_{20}ClNO_2$	$C_{14}H_{19}NO_4$	$C_{15}H_{23}NO_5S$	$C_{15}H_{21}NO_4$
Semi-developed formula	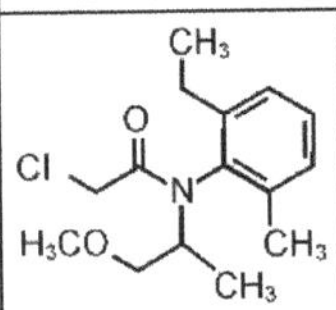	–	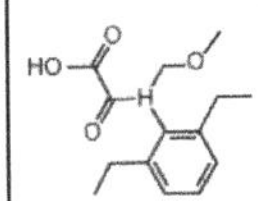	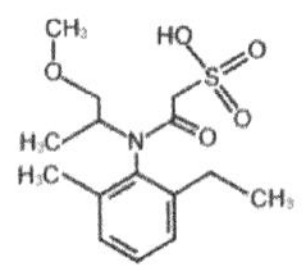	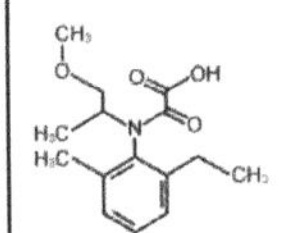
Log Kow (pH 7, 20°C)	3.13	3.52	–	0.6	<1
Log Koc	2.05	2.52	–	0.82	0.7
Water solubility	530 $mg \cdot L^{-1}$ (20°C)	240 $mg \cdot L^{-1}$ (25°C)	–	212,461 $mg \cdot L^{-1}$ (20°C)	8,500 $mg \cdot L^{-1}$ (20°C)

Table 10.7. *Characteristics of metolachlor, alachlor and metabolites (ESA and OXA)*

A priori, the characteristics of metolachlor ESA are very unfavorable for the adsorption onto AC (high solubility in water).

10.4.1.2.8. The efficiency of powdered activated carbon

The characteristics of the ACs tested are summarized in Table 10.8.

The tests were carried out on water drawn from the Garonne (France), with the following characteristics: UV 254 = 4.4 m^{-1}, DOC = 1.6 $mg \cdot L^{-1}$, metolachlor = 3.6 $\mu g \cdot L^{-1}$, metolachlor ESA = 1.6 $\mu g \cdot L^{-1}$, metolachlor OXA = 1.7 $\mu g \cdot L^{-1}$. The rate of AC varied between 10 and 30 $g \cdot m^{-3}$. The results obtained show that all the powdered carbons tested have a removal efficiency greater than 90% for metolachlor. For metabolites ESA and OXA, the efficiency is between 80 and 95%. Nevertheless, Norit SA Super AC offers a removal efficiency greater than 94% for the three pesticides (Figure 10.11).

Parameters	Powdered activated carbon					
	Aquasorb MP 23 (Jacobi)	Aquasorb MP 25 (Jacobi)	Aquasorb LP 17 (Jacobi)	SA Super (Norit)	W 35 (Norit)	Pulsorb WP240 (Chemviron)
Origins	Vegetal	Vegetal	Vegetal	Coal	Coal	Coal
d_{10} µm	15–35	15–35	15–35	10	10	20
d_{50} µm	20	20	20	15	15	30–35
Specific surface $m^2 \cdot g^{-1}$	820	1,100	950	1,150	780	950
Apparent density $kg \cdot m^{-3}$	260	480	500	250	330	370–430
Iodine index	760	1,100	800	1,050	850	900
pH	–	Alkaline	Alkaline	Alkaline	Alkaline	–

Table 10.8. *Characteristics of the powdered activated carbons tested*

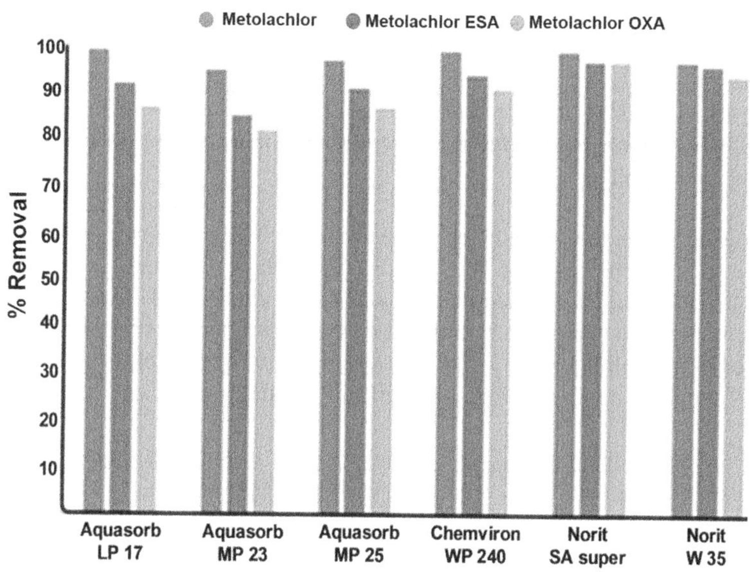

Figure 10.11. *Removal performance of metolachlor and its metabolites (ESA and OXA) with powdered activated carbon. For a color version of this figure, see www.iste.co.uk/gaid/watertreatment3.zip*

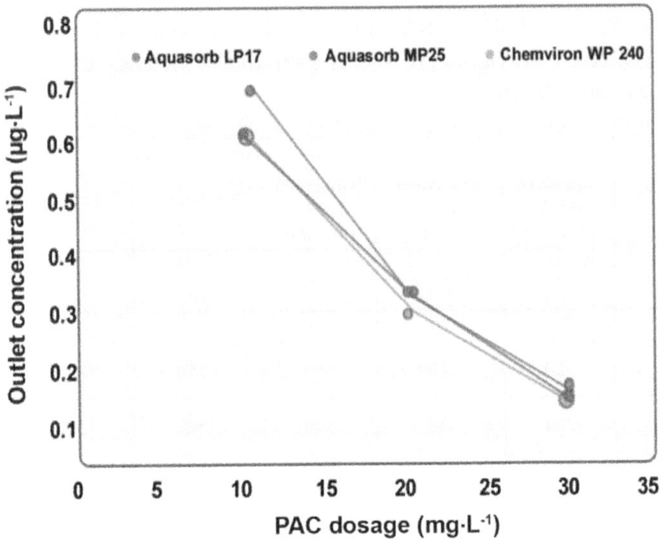

Figure 10.12. *Metolachlor ESA reduction tests on lake water. For a color version of this figure, see www.iste.co.uk/gaid/watertreatment3.zip*

Other trials have confirmed the effectiveness of PAC on lake water (Figure 10.11) with an initial metolachlor ESA concentration of 1.8 $\mu g \cdot L^{-1}$. A dose of 30–35 $g \cdot m^{-3}$ is required to reduce the concentration of treated water to 0.1 $\mu g \cdot L^{-1}$ with Chemviron WP 240 AC.

Knowing Freundlich parameters, as well as the MP's initial concentration, it is possible to infer the mass of PAC required to obtain an output concentration of 0.1 $\mu g \cdot L^{-1}$. We can thus notice that the adsorption capacity of PACs in relation to metolachlor is high (Table 10.9) and removes a large amount of this substance. This is related to its high log Kow value. The adsorption of metolachlor can be classified as type L, suggesting that AC has a high affinity with this pesticide.

On the other hand, ACs have a much lower adsorption capacity regarding metabolite metolachlor ESA, which is partly justified due to its low Kow. As a matter of fact, a low or negative log Kow value may translate a very weak adsorption for the molecule considered on its own, but its interaction with the different organic components of the water matrix may result in a higher removal than expected. An output concentration of 0.1 $\mu g \cdot L^{-1}$ can be reached provided that the PAC dosage is sufficient. With granular AC, it is the saturation time of the AC and its replacement which should vary between these two molecules.

Parameters	Aquasorb LP 27	Aquasorb MP 25	Chemviron WP 240		Aquasorb LP 27	Aquasorb MP 25	Chemviron WP 240
	Metolachlor ESA (Log Kow 0.6)			Metolachlor (Log Kow 3.13)			
Freundlich							
qe $mg \cdot g^{-1}$	0.047	0.049	0.057	qe $mg \cdot g^{-1}$	0.169	0.123	0.165
k $(mg \cdot g^{-1})$ $(L \cdot mg^{-1})1/n$	142	142	188	k $(mg \cdot g^{-1})$ $(L \cdot mg^{-1})1/n$	559	406	547
1/n	0.48	0.46	0.52	1/n	0.52	0.52	0.52
Langmuir							
qmax $mg \cdot g^{-1}$	0.123	0.137	0.213	qmax $mg \cdot g^{-1}$	0.380	0.233	0.263
b $mg \cdot L^{-1}$	5.91	5.41	2.09	b $mg \cdot L^{-1}$	8.19	10.59	18.36
qe $mg \cdot g^{-1}$	0.0457	0.0481	0.0372	qe $mg \cdot g^{-1}$	0.171	0.119	0.170
Temkin							
B $mg \cdot L^{-1}$	0.03	0.04	0.05	B $mg \cdot L^{-1}$	0.1	0.09	0.44
A	39.6	24.3	20.6	A	52.5	34.4	39.98
qe $mg \cdot g^{-1}$	0.0438	0.379	0.0334	qe $mg \cdot g^{-1}$	0.167	0.110	0.162
RL (Bayat)	0.99	1	1	RL (Bayat)	0.09	1.09	0.05

Table 10.9. *Adsorption parameters of activated carbons in relation to metolachlor and metolachlor ESA*

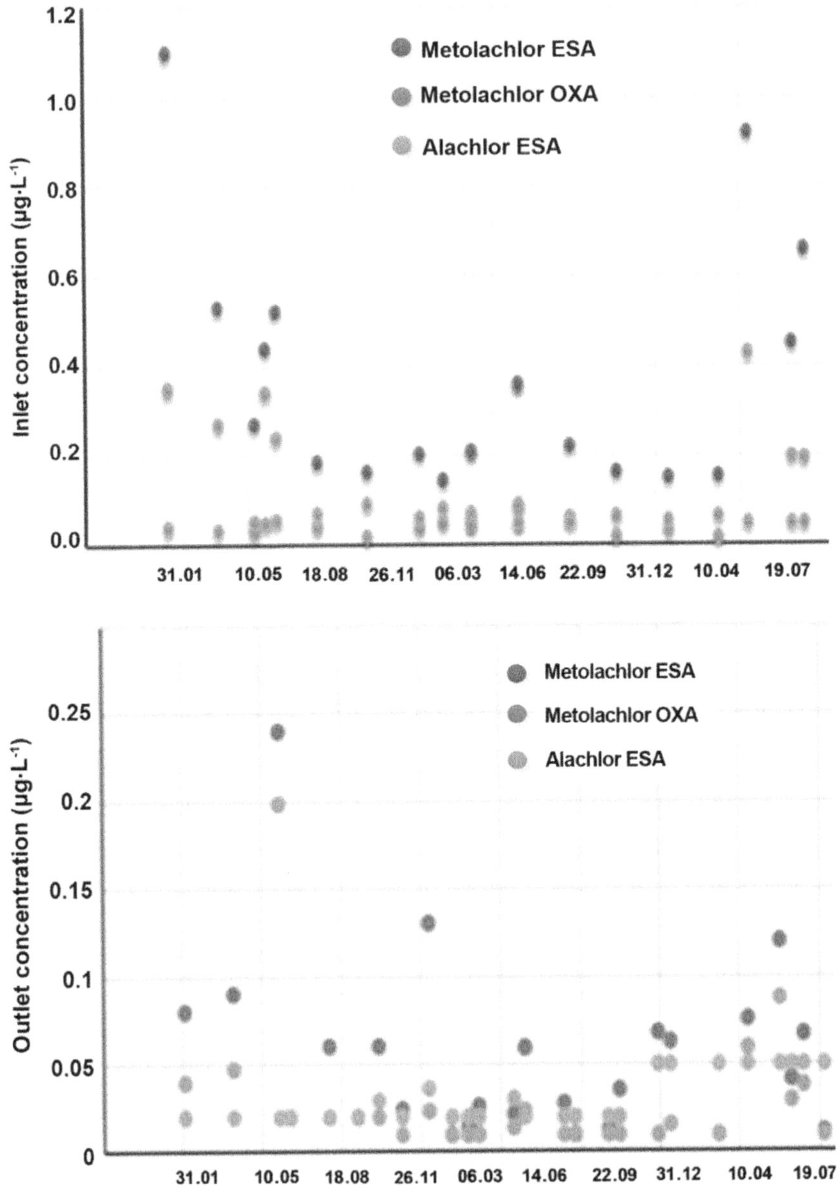

Figure 10.13. *(A) Inlet and outlet concentrations for metolachlor ESA and OXA, as well as alachlor ESA at Durtal plant with Actiflo® Carb (2016–2018). (B) Inlet and outlet concentrations for metolachlor ESA and OXA, as well as alachlor ESA at Durtal plant with Actiflo® Carb (2016–2018). For a color version of this figure, see www.iste.co.uk/gaid/watertreatment3.zip*

The removal performance of metabolites (ESA and OXA), as well as alachlor ESA at the Durtal plant (France) with Actiflo® Carb for the period 2016–2018, shows that PAC (Norit W 35) leads to residuals ≤ 0.1 μg·L^{-1} with a dosage of 15 g·m^{-3}.

Input concentrations oscillate between 0.05 and 0.4 μg·L^{-1} for metolachlor OXA, and between 0.18 μg·L^{-1} and 1.2 μg·L^{-1} for metolachlor ESA, whereas the concentrations for alachlor ESA are <0.15 μg·L^{-1}.

We can observe that in the last 3 years, the output concentration for metolachlor ESA only exceeded the standard of 0.1 μg·L^{-1} on three occasions, due to a mechanical PAC injection problem. However, it should be noted that during these same periods, the outlet concentrations for metolachlor OXA and alachlor ESA remained well below the standard.

According to previous calculations, for an inlet concentration of 0.4 μg·L^{-1} for metolachlor ESA, a 7 g·m^{-3} dosage of PAC (Norit W 35) is sufficient.

10.4.1.2.9. The efficiency of μgrain activated carbon

A settling reactor variant concerns the implementation of AC in a fluidized bed with μgrain AC (see Chapter 9). The characteristics and adsorption parameters of the μgrain AC tested are summarized in Table 10.10.

The results of Freundlich constants with μgrain activated carbon led to an adsorption capacity of: qe (mg·kg^{-1}) = 74 Cout$^{0.89}$. The dosages of μgrain activated carbon are higher than the ones obtained with powdered carbon, due to the larger particle size which reduces the exchange surfaces as well as intra-pore diffusion (Table 10.10 and Figure 10.14).

Nevertheless, the processes involving the use of μgrain activated carbon have adsorbent concentrations in the reactor between 200 and 350 g·L^{-1}. As a result, the removal kinetics of pesticides are still interesting even for variable AC dosages between 15 and 30 g·m^{-3}.

The results obtained on two Veolia sites in France confirm the removal performance of pesticides metolachlor ESA and OXA because of the use of μgrain activated carbon. Concentrations in treated water are lower than 0.1 μg·L^{-1} for a variable input concentration between 0.2 and 1.6 μg·L^{-1}.

Parameters	Micrograin activated carbon	Adsorption parameters	
	Microsorb 400 R chemviron	Metolachlor ESA	
Origins	Coal	**Freundlich**	
d_{10} µm	250	**qe, mg·g^{-1}**	0.00953
d_{50} µm	500	**K, mg·g^{-1} L·mg^{-1}**	74
Specific surface, m²·g^{-1}	860	**1/n**	0.89
Apparent density, kg·m^{-3}	450	**Langmuir**	
Iodine index	800	**qmax, mg·g^{-1}**	0.0125
		b, mg·L^{-1}	0.85
		qe, mg·g^{-1}	0.0098

Table 10.10. *Characteristics and adsorption parameters of micrograin activated carbon*

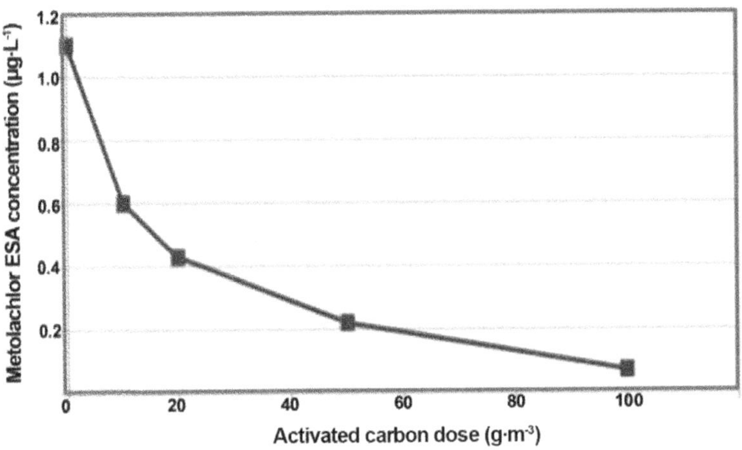

Figure 10.14. *Adsorption of Metolachlor ESA onto micrograin activated carbon. For a color version of this figure, see www.iste.co.uk/gaid/watertreatment3.zip*

10.4.1.2.10. Medicinal products

Table 10.11 summarizes the characteristics of some pharmaceutical residues found in water which make it possible to predict their adsorption onto AC. In particular, the parameters considered are molecular weight, log Kow, solubility as well as the presence of functional groups COOH, NH_2, SO_2.

As illustrated by the curve qe ($mg \cdot g^{-1}$) as a function of log Kow for pesticides, this same approach applied to pharmaceutical residues lets us predict that negative log Kow values are considered as having a negative impact, whereas very high values have a positive impact on adsorption. In the same way, low molecular weights (<200 $g \cdot mole^{-1}$) are judged to be negative and high molecular weights (>700 $g \cdot mole^{-1}$) are considered positive. Finally, despite the fact that MPs are, by definition, at trace levels, very high concentrations were judged as positive.

Thus, gabapentin and metformin have the lowest molecular weights and a negative log Kow. They also contain a COOH and NH_2 group, which decrease adsorption. They should therefore be less adsorbed than other MPs. Diclofenac, fenofibrate, estradiol, nonylphenol, triclosan, triclocarban and bezafibrate have very high log Kow values (>4) and are therefore expected to have a high affinity with AC. Triclosan has a high log Kow (4.76) with a symmetric and coplanar structure, and should probably adsorb well onto AC. Iohexol has a negative log Kow (–3.05), but a very high molecular weight is not likely to be well adsorbed, even if its solubility is not significant.

Adsorption results obtained with Actiflo® Carb onto PACs (Aquasorb MP23, Cabot-Norit SAE super and Pulsorb RD 30), for variable MP concentrations between 500 and 1,000 $ng \cdot L^{-1}$, confirm the predictions.

Iohexol and iomeprol have a high molecular weight, but due to their low log Kow and their size, they are poorly adsorbed. Gabapentin, iohexol and metformin are poorly adsorbed (<20%). Atenolol, tramadol, propanolol, carbamazepine, metoprolol and fenofibrate are well adsorbed (>70%), actually the best among all the other MPs. Their high log Kow provides them with a great affinity with PAC. Triclosan, which has a symmetrical, coplanar and relatively small structure and is characterized by a high log Kow, is very well adsorbed, indeed.

The PAC reactors developed by Veolia can obtain removal efficiencies for a large number of MPs with values oscillating between 60 and 95% and with variable doses of AC between 10 and 20 $mg \cdot L^{-1}$.

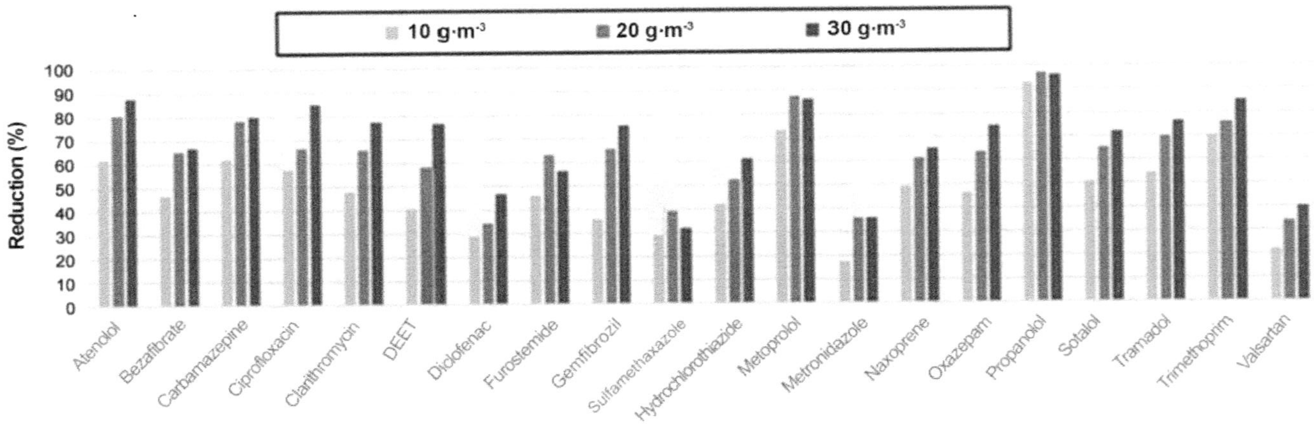

Figure 10.15. *Micropollutant removal performance with different dosages of PAC Pulsorb RD 90. For a color version of this figure, see www.iste.co.uk/gaid/watertreatment3.zip*

Pharmaceutical residue micropollutants – endocrine disrupting compounds	MW	log Kow	Solubility (ppm)	Freundlich parameters k 1/n $(mg{\cdot}g^{-1})(L{\cdot}\mu g^{-1})1/n$	
Atenolol	266.3	0.15	13,330	13	0.22
Bisphenol A	228.3	3.32	120	35.2	0.23
Bezafibrate	361.8	4.25	1	5.8	0.24
Carbamazepine	236.2	2.45	18	9.6	0.21
Caffeine	194.2	–0.07	21,600	10.1	0.21
Ciprofloxacin	331.3	0.38	30,000	12.5	0.23
Clarithromycin	748	3.16	1,693	30.8	0.22
Diclofenac	296.1	4.75	2.37	25	0.32
DEET	191.2	2.40	912	8	0.28
Erythromycin	733.9	3.06	2,000	33.1	0.26
Estradiol	272.4	4.01	3.6	21.1	0.30
17α Estradiol	272.4	4.01	0.39	20.5	0.28
Estriol	288.4	2.45	27.4	14.1	0.22
Ethinylestradiol	296.4	3.67	11.3	38	0.25
Fluoxetine	309.3	4.05	0.017	9.1	0.28
Fenofibrate	360.8	5.20	0.42	24.5	0.31
Fluvoxamine	318.3	2.80	7.34	14.9	0.23
Furosemide	330.7	2.03	73.1	12.8	0.21
Gabapentin	171.2	–1.10	4,490	1.4	0.28
Gemfibrozil	250.3	3.4	27.8	13	0.22
Hydrochlorothiazide	297.7	–0.07	722	10.8	0.28
Ibuprofen	206.4	3.97	21	33.1	0.27
Iohexol	821.1	–3.05	790	2.2	0.24
Iopromide	791.1	2.49	–	4.05	0.28
Iopamidol	777.1	–2.42	120,000	22	0.34
Ketoprofen	254.3	3.12	51	35	0.26
Metformin	129.1	–2.60	10	–	–
Metoprolol	267.3	1.88	16,041	12.4	0.205
Naxoprene	230.3	3.18	150	25	0.32
Nonylphenol	220.3	5.76	7	31.1	0.28
Omeprazole	345.4	2.23	359	12.5	0.23
Oxazepam	286.7	2.24	179	12.8	0.21

Pharmaceutical residue micropollutants – endocrine disrupting compounds	MW	log Kow	Solubility (ppm)	Freundlich parameters k 1/n $(mg\cdot g^{-1})(L\cdot \mu g^{-1})1/n$	
Ofloxacin	361.4	–0.39	10,800	11.2	0.31
Progesterone	272.4	4.01	3.60	25.6	0.29
Paracetamol	151.1	0.46	14,000	2.8	0.26
Primidone	218.2	0.91	500	3.5	0.29
Propanolol	259.3	3.48	61.7	41	0.19
Sotalol	272.3	0.24	782	8	0.22
Sulfamethoxazole	253.2	0.89	610	3.5	0.24
Tramadol	299.8	3.01	1,151	13	0.22
Triclocarban	315.6	4.34	0.11	23.3	0.31
Triclosan	289.5	4.76	10	17.2	0.38
Trimethoprim	290.3	0.91	400	35	0.23
Valsartan	485.5	1.49	8,210	7	0.19
Venlafaxine	277.4	3.20	267	12.8	0.21

Table 10.11. *Freundlich parameters (k and 1/n) for some drugs*

The performances of the PAC reactor are evaluated considering the total pesticides and other MPs present in the water to be treated. This is what differentiates this application from that of NOM, for which there was only one parameter (DOC), or at the most two parameters (color and DOC), to take into account.

PAC dosages can be increased depending on the DOC concentration in the water, which is an important factor because it influences the removal of MPs due to competitive adsorption.

10.4.1.2.11. PAC and ultrafiltration membrane (UF) combination

A variant of conventional reactors associates PAC with an ultrafiltration membrane. The PAC-membrane combination featured by the Opaline® C process (Figure 10.16) is useful when the PAC concentration to be injected does not require high doses and the purpose is mainly to ensure an initial removal of microorganisms. The PAC contact tank is designed in an equivalent manner to conventional reactors, with an equivalent time (10 min).

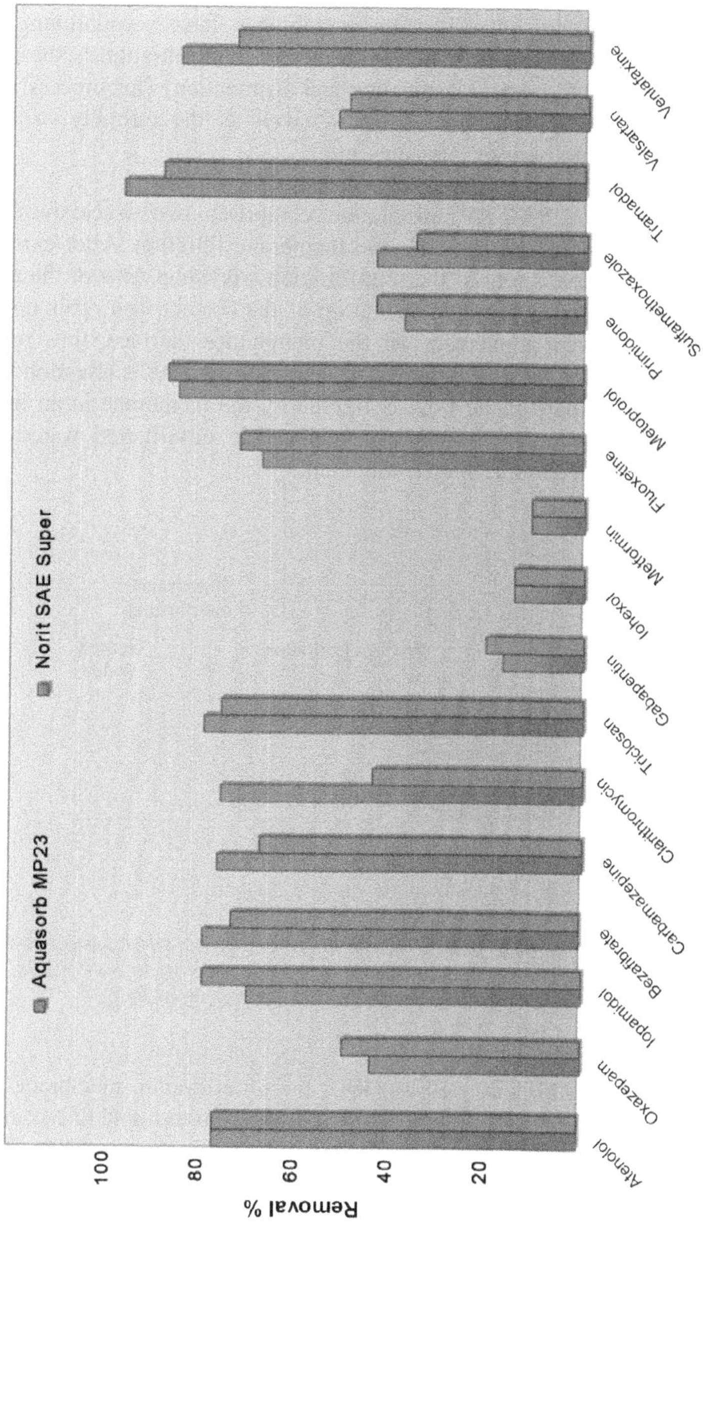

Figure 10.16. *Average performance with Aquasorb MP23 and Norit SAE Super. For a color version of this figure, see www.iste.co.uk/gaid/watertreatment3.zip*

Membranes are classified by solute exclusion size, which is sometimes referred to as pore size. The molecular weight cut-off is the degree of exclusion of a known solute. Ultrafiltration and microfiltration membranes have a minimum solute rejection size of 0.01 μm and 0.10 μm, respectively. Ultrafiltration is the process that is most often used for water clarification and disinfection. This process enables a very good removal of suspended solids as well as the turbidity of various microorganisms (bacteria, viruses, parasites, etc.).

The operation of a PAC-UF installation comprises two successive stages: adsorption onto PAC in a specific tank and membrane filtration. After exiting the contact tank, the mixture is sent to the ultrafiltration workshop. Part of the mixture (5–20%) containing PAC is recycled at the top of the contact tank. This operation mode creates a turbulent movement on the membranes' surface, thus reducing clogging. Recirculation also concentrates and increases the PAC's retention time in the system. The other part of the flow is filtered by the membrane in an internal-external direction, operating in tangential mode. The ultrafiltered water, called permeate, is then directed to the filtered water tank.

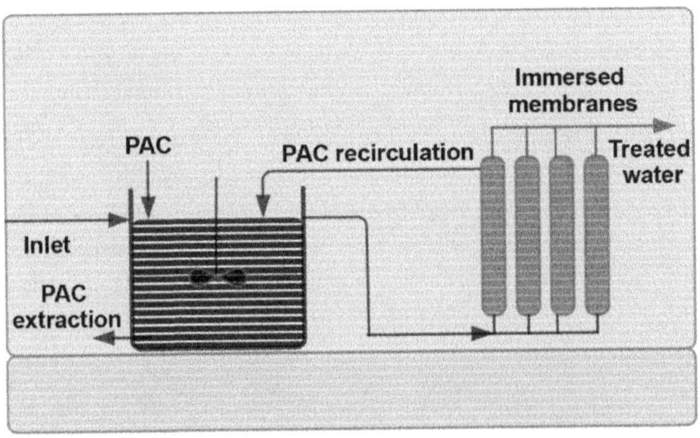

Figure 10.17. *Opaline® C process: combination of powdered activated carbon with a pressurized ultrafiltration membrane. For a color version of this figure, see www.iste.co.uk/gaid/watertreatment3.zip*

While AC removes MPs as well as DOC, the ultrafiltration membrane has no effect on the reduction of MPs. Figure 10.17 illustrates an example of MP removal by means of the Opaline® C process.

The results obtained with the Opaline® C process are identical to the ones presented for PAC reactors, insofar as the operating provisions of PAC are similar (both in terms of the AC concentration in the reaction tank and on the previously mentioned types of AC).

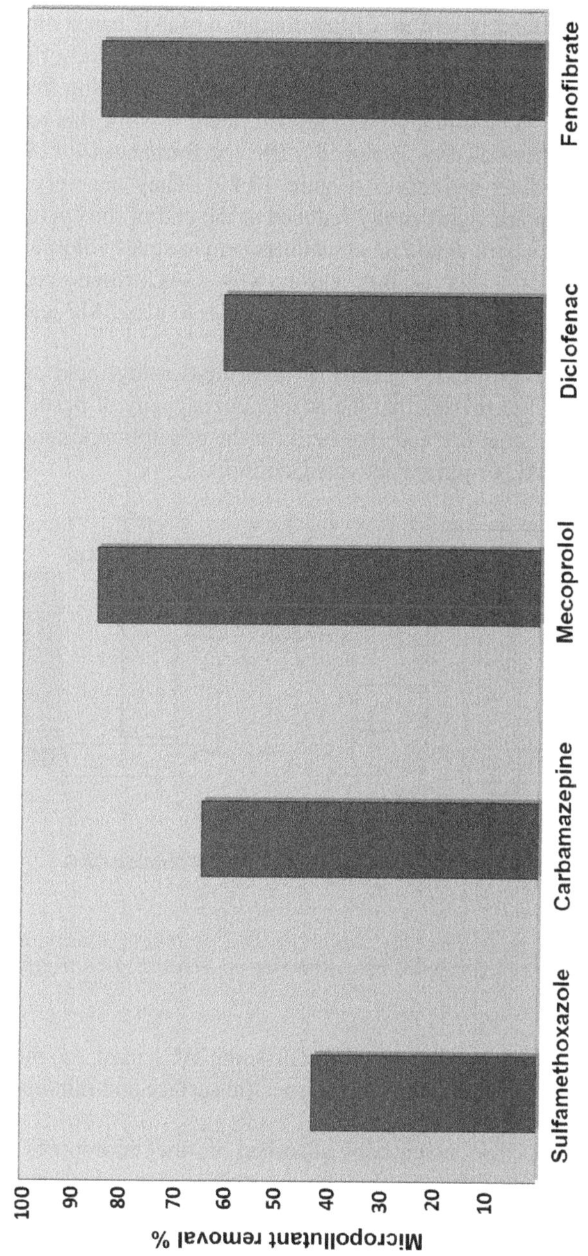

Figure 10.18. *Removal of micropollutants with the Opaline® C process (Veolia)*

10.4.1.2.12. Granular activated carbon

The water to be treated is sent to a filter containing GAC based on a downward flow. During its passage, organic substances and MPs are adsorbed onto the material. A media adsorption front is thus created, gradually moving from the filter's inlet to its outlet. The AC gradually becomes saturated. It is for this reason that the GAC needs to be periodically replaced. The performances of GAC evolve depending on the media's saturation (Figure 10.19). They are very good at the cycle's beginning, then are significantly reduced at the end of the cycle, because the unsaturated GAC that is still available constitutes only a small volume compared to the total load capacity. This is the reason why GAC filters require regular monitoring and control of the media's capacity in order to avoid MP leaks.

The adsorption of pesticides (atrazine, atrazine-desethyl and triflusulfuron-methyl) onto a GAC filter reveals that the adsorption capacity of pesticides depends on the media's porous structure and size because the adsorption kinetics are slower with GAC than with PAC or µgrain activated carbon.

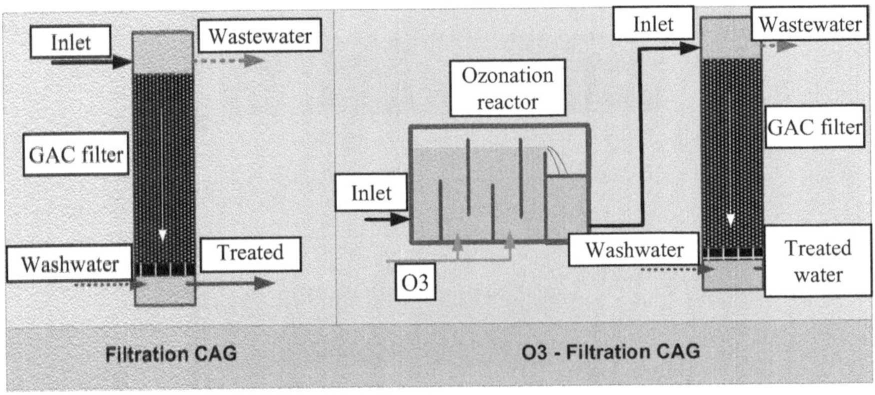

Figure 10.19. *Removal of micropollutants on GAC or in combination with ozone. For a color version of this figure, see www.iste.co.uk/gaid/watertreatment3.zip*

Table 10.12 presents the characteristics of some ACs used for the removal of MPs. We notice the correlation between the specific surface and microporosity.

Pesticide molecules are essentially adsorbed in the micropores without the resistance of intraparticle diffusion through the macropores, which is generally the controlling stage for GACs.

Parameters	Granular activated carbon			
Manufacturer	Jacobi	Jacobi	Cabot Norit	Chemviron
Name	Aquasorb NC 60	Aquasorb 2000	Norit Row 0.8	12X40
Origin	Coconut	Coal	Peat	Coal
Grain diameter (mm)	1.0	1.0	0.9	1.1
Specific surface ($m^2 \cdot g^{-1}$)	1,200	1,000	915	1,100
Total pore volume ($cm^3 \cdot g^{-1}$)	0.590		0.648	0.646
Micropore volume ($cm^3 \cdot g^{-1}$)	0.513		0.388	0.466
Macropore volume ($cm^3 \cdot g^{-1}$)	0.242		0.375	0.646
Microporosity (%)	87		59.9	72.2
Average pore diameter (nm)	0.59		0.55	0.58
Density ($kg \cdot m^{-3}$)	0.783		0.646	0.579
Apparent density ($kg \cdot m^{-3}$)	0.97	0.95	0.85	0.92

Table 10.12. *Characteristics of some granular activated carbons used for the removal of micropollutants*

10.4.1.2.13. The case of metolachlor and its metabolites

As previously described, two diffusion processes within the organic substance complement one another: external (surface) diffusion and internal diffusion (diffusion within the adsorbent's pores). In a fixed bed of GAC, the limiting diffusion mechanism is internal diffusion. In the internal diffusion process, organic molecules move into macropores, mesopores or micropores, depending on their size. This is the molecular diffusion process. During external diffusion, the molecules are adsorbed onto the solid's surface.

The removal of these pesticides and metabolites using two GACs (whose characteristics are summarized in Table 10.13) is performed on raw water containing a variable metolachlor concentration of 0.9–3.5 $\mu g \cdot L^{-1}$, and a concentration of metabolites (ESA and OXA) ranging between 0.5 and 0.9 $\mu g \cdot L^{-1}$ for each molecule. The contact times vary between 10 and 12 min.

The results obtained (Figure 10.20) show that the metolachlor present without water was well adsorbed onto Aquasorb 2000, considering that after 10 months of operation no leakage (concentration limit 0.1 $\mu g \cdot L^{-1}$) was observed with this pesticide onto Aquasorb 2000. On the other hand, a leak appeared as from the fifth month with Filtrasorb 400 activated carbon. ESA and OXA metabolites present in the water were well adsorbed onto Aquasorb 2000 activated carbon.

Parameters	Granular activated carbon	
	Aquasorb 2000 (Jacobi)	**Filtrasorb 400 (Chemviron)**
Origin	Vegetal	Coal
d_{10} µm	850	650
d_{50} µm	1,300	1,020
Specific surface $m^2 \cdot g^{-1}$	1,000	1,050
Apparent density $kg \cdot m^{-3}$	500	425
Iodine index	950	1,050
pH	8–11	–
CU	1.8	1.9

Table 10.13. *Characteristics of granular activated carbons*

There was a leak above 0.1 $\mu g \cdot L^{-1}$ in metolachlor OXA with Filtrasorb 400 carbon after 4 months of operation (Figure 10.21). In contrast, Aquasorb 2000 carbon remained below 0.1 $\mu g \cdot L^{-1}$ after 6 months of operation. The adsorption capacity of Aquasorb 2000 carbon is better than that of Filtrasorb 400 for these molecules.

In general, it is accepted that adsorption onto GAC is an interesting stage when the results obtained in laboratory column tests lead to volumes of treated water of at least 15,000 BV. This is the result obtained for metolachlor metabolites ESA and OXA with Aquasorb 2000 carbon. With metolachlor and alachlor, volumes exceed 30,000 BV for this same AC. The adsorbed quantities of pesticides deduced from these tests lead to respective values greater than 100 $mg \cdot kg^{-1}$ of GAC for metolachlor, and around 33 $mg \cdot kg^{-1}$ of GAC for ESA and OXA metabolites.

We can consider that GAC must be able to treat more than 100,000 BV for surface water containing pesticides such as metolachlor and alachlor. In the presence of metolachlor ESA and OXA and alachlor ESA, volumes of treated water should be between 30,000 and 45,000 BV for a contact time of 10–12 min and for inlet concentrations below 0.7 $\mu g \cdot L^{-1}$.

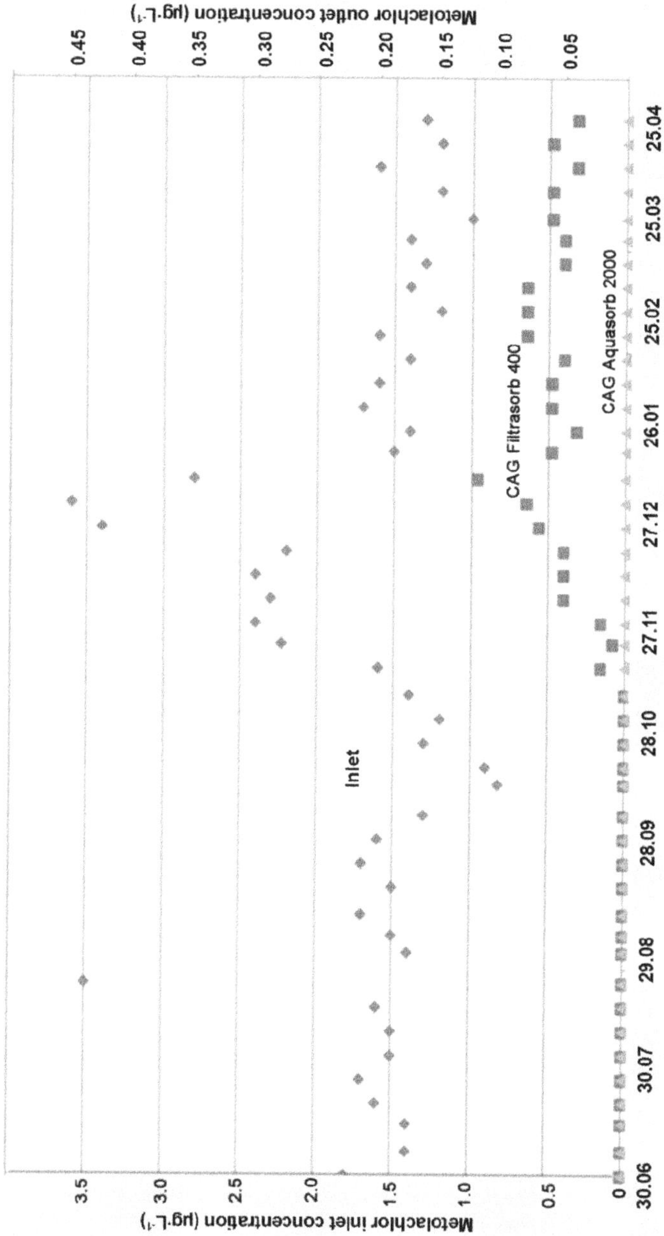

Figure 10.20. *Adsorption of metolachlor onto two granular activated carbons (Aquasorb 2000 and Filtrasorb 400). For a color version of this figure, see www.iste.co.uk/gaid/watertreatment3.zip*

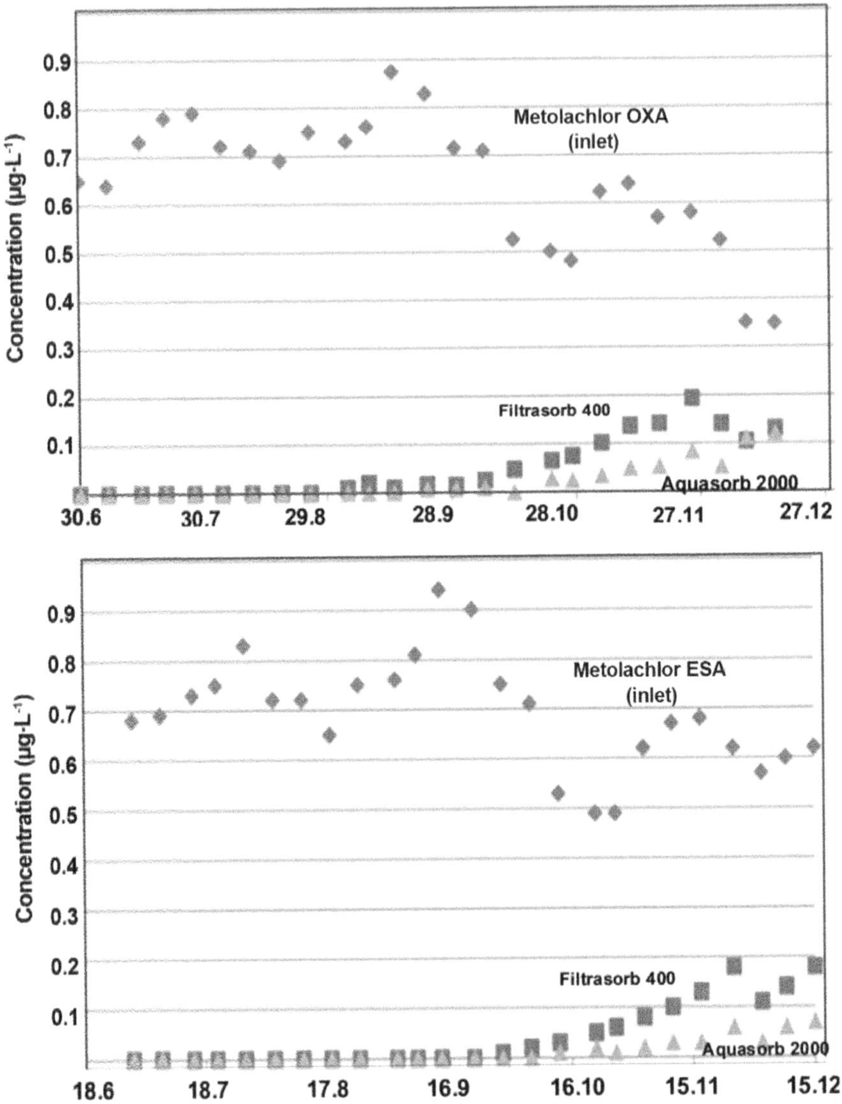

Figure 10.21. *Adsorption of metolachlor OXA and ESA onto two granular activated carbons. For a color version of this figure, see www.iste.co.uk/gaid/watertreatment3.zip*

10.4.1.2.14. Adsorption of pharmaceutical residues and other endocrine-disrupting compounds

The water feeding GAC filters is most often water which has been coagulated, flocculated, settled and filtered. There are still residues of dissolved organic substances (DOC), in addition to the MPs which were not removed during the previous stages. This leads to an adsorption competition which can hamper the removal capacity of AC in relation to MP. The compounds present in NOM extracted from settled water are of the acid type with low molecular weight and/or neutral and can compete with MP in terms of adsorption.

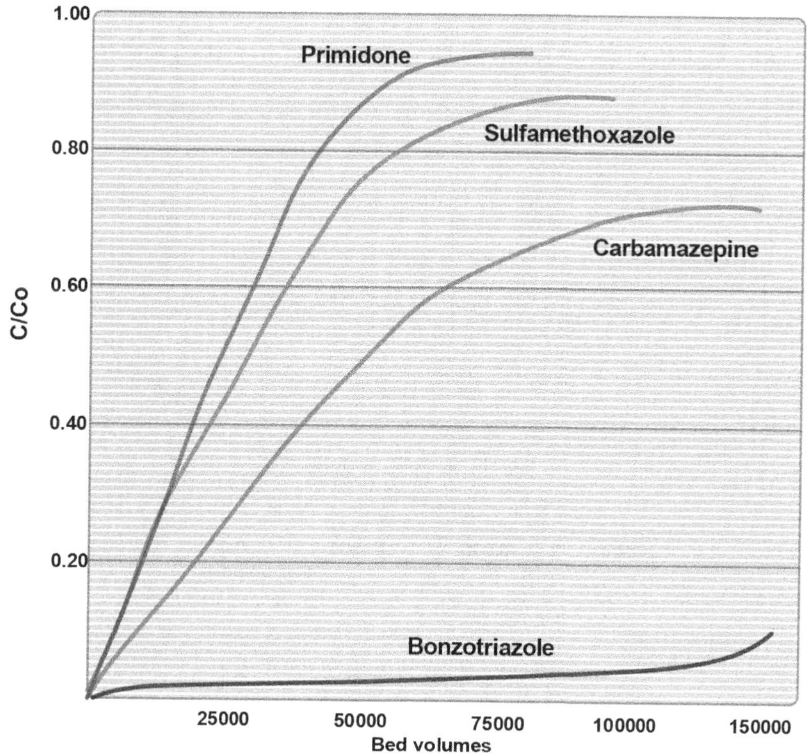

Figure 10.22. *Adsorption of four drugs on a GAC filter (Aquasorb 2000). For a color version of this figure, see www.iste.co.uk/gaid/watertreatment3.zip*

Figure 10.22 illustrates the results of adsorption onto a GAC filter of three pharmaceutical residues (carbamazepine, primidone and sulfamethoxazole) and industrial waste (benzotriazole).

The GAC filter is made of Aquasorb 2000 activated carbon, with a contact time of 10 min, operating at a speed of 8 $m \cdot h^{-1}$. Input MP concentrations vary between 5 and 10 $\mu g \cdot L^{-1}$. It appears that benzotriazole is best adsorbed, whereas GAC saturates with primidone after 75,000 BV. Sulfamethoxazole behaves almost similarly to primidone. Carbamazepine saturates GAC up to 50% after a passage of 50,000 BV.

The ability of carbon to adsorb the MPs present in water depends on the initial MP concentrations: the lower the initial concentration in natural water, the lower the observed adsorption capacity.

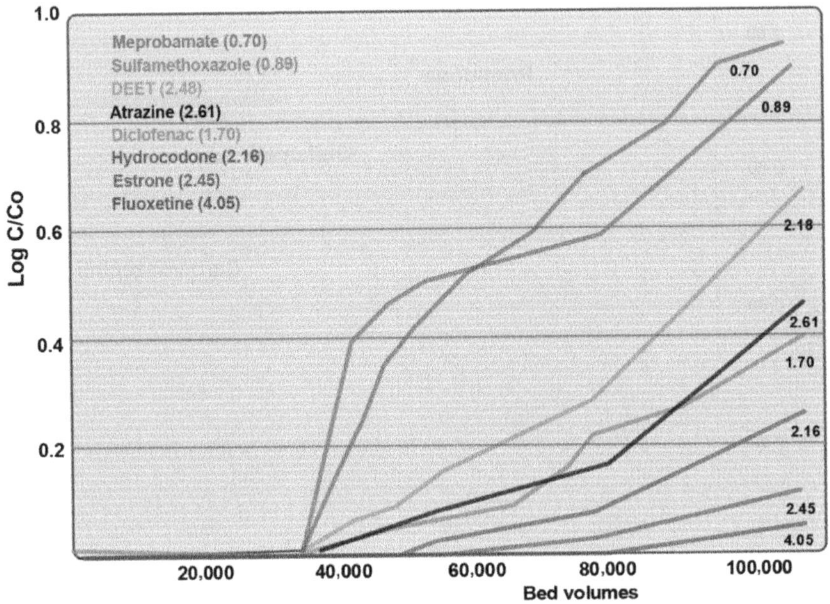

Figure 10.23. *Efficiency of GAC against various medicinal products depending on their log Kow. For a color version of this figure, see www.iste.co.uk/gaid/watertreatment3.zip*

Figure 10.23 shows the efficiency of Cabot Norit activated carbon (12 × 40) against various medicines with different log Kow values. These results confirm that a molecule with a high log Kow is better adsorbed than another molecule with a low to negative log Kow. This is a trend, but does not constitute a general rule.

Well adsorbed pharmaceutical residues		Moderately adsorbed pharmaceutical residues	Poorly adsorbed pharmaceutical residues
Amlodipine	Levonorgestrel	Bisphenol A	Caffeine
Atenolol	Losartan	Ciproflacin	Fosfomycin
Bezafibrate	Metoprolol	Cyclophosphamide	Gabapentin
Carbamazepine	Naxoprene	Doxycycline	Hydroxycarbamide
Chlorambucil	Nonylphenol	Fosfamide	Hydrochlorothiazide
Clarithromycin	Octylphenol	Heptaminol	Iobitridol
DEET	Omeprazole	Meprobamate	Iohexol
Diclofenac	Oxazepam	Paracetamol	Iopamidol
Erythromycin	Oxylbenzone	Sulfamethazine	Levodopa
Estradiol	Pravastatin	Sulfamethoxazole	Metformin
17α Estradiol	Primidon		Ofloxacin
Estriol	Propanolol		Ranitidine
Ethinylestrone	Progesterone		
Fenofibrate	Sotalol		
Fluoxetine	Testosterone		
Fluvoxamine	Tilosin		
Furosemide	Tramadol		
Galaxolid	Triclocarban		
Gemfibrozil	Triclosan		
Ibuprofen	Trimetropine		
Iopromide	Valsartan		
Ketoprofen	Venlafaxine		

Table 10.14. *Adsorption conditions of some pharmaceutical residues onto activated carbon*

General discussion on the effectiveness of activated carbon in relation to pesticides and other micropollutants

Many parameters influence the adsorption mechanism, such as the pesticide's solubility and structure. In broad terms, the main adsorption force is placed between the scattering forces of π electrons (which are part of the pesticide's structure) and π electrons present on the surface of AC. For example, the aromatic nucleus of the molecules of metolachlor, alachlor and their metabolites tends to promote interactions between AC and these molecules due to the delocalization of π electrons around the aromatic nucleus. The branched alkyl substituent for metolachlor's and alachlor's aromatic nucleus is involved in the proper adsorption of these pesticides, because it gives them a high hydrophobicity (high log Kow).

On the other hand, however, a loss of the alkoxy alkyl link or the replacement of metolachlor's Cl by SOO-OH (ESA) or by CO-OH (OXA) results in an increase in the solubility of these molecules in water and a decrease in the affinity of AC with metabolite molecules. The same is true for alachlor, where the replacement of the -CH_2Cl branch by -CO-OH (OXA) tends to increase the metabolite's solubility.

In addition, the carbonyl group (CO) has a larger dipole moment than the carboxylic acid group, and therefore acts as a stronger electron donor. This explains why metolachlor and alachlor are better adsorbed than their OXA metabolites.

As far as their metabolites are concerned, it is possible for the adsorption mechanisms of metolachlor and alachlor to be different in reference to Freundlich or Langmuir isotherms. For example, there could be a stronger competition for metabolites from adsorption sites.

In a water matrix containing NOM, the adsorption behavior of alachlor, metazachlor and their metabolites is influenced by competitive adsorption. Although pesticides are adsorbed in the micropores due to their low molecular weight, organic substances with a mean molecular weight which integrate mesopores interfere with the adsorption kinetics. Therefore, a good preliminary removal of NOM is recommended in order to reduce the dosages of powdered or µgrain activated carbon, or to increase the lifetime of GAC. GAC treating water having a DOC concentration <1 $mg \cdot L^{-1}$ has a better adsorption capacity for ESA and OXA metabolites compared to water containing 2 $mg \cdot L^{-1}$ DOC due to this competitiveness.

In fixed beds, an increase in contact time (beyond the conventional 10 min), and consequently the GAC volume, leads to much larger volumes of treated water, and the kinetic effects of competing substances are attenuated.

Through these few non-exhaustive examples, Veolia has tools, a database and feedback from the fieldwork offering invaluable insight on the decontamination of water polluted by pesticides, namely alachlor, metazachlor, metolachlor and their metabolites.

This also makes it possible to choose the most suitable AC for this type of contaminant. In the light of its performance against these pesticides, Aquasorb 2000 has shown that it is an interesting adsorbent, in virtue of its chemical characteristics and porous distribution. Because even if pesticides such as metolachlor ESA have a lower log Kow, it is the longer contact time in the filter which makes it possible to obtain high performances. Instead of the conventional 10–12 min, we operate with contact times in the range of 20–25 min, especially considering that the interactions between humic substances and these metabolites can favor the adsorption mechanisms inside the porous structure (either in monolayer or in multilayer).

The presence of DOC can reduce this efficiency and oblige the operator to increase the PAC doses or to reduce the life expectancy of filters. But, since DOC is found in concentrations up to 1,000 – 10,000 times higher than MPs, it is this parameter which guides the design of structures so that the quality of the outlet water meets the standards both for DOC and standardized organic MPs.

PAC or µgrain reactors offer the advantage of continuously supplying fresh carbon and ensuring the quality of treated water, despite variations in raw water. These reagents are periodically reduced, depending on the seasons when DOC is lower.

GAC can be very effective provided that its operation is continuously monitored and that the operator is ready to renew it in case there are DOC and/or MP leaks. It also appears that steam-activated GAC often exhibits a higher adsorption capacity and involves longer turnover times.

In any case, the selection and evaluation of GAC-type are fundamental and, often, only pilot tests can help in the best choice of AC to remove these pesticides and their metabolites due to a lack of feedback and know-how.

This is because a direct correlation with the characteristics of the adsorbent material (specific surface, iodine index, etc.) is not sufficient. It is also important to consider the pore distribution associated with the physicochemical characteristics of pesticides (Kow, ionization form, etc.) as well as the water's organic matrix.

In all cases involving the use of AC (powder, µgrain or grain), it is advisable to ensure that the characteristics of the samples which made it possible to carry out laboratory or pilot tests are identical to the batches supplied at the treatment station. Differences were often observed between batches for the same AC.

10.4.2. *Ozonation*

Ozone exists as a gas at room temperature. The gas is colorless, with an easily detectable pungent odor at concentrations as low as 0.02–0.05 ppm (by the volume it is dissolved in). This is still below concentrations which may be hazardous for health. Ozone gas is highly corrosive and toxic. Ozone is a molecule composed of three oxygen atoms (in the gaseous state, under normal temperature and pressure conditions), acting as a powerful oxidant that can transform and/or degrade many organic substances.

Among the chemicals generally used in water treatment, ozone is a powerful oxidant, ranking barely behind the hydroxyl free radical (OH•) in terms of oxidation. It oxidizes many organic and inorganic compounds. The reactions with these compounds trigger a demand for ozone in treated water, which must be satisfied before developing a measurable residual.

Ozone is poorly soluble in water. At 20°C, the 100% solubility of ozone is only 570 $mg \cdot L^{-1}$. Ozone is more soluble than oxygen, but chlorine is 12 times more soluble than ozone. Ozone concentrations used in drinking water treatment oscillate between 0.5 and 10 $mg \cdot L^{-1}$, limiting the driving force for the mass transfer of ozone gas into water.

In water, ozone decomposes spontaneously by means of a complex mechanism involving the generation of hydroxyl free radicals. Hydroxyl free radicals are among the most reactive oxidizing agents in water, with reaction rates in the range of 10^{10}–10^{13} $M^{-1} \cdot s^{-1}$. Due to the short half-life of hydroxyl free radicals (only a few microseconds range), their concentrations can never reach levels greater than 10^{-12} M.

Figure 10.24 shows that ozone reacts following two modes of action:

– direct oxidation of compounds by the O_3 molecule;

– oxidation of compounds via the OH• free radicals resulting from the decomposition of O_3;

– these two modes of action can take place together or in parallel.

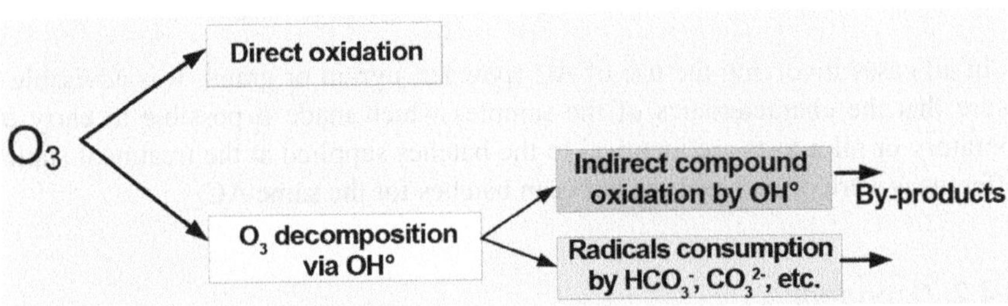

Figure 10.24. *Decomposition and reactions of ozone in water. For a color version of this figure, see www.iste.co.uk/gaid/watertreatment3.zip*

The two oxidation pathways are in competition in relation to the compounds to be oxidized. Direct oxidation with ozone is relatively slow compared to the oxidation of hydroxyl free radicals, even if the concentration of ozone in water is relatively high. On the other hand, the reaction of hydroxyl radicals is quick, but their concentration under normal ozonation conditions is relatively low.

In an acid medium, direct oxidation with molecular ozone predominates. At high pH levels, the production of hydroxyl free radicals is favored and the addition of H_2O_2 or UV makes their oxidation kinetics predominate. This mode of action is notably used as an advanced oxidation process.

The spontaneous decomposition of ozone takes place throughout a series of stages. Hydroxyl radicals are formed as one of the intermediate products and can react directly with compounds in water. The decomposition of ozone into pure water is prompted by the hydroxyl free radicals produced as an intermediate product of ozone decomposition, resulting in the net production of 1.5 moles of hydroxyl free radicals per mole of ozone.

Ozone demands are associated with the following elements:

– reactions with NOM in water: the oxidation of NOM leads to the formation of aldehydes, organic acids and aldo and keto acids;

– organic oxidation by-products: organic oxidation by-products are generally better suited for biological degradation and can be measured as assimilable organic carbon (AOC) or biodegradable dissolved organic carbon (BDOC);

– synthetic organic compounds (SOCs): some SOCs can be oxidized and mineralized under favorable conditions. To obtain total mineralization, the oxidation of hydroxyl radicals is generally the dominant route, as obtained in advanced oxidation processes;

– bromide ion oxidation: bromide ion oxidation results in the formation of hypobromous acid, hypobromite ion, bromate ion, brominated organic compounds and bromamines;

– bicarbonate or carbonate ions, usually measured in alkalinity, scavenge hydroxyl radicals and form carbonate radicals. These reactions are important for advanced oxidation processes where the radical oxidation pathway is predominant.

It is notably used on a large scale for the removal of tastes, odors and colors as well as for the disinfection of drinking water. Ozone is electrophilic and reacts rapidly and selectively, primarily with amines, phenols, aromatic compounds and double bonds.

The hydroxyl radical (OH•) reacts very quickly – in a non-selective manner – with most organic molecules. The formation rate of OH• depends on the properties of the water matrix: pH, alkalinity, type and content of NOM. It significantly improves the oxidation of certain MPs. The reactivity of a substance with ozone depends on its molecular structure and can be characterized by its second-order reaction rate constants kO_3 (with ozone) and kOH• (with the hydroxyl radical). The reaction kinetics accompanying the mechanisms of chemical oxidation with ozone take place in ozonation tanks (two to four chambers) into which O_3 is injected (Figure 10.25). Contact times in each chamber may vary between 4 and 8 min. The reactor's design and the composition of the water to be treated are important parameters, which influence the performances of oxidation by O_3.

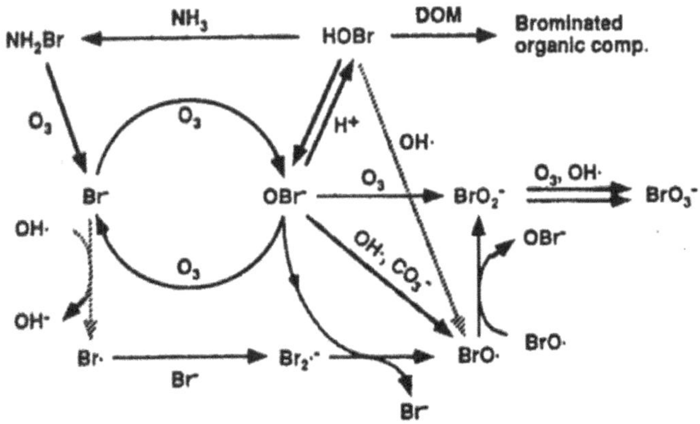

Figure 10.25. *Reaction of ozone with bromides for the production of bromates and organobromine compounds*

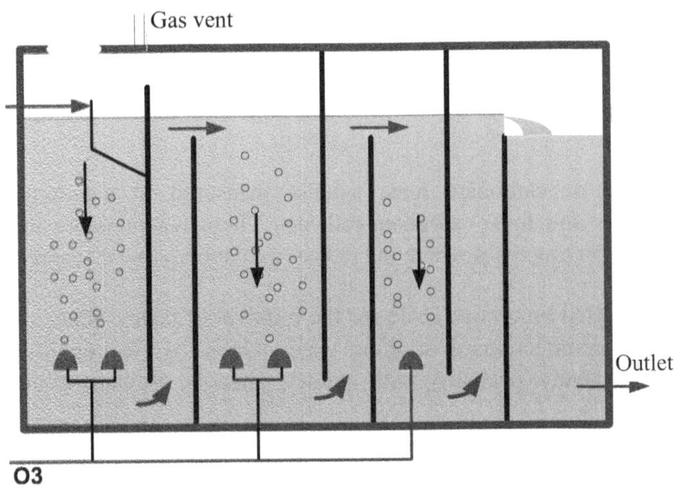

Figure 10.26. *Ozonation reactor with three tanks*

The higher the ozone concentration and the exposure time, the better the removal of pollutants. Ozone reacts not only with MPs, but also with organic matter in general, with nitrites and other inorganic compounds. But this can risk the formation of undesirable reaction by-products such as nitrosamines, bromates, and so on, because many oxidized substances are generally not completely mineralized, and therefore, by-products are formed.

10.4.2.1. *The case of pesticides*

Ozone reacts rapidly with double bonds, deprotonated amines and activated aromatic compounds. These types of bonds or organic functions are often part of the structure of pesticides and other MPs, such as pharmaceutical residues and synthetic substances. At $pH > 8$, ozonation is considered an advanced oxidation process due to the breakdown of O_3 molecules into OH• radicals, whose high concentration involves interactions with organic molecules.

Pesticide reaction rate constants with ozone vary between $<0.04\ M^{-1}{\cdot}s^{-1}$ for lindane and $4.4 \times 10^4\ M^{-1}{\cdot}s^{-1}$ for aldicarb. The reaction with lindane is very slow, whereas aldicarb reacts within seconds.

Reactions with pesticides are varied, depending on the molecules' chemical properties. Their oxidation does not lead to complete mineralization, but to the formation of oxidation by-products through protonation, dimerization, isomerization and tautomerization reactions. The rate constants of atrazine, cyanazine and deisopropylatrazine (DIA) are very close to one another. Atrazine has less than 50% removal with an ozone dose of 2.5 $mg{\cdot}L^{-1}$, but this removal is significantly improved when using advanced oxidation processes (UV/O_3, O_3/H_2O_2) (see section 10.4.4). Desethylatrazine (DEA) and DIA, which are by-products of the chemical oxidation of atrazine, and the cyanazine herbicide are fairly recalcitrant to oxidation. This is because they do not contain an aromatic function but rather an electron-withdrawing group (chloro double bond), which produces this low reactivity in relation to ozone. In general, pesticides are more difficult to oxidize than pharmaceutical residues. See, for example, the reduction of less than 65% for cyanazine, despite a Ct of 13.2 $mg{\cdot}min{\cdot}L^{-1}$. The reductions for DEA and DIA are around 38–50% for this same Ct.

The ozonation of river water having undergone coagulation–flocculation–settling and sand filtration, containing about 2–2.5 $mg{\cdot}L^{-1}$ DOC and 3–4 $\mu g{\cdot}L^{-1}$ of atrazine, reveals a complete degradation of this pesticide for a dose of 4 $mg{\cdot}L^{-1}$ with a 6 min contact time. For a 2 $mg{\cdot}L^{-1}$ dose of O_3, the atrazine residual is around 0.8 $\mu{\cdot}L^{-1}$. DEA and DIA by-products are rapidly present for an O_3 dose lower than 0.5 $mg{\cdot}L^{-1}$ (Figure 10.27). This means that an additional downstream removal of atrazine and its by-products should be contemplated if the water is seriously contaminated.

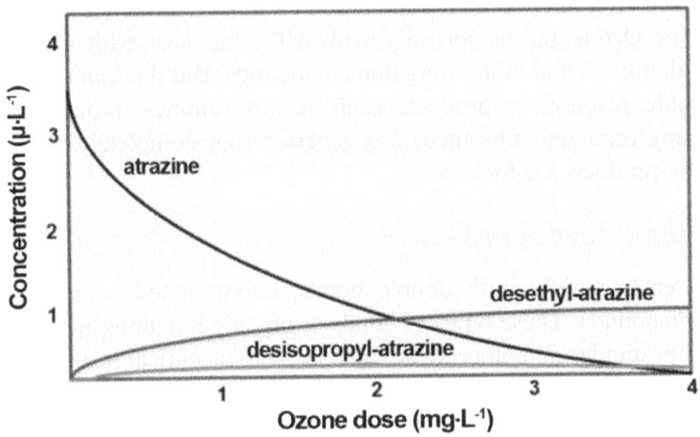

Figure 10.27. *Ozonation of atrazine and its by-products using different ozone doses. For a color version of this figure, see www.iste.co.uk/gaid/watertreatment3.zip*

As pesticides react well with O_3, an additional reaction with the OH• radicals is not necessary. The reaction time of these substances with ozone is faster than the time required for the formation of free radicals. On the other hand, for pesticides whose oxidation kinetics are slower, the reaction with OH• radicals during ozonation plays a more prominent role in relation to their oxidation. For example, mecoprop, which is a difficult pesticide to oxidize, requires a reaction time of more than 40 min and more than 1.1 g $O_3 \cdot g^{-1}$ DOC to obtain a reduction of around 70%. The injection of H_2O_2 makes it possible to increase this efficiency and to reduce reaction times.

Ozone oxidation is accompanied by the formation of metabolites whose toxicity may be related to the parent molecule. Thus, dichlorine pesticides such as diuron (Figure 10.28) and linuron are oxidized to by-products such as 3 (3.4 dichlorophenyl)-1, methylurea) and 3.4 dichloroformamide.

Although alachlor is actively oxidized by ozone in drinking water systems, many by-products persist (Figure 10.29). For example, alachlor's various chemical degradation pathways include oxidation of the arylethyl group, N-dealkylation, cyclization and cleavage of the benzene ring. Intermediate reactions give rise to numerous by-products, including an aromatic nucleus in their structure. The ozonation of alachlor leads to the following by-products: N(2,6 diethylphenyl)-methyleneamide, 8-ethyl-3,4-dihydro-quinoline, 8-ethyl-quinoline, 1-chloroacetyl-2-hydro-3-ketone-7-acetyl-indole, 2-chloro-2′,6′-diacetyl-N-(methoxymethyl)-acetanilide, 2-chloro-2′-acetyl-6′-ethyl-N-(methoxymethyl)-acetanilide, two isomers of hydroxylated alachlor. As well as 1-chloroacetyl-2,3-dihydro-7-ethyl-indole,

2-chloro-2′,6′-ethyl-acetanilide, 2-chloro-2′,6′-acetyl-acetanilide, 2-chloro-2′-ethyl-6′-acetyl-N-(methoxymethyl)-acetanilide.

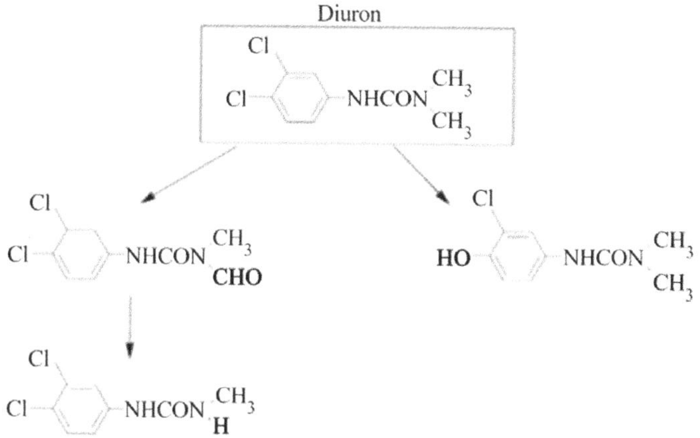

Figure 10.28. *Advanced ozonation by-products (O_3/H_2O_2) of diuron*

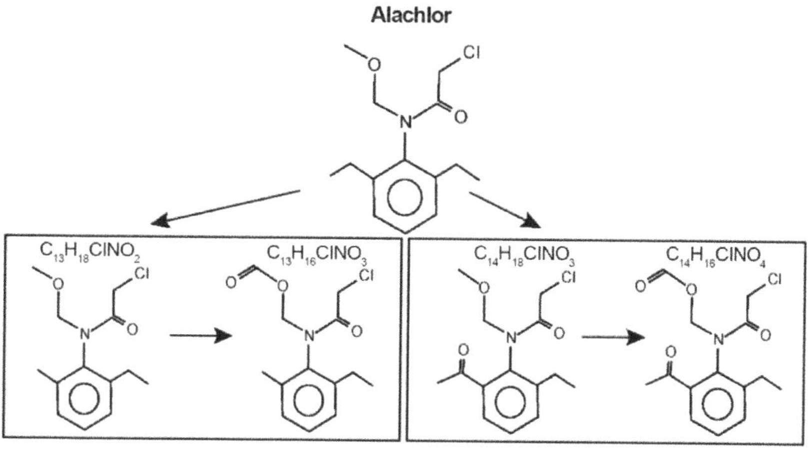

Figure 10.29. *Alachlor oxidation by-products*

The case is similar for metalochlor, whose main products obtained are summarized in Figure 10.30.

The main by-product generated by ozonation is oxo-metolachlor, resulting from the oxidation of the ethyl group's carbon. Tests carried out by Veolia have shown the chemical degradation of the parent molecule, which is replaced by oxidation by-products (Figure 10.30). Although some reactions carried out in the laboratory could lead to final products such as acetic, chloroacetic, formic, propionic and oxalic acids, they could only be obtained with high dosages of ozone and very long contact times.

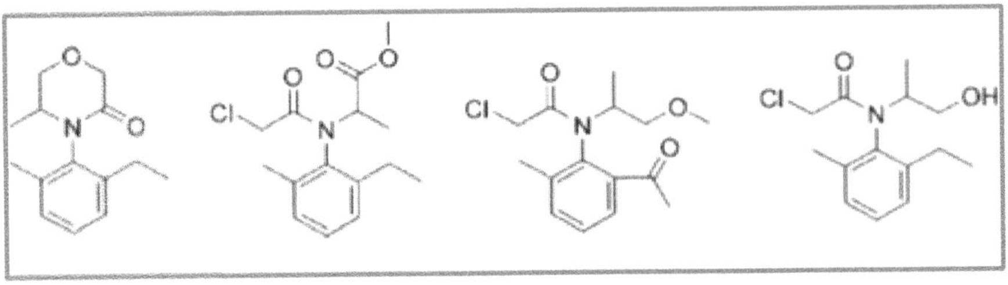

Figure 10.30. *Ozonation by-products of metolachlor*

Admittedly, the ozonation conditions in drinking water plants (6 min contact time and 2–3 $gO_3 \cdot m^{-3}$) do not make it possible to produce a sufficient amount of OH• radicals, nor to completely oxidize metolachlor, for instance (70% with 3 $g \cdot m^{-3}$), nor to degrade the intermediate products until the formation of acids. The potential for the formation of bromates (standard at 10 $\mu g \cdot L^{-1}$) should not be overlooked, since it can prove to be a limiting parameter for the use of ozone.

10.4.2.2. *The case of pharmaceutical residues*

The ozonation mechanisms of pharmaceutical residues are similar to those of pesticides, and depend on the molecule's structure, its concentration and the ozonation conditions. But ozone obviously has a high potential for the oxidation of certain pharmaceuticals and EDCs in drinking water plants.

An oxidation efficiency of 92% is obtained for diclofenac with an ozone dose of 0.21 $gO_3 \cdot g^{-1}$ DOC, and 99% with an ozone dose of 0.41 $gO_3 \cdot g^{-1}$ DOC. It is this last dose that should be injected to obtain a removal of carbamazepine and atenolol over 90% (Figure 10.31). The removal kinetics of medroxy-progesterone is a bit slower because only a 57% removal is obtained with a Ct of 1.03 $mg \cdot min \cdot L^{-1}$, whereas norethindrone and levonorgestrel reach a removal percentage over 90%. A Ct of 1.85 $mg \cdot min \cdot L^{-1}$ is required to achieve a caffeine removal over 80%.

Norethindrone and levonorgestrel contain ethynyl groups with a higher reactivity to ozone than substances such as caffeine or progesterone.

Substances reacting quickly to O_3 such as clindamycin, erythromycin, mefenamic acid, naproxen, sulfapyridine, trimethoprim and sotalol are oxidized over 90% with a dose of 0.21 $gO_3 \cdot g^{-1}$ DOC.

Trimethoprim is oxidized with an ozone dose of 0.3 $mg \cdot L^{-1}$, whereas a 0.41 $gO_3 \cdot g^{-1}$ DOC dose is required for the removal of gemfibrozil and estrogenic steroids (estrone, 17β-estradiol, estriol and 17α-ethynyl estradiol).

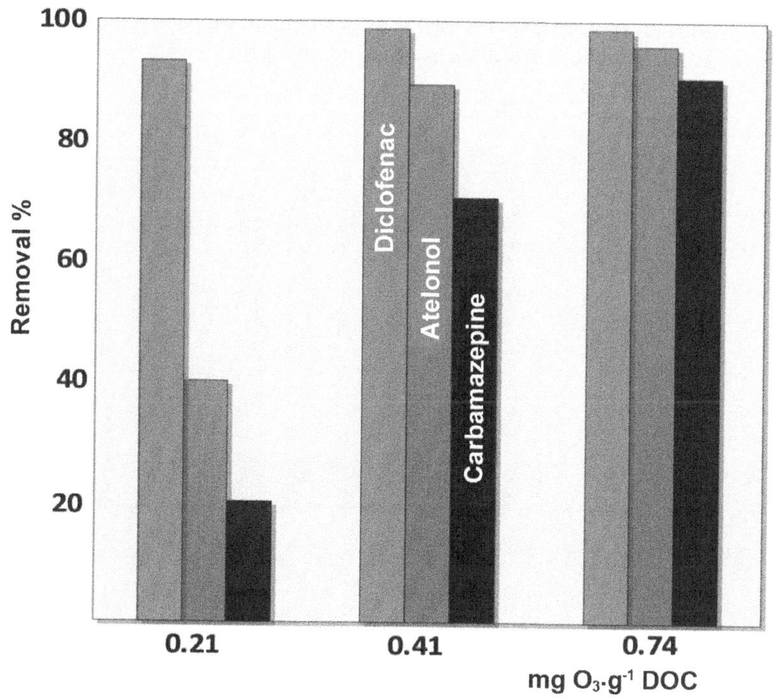

Figure 10.31. *Drugs percent removal with different ozone doses.*
For a color version of this figure, see www.iste.co.uk/gaid/watertreatment3.zip

Slow reacting substances such as ibuprofen, paracetamol, sulfamethazine, sulfadiazine and phenazone are oxidized at more than 90% for a dose of 0.81 $gO_3 \cdot g^{-1}$ DOC, whereas a dose of at least 0.60 $gO_3 \cdot g^{-1}$ DOC is required for beza-fibrate. It appears that ozone doses applied during drinking water treatment (2–3 mg $O_3 \cdot L^{-1}$) generally lead to reductions over 95% for many pharmaceutical residues, with 2–3 min contact times in ozonation tanks.

A ratio greater than 1.2 $gO_3 \cdot g^{-1}$ DOC is recommended for removing more than 80% of MPs in water containing 6.9 $mg \cdot L^{-1}$ of DOC. On the other hand, a ratio over 1.5 g $O_3 \cdot g^{-1}$ DOC leads to an efficiency over 90%, even for refractory MPs.

Figure 10.32 shows the importance of the O_3 dose in order to obtain high efficiency in water heavily loaded with NOM. This is due to a high O_3 demand related to the presence of organic matter. It could thus require high dosages, even if the MP concentration to be removed is low. When the water to be treated contains a high concentration of DOC, this leads to an immediate demand for O_3, which can result in a decrease in the performances of O_3 against MPs.

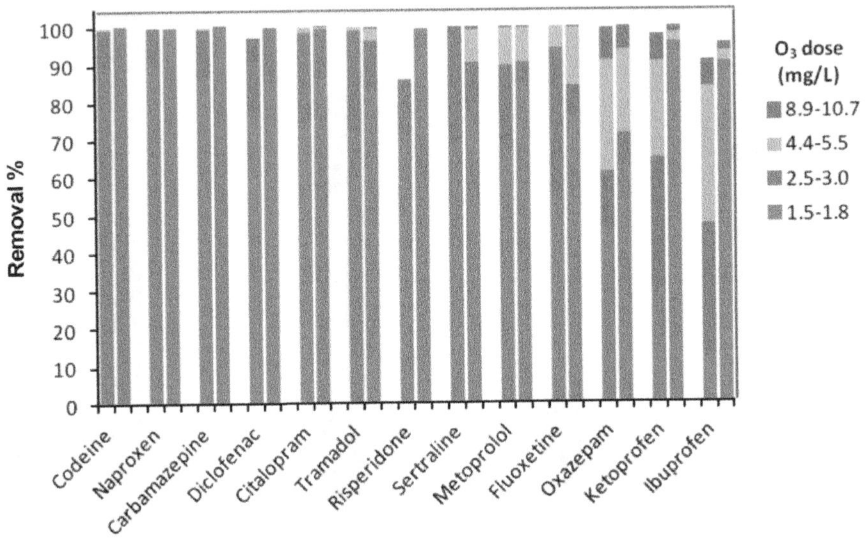

Figure 10.32. *Ozonation performances at pH 6.0 for different O_3 dosages (DOC 6.9 $mg \cdot L^{-1}$). For a color version of this figure, see www.iste.co.uk/gaid/watertreatment3.zip*

Figure 10.33 shows the logarithm for the ozone and hydrogen peroxide (H_2O_2) residual concentration from oxidation for some MPs (ethynylestradiol, sulfamethoxazole, carbamazepine and ibuprofen) at pH 8, depending on the oxidant's doses.

Both substances (ethynylestradiol and sulfamethoxazole) are characterized by active aromatic rings (respectively, Ar–OH and Ar–NH_2) whose ozonation kinetics are the fastest and requiring the lowest O_3 dosage in comparison with the other two drugs.

Linear ratios are observed between the OH• doses injected and MP reduction, the kinetics always being favorable to ethynylestradiol. It appears that O_3 oxidizes all the MPs having an electron-rich configuration, but whose reactivity also depends on the presence of natural organic molecules (DOC) which contribute to the demand for ozone.

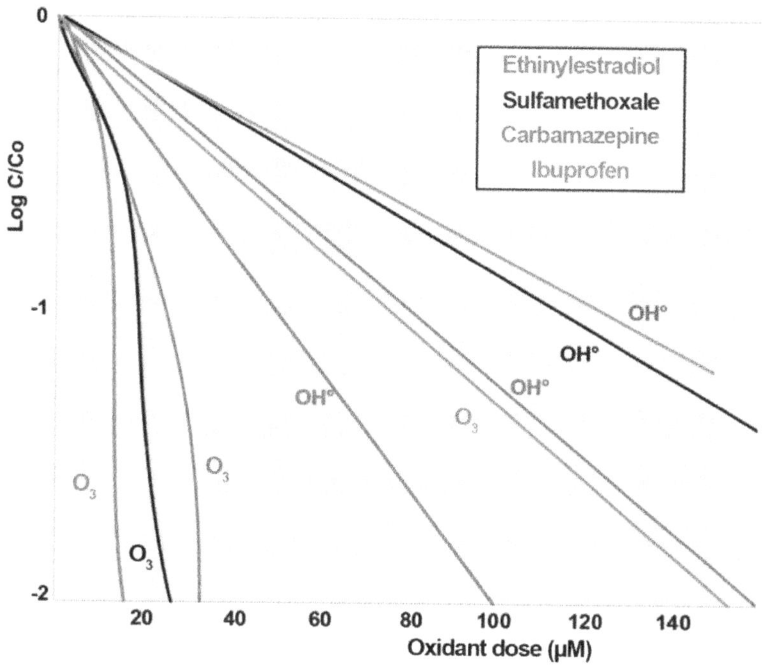

Figure 10.33. *Logarithm of the residual concentration for some drugs as a function of the oxidant's dose (O_3 or H_2O_2) at pH 8. For a color version of this figure, see www.iste.co.uk/gaid/watertreatment3.zip*

For many medicinal products with a good reaction with O_3, an additional reaction with the OH• radicals is not necessary. Because the reaction time of these substances with ozone is faster than the time required for the formation of free radicals.

The results obtained (Table 10.15) clearly illustrate the high efficiency of chemical oxidation with ozone under the following operating conditions: contact time 15 min, O_3 dose: 8–10 $g \cdot m^{-3}$. It appears that, apart from gabapentin (66.6%), all the other molecules are removed by ozone over 80%, with an efficiency reaching

99% for carbamazepine, diclofenac, hydrochlorothiazide and sulfamethoxazole (98%).

Substances	Ozonation ($6\ mg\cdot L^{-1}\ O_3$)		Substances	Ozonation ($6\ mg\cdot L^{-1}\ O_3$)	
	Cin mean value ($ng\cdot L^{-1}$)	**Average removal %**		**Cin mean value ($ng\cdot L^{-1}$)**	**Average removal %**
Atenolol	500	95.0	**Levetiracetam**	81	94.0
Enolic acid	380	90.0	**Methyl-benzotriazole**	3,267	96.0
Benzotriazole	3,633	77.6	**Metoprolol**	313	97.0
Carbamazepine	390	99.0	**Primidone**	103	90.2
Clarithromycin	153	97.0	**Sulfamethoxazole**	297	98.0
Diclofenac	813	99.0	**Trimethoprim**	29	82.5
Gabapentin	1,017	66.6	**Valsartan**	81	93.7
Hydrochlorothiazide	1,213	99.0	**Venlafaxine**	267	96.0

Table 10.15. *Ozonation performances on pharmaceutical residues*

The ozonation by-products from medicines have been identified on many occasions. The ozonation of diclofenac (Figure 10.35) leads to intermediate products such as 2-{2-[(2,6-dichlorophenyl)amino]-5-hydroxyphenyl}acetic acid (D1) and 2,6-dichloro-4-hydroxybenzenamine (D8).

Similarly, the ozonation of paracetamol and ethyl estradiol leads to numerous reaction by-products, as illustrated by Figures 10.36 and 10.37.

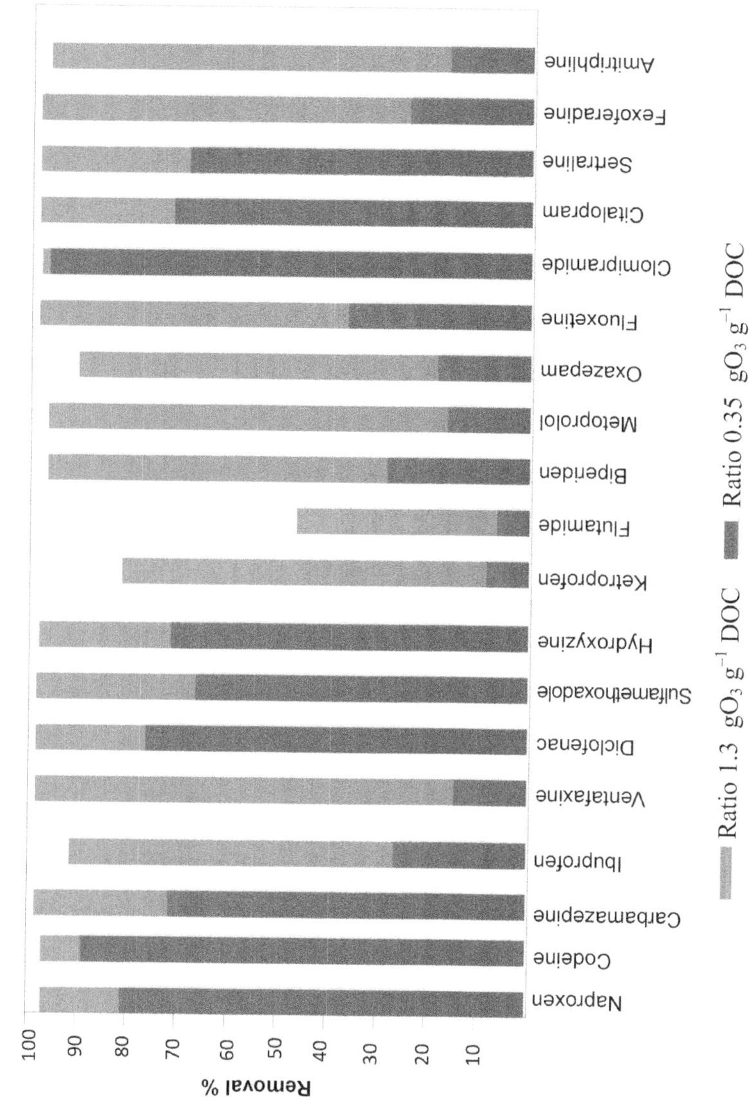

Figure 10.34. *Micropollutant ozonation for various $gO_3{\cdot}g^{-1}$ DOC ratios (DOC input 5 $mg{\cdot}L^{-1}$). For a color version of this figure, see www.iste.co.uk/gaid/watertreatment3.zip*

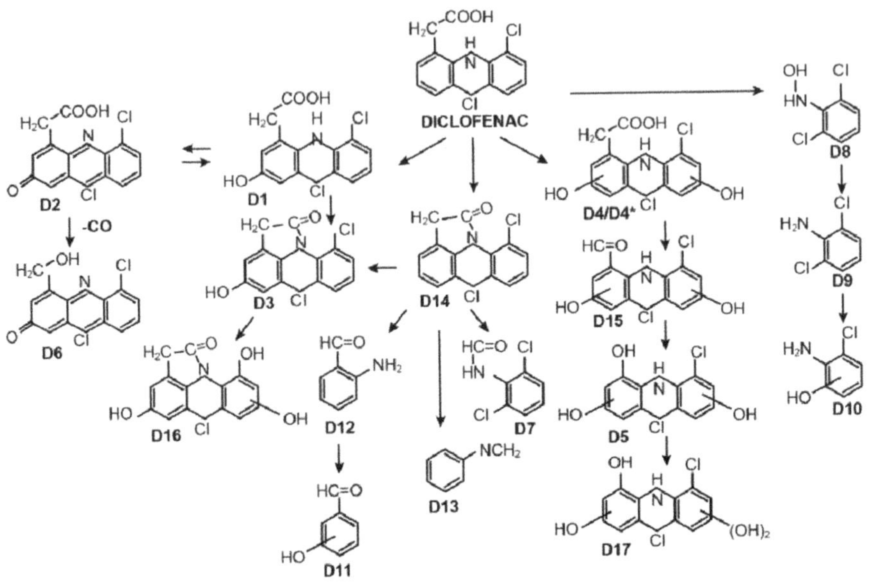

Figure 10.35. *Demonstration of diclofenac ozonation by-products*

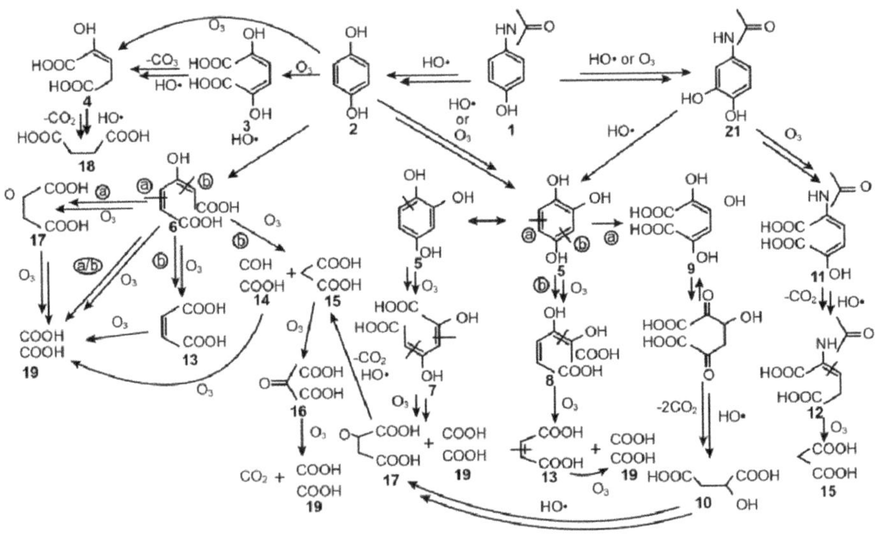

Figure 10.36. *Demonstration of paracetamol ozonation by-products*

Formulae and molecular weight	Developed formulae of ethyl estradiol ozonation by-products
$C_{18}H_{20}SO_7$ MW 379.08	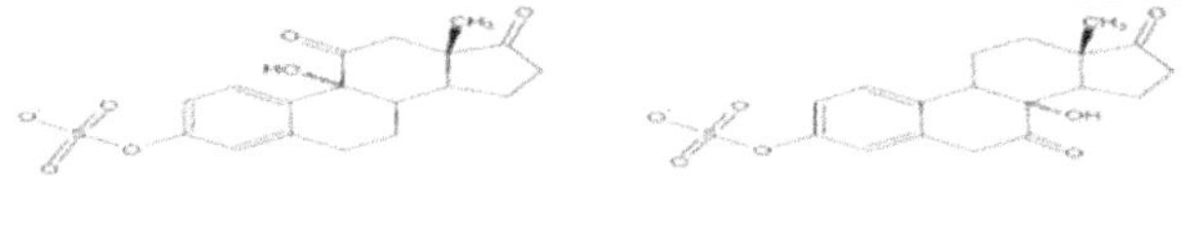
$C_{18}H_{22}SO_6$ MW 365.10	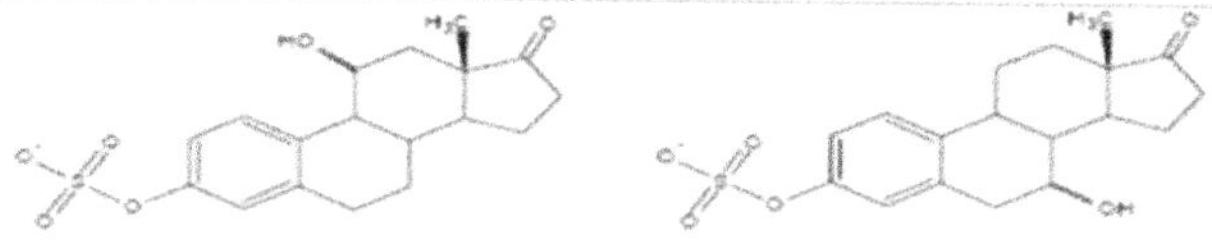
$C_{18}H_{22}SO_6$ MW 365.10	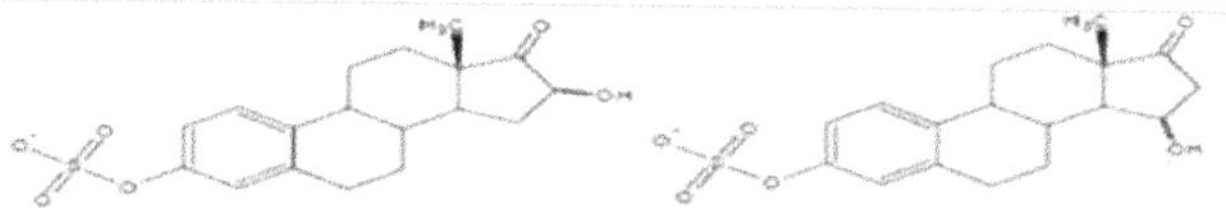
$C_{18}H_{20}SO_6$ MW 363.09	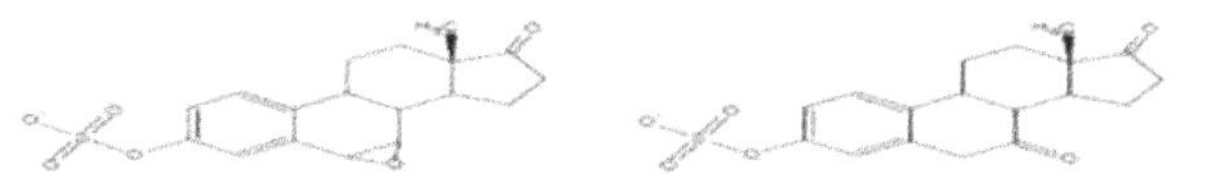

Figure 10.37. *Demonstration of ethyl estradiol ozonation by-products (source: Veolia)*

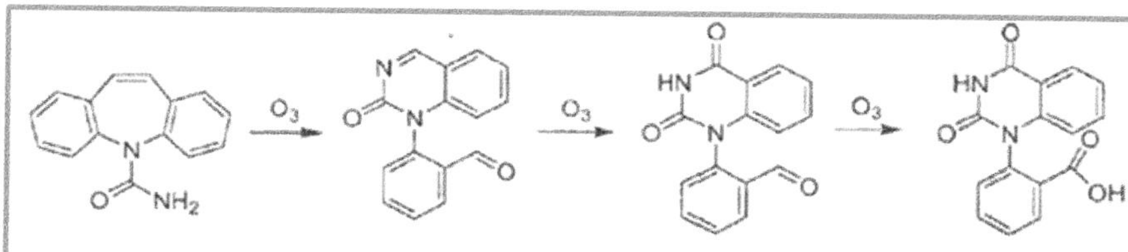

carbamazepine BQM BQD BaQD

Figure 10.38. *Demonstration of carbamazepine ozonation by-products*

The ozonation of carbamazepine (Figure 10.38) leads to by-products such as 1-(2-benzaldehyde)-4-hydro-(1H,3H)-quinazoline-2-one (BQM), 1-(2-benzaldehyde)-(1H,3H)-quinazoline-2,4-dione (BQD) and 1-(2-benzoic acid)-(1H,3H)-quinazoline-2,4-dione (BaQD).

10.4.3. *Ozone-activated carbon combination*

Ozone is known as an effective MP oxidation solution, but has the disadvantage of forming by-products which could potentially become EDCs. Like the parent substance, the O_3–PAC combination is a solution for preventing the formation of by-products, while improving MP removal performances. This solution comes down to blowing the ozone directly into the AC reactor itself, thus preventing the formation of by-products while ensuring a significant increase in removal efficiency. The underlying principle is to make O_3 act upon the adsorbent material, but not on the MP to be removed.

10.4.3.1. *Ozone-PAC*

Veolia has developed a process that cleverly combines a PAC reactor with ozone. The in situ injection of O_3 into the AC reactor leads to reduction rates over 90% for the vast majority of MPs. The introduction of ozone in direct contact with the AC leads to a modification of the AC's surface chemistry by creating surface functions, which contribute to an increase in the adsorption capacity of the adsorbent media. In addition to pore diffusion, the MPs become bound to the surface functions. This results in a better removal of MPs with an equal, or even lower, dose of AC. It is thus possible to significantly reduce PAC doses because of the implementation of ozone, while maintaining identical procedure performances. There are no organic by-products or bromates formed.

Substances	O_3–PAC (Aquasorb MP 23)			O_3–PAC (Norit SAE Super)		
	Inlet	Outlet	removal %	Inlet	Outlet	removal %
Diclofenac, $\mu g \cdot L^{-1}$	0.839	0.149	82	1.093	0.034	97
Carbamazepine, $\mu g \cdot L^{-1}$	0.982	0.024	98	1.288	0.014	99
Hydrochlorothiazide, $\mu g \cdot L^{-1}$	2.333	0.077	97	3.658	0.146	95
Metoprolol, $\mu g \cdot L^{-1}$	0.033	0.010	69	0.050	0.010	76
Sulfamethoxazole, $\mu g \cdot L^{-1}$	0.195	0.045	77	0.256	0.030	88
Atenolol, $\mu g \cdot L^{-1}$	0.150	0.013	90	0.142	0.018	82
Ethylesestradiol, $\mu g \cdot L^{-1}$	5.211	0.108	98	11.075	0.301	97

Table 10.16. *Performance of two PACs in combination with ozone*

The results obtained with 1–2 $mg \cdot L^{-1}$ dosage of O_3, PAC contact time of 10 min and PAC dosages of 5–10 $mg \cdot L^{-1}$ have shown a gain between 10 and 15% in the average removal performance of various MPs (Table 10.16 and Figure 10.36), with removal mean values between 85 and 92%, instead of the 70–80% observed with PAC.

It is notable that MP removal performances are quite close to those of pharmaceutical residues.

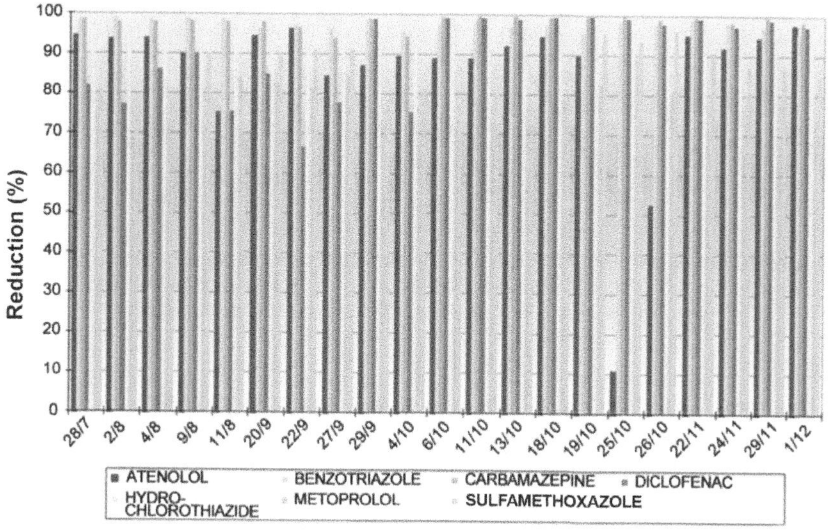

Figure 10.39. *Summary of Actiflo® Carb performances with the O_3–PAC reactor coupling. For a color version of this figure, see www.iste.co.uk/gaid/watertreatment3.zip*

As a result, the optimal removal of MPs is obtained because of this O_3–PAC reactor combination, without releasing any by-products (Figure 10.39) into the receiving environment. The results obtained on a second site confirm the effectiveness of this combination applied under the same operating conditions as described earlier. An efficiency over 80% was obtained for all the molecules analyzed, with values over 98% for carbamazepine, diclofenac, hydrochlorothiazide and sulfamethoxazole (Figure 10.40).

Compounds which are poorly or moderately removed by PAC alone (particularly due to the low doses of AC) such as gabapentin (10%), atenolol (40%) and valsartan (40%) reach an efficiency removal over 80% in the presence of O_3 in situ. This efficiency can be significantly improved by intervening on the dosages of the two reagents while controlling the ozone dosage in order to prevent the formation of by-products.

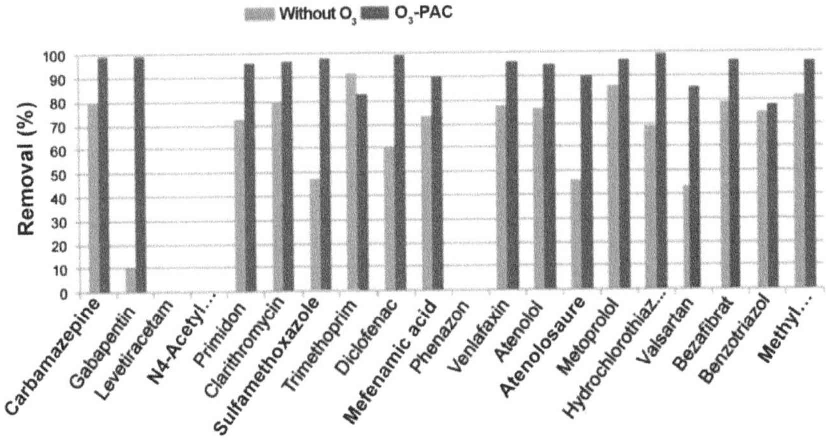

Figure 10.40. *Removal of micropollutants with the combination of in situ O_3–PAC (O_3: 2 g·m^{-3}, PAC (Aquasorb MP23): 10 g·m^{-3}). For a color version of this figure, see www.iste.co.uk/gaid/watertreatment3.zip*

The mass reduction of all the molecules analyzed increases to over 91% (compared to 75% without ozone). The average reduction of all the molecules analyzed is 89% (compared to 77% without ozone).

Even though the 0.5–0.6 mg O_3·g^{-1} DOC ratios applied comply with those previously indicated for oxidizing MPs by O_3, the mechanisms of O_3–PAC combined are different because O_3 is no longer injected to play an ozonation role, but rather to modify the AC's surface.

Ozone directly injected into the PAC contact tank whose concentration is between 1.5 and 2 g·L^{-1} triggers a direct reaction between the oxidant (O_3) and the reducer (PAC) to the detriment of organic matter and MPs. This is the reason why the ozonation reaction does not modify the MPs, since the O_3–PAC reaction kinetics are much faster. In addition, the residual ozone was always nonexistent in the interstitial water. As a complement to MP analyses, SEC-UVD chromatograms (with and without ozone) have shown that there is an actual removal of humic substances due to the addition of AC (Figure 10.41). The O_3 injection into the PAC reaction tank results in a systematic reduction of the humic substance peaks, compared to the performances obtained with Actiflo® Carb (without O_3). The sample treated in this manner contains a lower concentration of organic carbon. This can be explained by the better adsorption efficiency of PAC after being oxidized by the ozone in situ, without generating the appearance of new peaks, something which could have been assimilated to oxidation by-products.

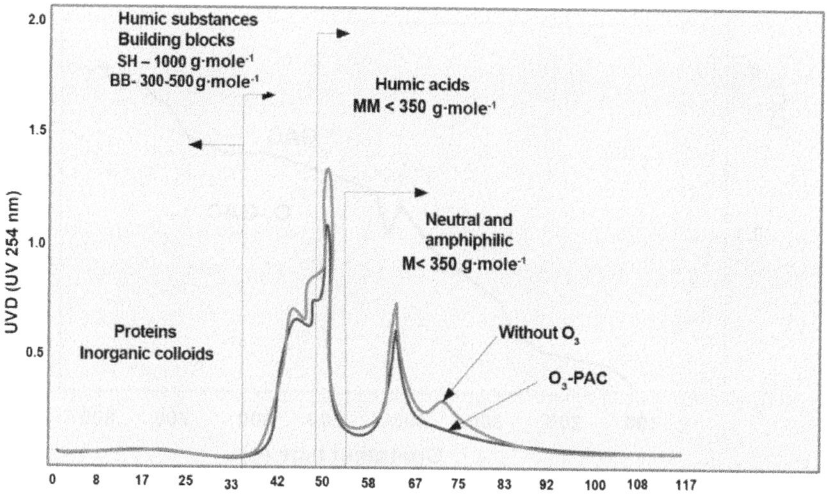

Figure 10.41. *LC UVD of samples treated with Actiflo® Carb with and without activated carbon ozonation. For a color version of this figure, see www.iste.co.uk/gaid/watertreatment3.zip*

10.4.3.2. O_3–granular activated carbon

Ozone doses of 0.8–1.2 $gO_3 \cdot g^{-1}$ DOC and an HRT of 12–20 min are sufficient to remove a wide range of MPs. Most MPs are transformed into more hydrophilic substances with similar or lower molecular weights such as aldehyde, ketone and carboxylic acid-type compounds, for example.

The conventional goal of an ozone-GAC filter combination is to initiate and promote the biological degradation of organic substances generated during the ozonation stage.

Figure 10.42 illustrates the lifetime evolution of GAC supplied with surface water (DOC: 2.4–3.0 $mg \cdot L^{-1}$), atrazine (0.5–0.7 $\mu g \cdot L^{-1}$), and operating with a contact time of 10 min.

The same water is also treated with an O_3–GAC combination and an ozonation contact time of 6 min. An atrazine leakage over 0.1 $\mu g \cdot L^{-1}$ is found after 400 days of operation with GAC. It is highly likely for the organic matter present to have blocked some adsorption sites, thus preventing atrazine from being retained any longer.

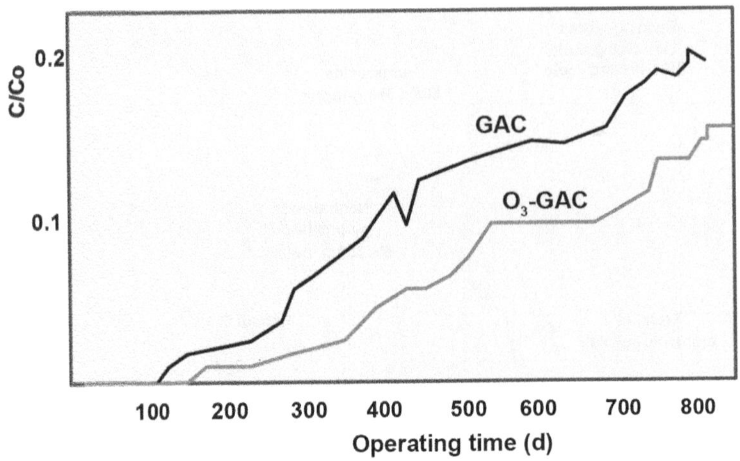

Figure 10.42. *Evolution of atrazine removal with ozonation alone and with the combination of O_3–GAC. For a color version of this figure, see www.iste.co.uk/gaid/watertreatment3.zip*

With the O_3–GAC combination, the atrazine leak is only observed after 700 days of operation. The oxidation of DOC by O_3 and its biodegradation onto GAC has undoubtedly released adsorption pores, leading to a greater retention of atrazine. Furthermore, atrazine oxidation by-products such as DEA and DIA are more biodegradable and less adsorbable than atrazine itself. It is quite difficult to predict the escape time of these two metabolites. Especially since depending on water temperature, the biodegradation function operates more effectively during hot periods than in colder periods.

The results obtained with a combination of O_3–GAC and GAC essentially operating in a biological mode (Figure 10.43) show that the formation of biodegradable by-products requires an O_3/DOC ratio greater than 1 for a large number of compounds.

GAC biological filtration based on a contact time of 15 min makes it possible to obtain an efficiency of 80% both on the mother molecule and on its by-products. Biological filtration onto GAC removes a large number of pharmaceutical residues over 80% provided that their concentration is at least 10 times their LOQ (limit of quality) before GAC filtration.

On the contrary, gabapentin is only removed at 53%, justified by its highly hydrophilic nature (log Kow: –1.37). GAC biological filtration can reduce between 25 and 35% of DOC for a temperature above 15°C.

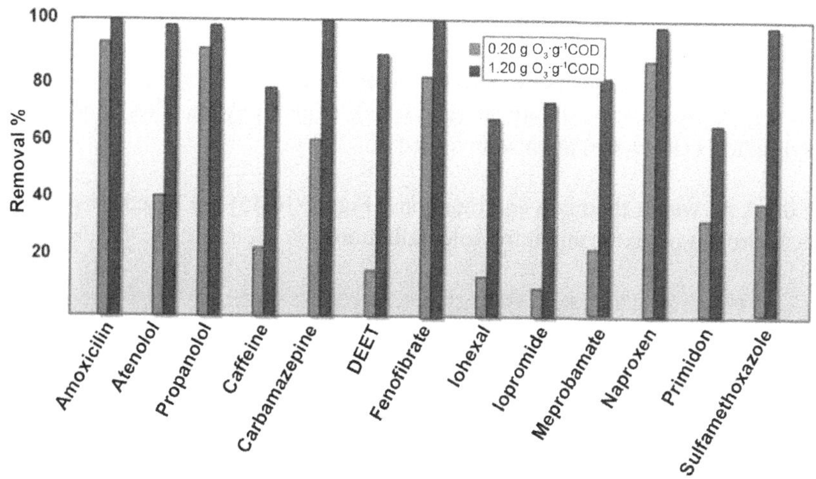

Figure 10.43. *Removal of some drugs with O_3–GAC depending on two O_3/DOC ratios. For a color version of this figure, see www.iste.co.uk/gaid/watertreatment3.zip*

10.4.4. *Advanced chemical oxidation*

When ozonation kinetics are slow, the way to accelerate ozone oxidation reactions is to associate them with hydroxyl radicals (OH•).

Advanced oxidation processes, known as AOPs, constitute an effective solution for degrading refractory compounds.

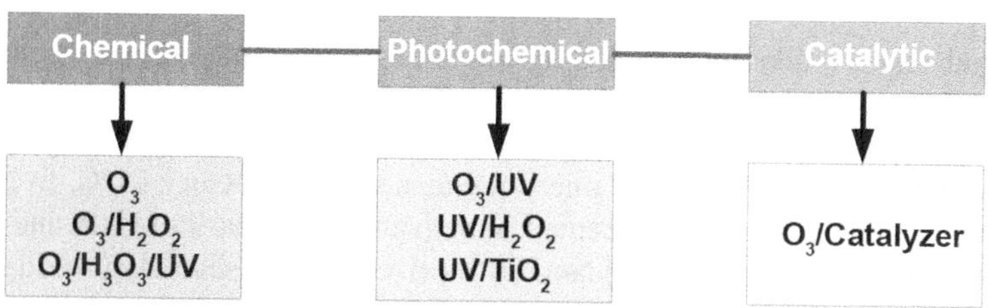

Figure 10.44. *Some advanced oxidation solutions. For a color version of this figure, see www.iste.co.uk/gaid/watertreatment3.zip*

AOPs are characterized by a variety of free radical reactions that involve combinations of chemical agents such as ozone (O_3), hydrogen peroxide (H_2O_2), transition metals and metal oxides, as well as auxiliary energy sources such as the visible ultraviolet (UV) and ultrasound. AOP examples include O_3–H_2O_2, O_3–UV, O_3–H_2O_2–UV, H_2O_2–UV, Fenton (Fe^{2+}–H_2O_2), or heterogeneous photo-oxidation using titanium dioxide (TiO_2). Hydroxyl radicals (OH•) are the main oxidant in AOPs, whereas other radical and active oxygen species such as superoxide radical anions ($O_2•^-$), hydroperoxyl radical (H–O–O•), triplet oxygen ($^3O_2$) and organic peroxyl radicals (R–O–O•) are also involved.

In drinking water, the main combinations (Figure 10.45) are based on the use of ozone, hydrogen peroxide and ultraviolet radiation.

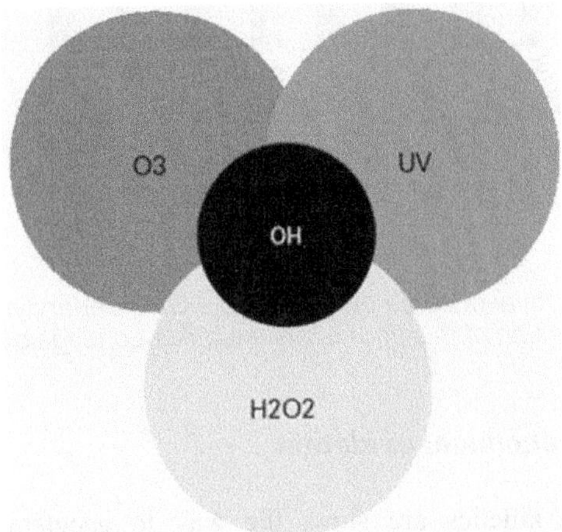

Figure 10.45. *Main combinations used for drinking water. For a color version of this figure, see www.iste.co.uk/gaid/watertreatment3.zip*

10.4.4.1. *Ozone-hydrogen peroxide coupling (O_3–H_2O_2)*

The advanced oxidation obtained by combining ozone and hydrogen peroxide is often used in an industrial environment to oxidize color or organic substances, which are difficult to biodegrade. This solution is proposed for the production of drinking water in order to oxidize micro-pollutants and refractory DOC. In general, this chemical activation is implemented to activate the primary oxidant and induce the formation of a secondary oxidant. In most cases, the secondary oxidant is a hydroxyl radical (OH•). The oxidation potentials of molecular ozone and hydroxyl radicals are 2.07 and 2.80 V, respectively, indicating that they are very strong oxidants.

The reaction is the following:

$$H_2O_2 + 2O_3 \rightarrow HO\bullet + HO\bullet + 3O_2$$

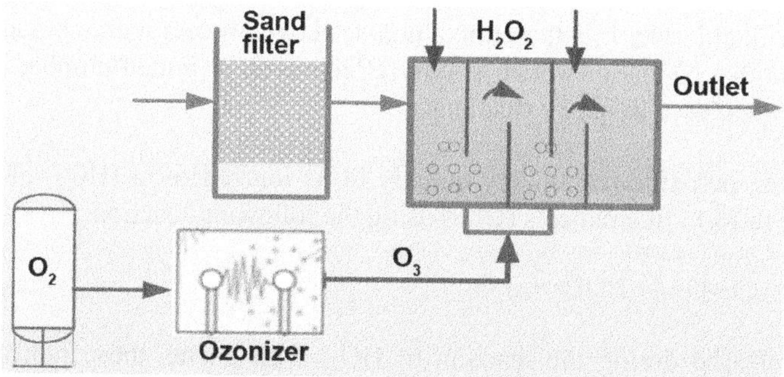

Figure 10.46. *Advanced chemical oxidation with O_3–H_2O_2. For a color version of this figure, see www.iste.co.uk/gaid/watertreatment3.zip*

Hydrogen peroxide is injected at the level of the ozonation chambers (Figure 10.46). In France, due to the potentially formed by-products, AOPs (except ozonation) are prohibited for drinking water. However, they are authorized in other northern European countries (e.g. the Netherlands) and the United States. The addition of H_2O_2 in the presence of ozone reinforces the oxidizing power of ozone in relation to organic matter (TOC, MPs). Hydrogen peroxide (H_2O_2) causes the extremely rapid (instantaneous) decomposition of ozone, giving rise to highly oxidizing and not very selective radical species. Among these, there is the majority formation of OH• hydroxyl radicals whose redox potential (2.8 V) is clearly higher than that of ozone (2.07 V).

Oxidation reactions involving hydroxyl radicals in the presence of organic substrates (in aqueous medium) are mainly electrophilic addition reactions and hydrogen abstraction reactions. The equations proposed to describe these addition and abstraction reactions are the following:

– first case, addition of the OH• radical to the organic compound R:

$$R + OH\bullet \rightarrow (R\text{- }OH\bullet) \rightarrow \text{hydroxylated products}$$

– second case, removal of a hydrogen atom:

$$RH_2 + OH\bullet \rightarrow (R\text{- }H) + H_2O \rightarrow \text{oxidized products}$$

In both cases, organic radicals are formed, which can subsequently react with other radicals (termination reaction) or even react with another molecular oxidant in solution (propagation reaction). The initial activation of the molecular oxidant leads to the generation of hydroxyl radicals (initiation reaction).

The combined oxidation of O_3-H_2O_2 degrades all types of MPs, except those saturated with chlorine. It is therefore a non-selective process with a broad spectrum of action, unlike ozonation which is only effective on a limited number of organic molecules when used in pure oxidation.

Hydrogen peroxide reacts very quickly in its ionized form (HO_2^- pKa = 11.6) with ozone to form free radicals (OH•) using the following reaction:

$$2\ O_3 + H_2O_2 \rightarrow 2\ OH\bullet + 3\ O_2$$

Once released during the reaction of HO_2^- with ozone, these radicals initiate other radical mechanisms involving the decomposition of hydrogen peroxide.

The reaction rate of hydroxyl radicals on organic molecules may be 1,000 to 1,000,000 times greater than the one obtained with molecular ozone. Therefore, the contact times required for the combined oxidation of O_3-H_2O_2 are shorter.

The efficiency of the process depends on the ozonation rate, the contact time, the injected H_2O_2–injected O_3 ratio, alkalinity, pH, TOC and temperature.

The minimum contact time to be applied in the O_3-H_2O_2 oxidation reactor is 6 min with a conventional bubble ozonation column. The standard ratio of reagents to be injected oscillates between 0.35 and 0.45 $g{\cdot}g^{-1}$ of injected H_2O_2/injected gO_3.

Site	T	TOC	pH	Alk	O_3 rate	g H_2O_2 /g O_3	Contact time	residual H_2O_2	Atrazine Co reduction	
	(°C)	($mg{\cdot}L^{-1}$)		($mg{\cdot}L^{-1}$ $CaCO_3$)	($mg{\cdot}L^{-1}$)	Optimal	(mn)	($mg{\cdot}L^{-1}$)	($\mu g{\cdot}L^{-1}$)	(%)
Neuilly s/Marne	18–22	1.5	7.8	200	2.5, 3, 4	0.45	8	0.05–0.2	500–2,000	80 90 92

Table 10.17. *Operating conditions and oxidation of atrazine with an H_2O_2/O_3 ratio*

When the H_2O_2/O_3 ratio is adjusted to its optimum between 0.35 and 0.45 $g{\cdot}g^{-1}$, and the minimum contact time of 6 min is respected, the residual ozone in the water leaving the oxidation reactor is very low or nonexistent (0.05–0.2 $mg{\cdot}L^{-1}$ max).

It is preferable to remove this residual H_2O_2 before the final disinfection with chlorine or ClO_2 because H_2O_2 consumes Cl_2 and ClO_2, resulting in an unnecessary overconsumption of these two disinfectants.

Although the consumption ratio is theoretically equal to 2.09 mg of Cl_2/mgH_2O_2, in practice, ratios of 3.00 mg Cl_2/mgH_2O_2 have been reached due to a chlorine demand by organic matter or other mineral elements. Other solutions are implemented in order to reduce the H_2O_2 residual, implementing reducing agents such as SO_2, $NaHSO_3$ or Na_2SO_3.

Consumption of reducing products:

– 1.33 g of sulfur dioxide (sulfur gas or SO_2) per g H_2O_2;

– 2.16 g of sodium bisulfite ($NaHSO_3$) per g H_2O_2;

– 2.62 g of sodium sulfite (Na_2SO_3) per g H_2O_2;

Compounds like trichlorethylene and perchlorethylene are well reduced by this combination via OH• free radicals.

10.4.4.2. *O_3/H_2O_2-GAC filter coupling*

The carbon filter has several roles:

– As a reducer, it removes any possible H_2O_2 residual after oxidation, whereas the assimilable organic carbon formed during the oxidation stage is removed biologically. It also removes pesticide residuals, as well as oxidation by-products (insofar as they are adsorbable and/or biodegradable).

– With an O_3–H_2O_2 treatment upstream, the carbon filter's lifespan is theoretically multiplied by a factor of 4–5. The implementation of this solution includes the ozonation stage (3–4 compartments) with a retention time of 6–10 min. The H_2O_2/O_3 ratio is adjusted to its optimum between 0.35 and 0.45 $g{\cdot}g^{-1}$, with a dosage of O_3 ranging from 2 to 2.5 $g{\cdot}m^{-3}$.

The high reactivity toward antibiotics is related to the presence of one or more functional groups in these molecules, such as amino nitrogen, sulfur, the activated aromatic ring and the carbon–carbon double bond. Acetylated sulfonamide metabolites, such as N4-acetylsulfamethoxazole, are more resistant to ozonation than parent compounds because acetylation blocks the amino aniline group, which is one of the targets of molecular ozone reactions. In contrast, the enhanced oxidation of antibiotics is more easily achieved by means of O_3–H_2O_2. However, the identity of the degradation by-products resulting from this oxidation is not completely known for a large number of molecules.

Reactivity for anticonvulsant and anti-anxiety groups show different reactivity between ozonation and the O_3–H_2O_2 combination. This is mainly due to their structural diversity and lack of common functional groups. Thus, carbamazepine is easily degraded by molecular ozone or hydroxyl radicals. On the other hand, molecules like clofibric acid, diazepam, ibuprofen, diazepam and primidone are relatively resistant to ozonation, though more easily oxidized with O_3–H_2O_2. The (O_3–H_2O_2) combination is generally more efficient than ozonation alone, due to the increased generation of hydroxyl radicals and the photon-triggered carbon-halogen bond cleavage. It is recommended for the treatment of ozone-refractory molecules.

10.4.4.3. *O_3–UV and O_3-H_2O_2–UV combinations*

The reactions between ozone and UV are summarized as follows:

$$O_3 + UV \rightarrow O_2 + O^1(D)$$

$$O^1(D) + H_2O \rightarrow HO\bullet + HO\bullet \rightarrow H_2O$$

where:

– $O^1(D)$: excited O_2 atoms. The (^1D) symbol is a spectroscopic notation used for specifying the atomic and molecular configuration (also known as singlet oxygen);

– HO•: hydroxyl radical.

Since the photolysis of ozone in the presence of water tends to form free radicals, the use of O_3 is squandered for this application.

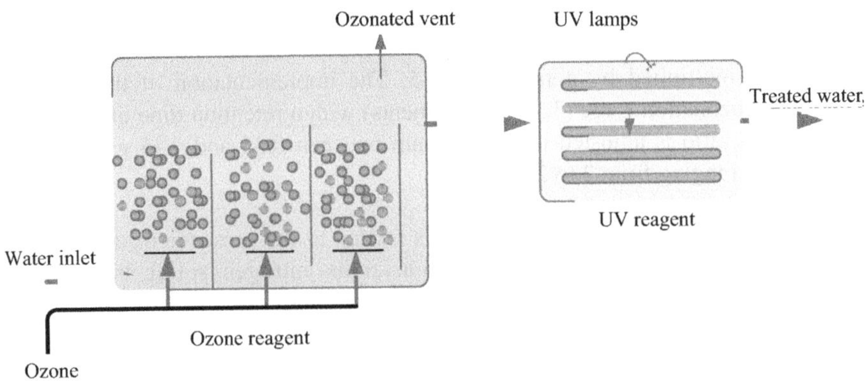

Figure 10.47. *Schematic representation of an AOP with O_3–UV. For a color version of this figure, see www.iste.co.uk/gaid/watertreatment3.zip*

The O_3–H_2O_2 combination can effectively degrade a large number of MPs. A low dose of ozone enables a first oxidation with a highly limited formation of bromates, whereas a complementary oxidation is obtained by UV–H_2O_2, because there is a residual of H_2O_2 at the ozone stage outlet which acts together with the UV.

The advantage of such a combination is that part of the ozone reacts with the DOC present in the water. This increases UV transmission and improves the UV–H_2O_2 process in terms of energy consumption and the formation of by-products.

The reaction rate constants of the hydroxyl radicals produced by O_3–H_2O_2 and UV–H_2O_2 are between 10^6 and 10^9 $m^{-1} \cdot s^{-1}$. For example, the reaction rate constant of atrazine is 2.7×10^9 $m^{-1} \cdot s^{-1}$.

10.4.4.4. *UV–H_2O_2 coupling*

Degradation by UV photolysis depends on the absorption of UV photons and the quantum yield. UV photolysis is a selective process. Besides the radical oxidation mechanism provided by H_2O_2, UV radiation itself can also degrade compounds by means of direct photolysis. The effectiveness of this process depends on the spectrum of the UV lamp used and the absorption spectra of the compounds present in the water to be treated.

Free radicals are formed when H_2O_2 is prompted to react with the UVs (200–280 nm):

$$H_2O_3 + UV \rightarrow HO\bullet + HO\bullet$$

The operating parameters are the UV dose, H_2O_2 dose and the pH level of water.

The electrical energy required for oxidation is expressed in EE/O units, corresponding to the electrical energy per volume and per log reduction. The ratio is as follows:

$$EE/O = \frac{EE}{V[\log\left(\frac{Cin}{Cout}\right)]}$$

where:

– EE: electrical energy in kWh;

– V: water volume in the reactor in m^3;

– Cin and Cout: compounds initial and final concentrations in $ng \cdot L^{-1}$.

Note that 1 log reduction (90%) of NDMA requires between 21 and 265 kWh·10^{-3} m^{-3} with 5–6 mg·L^{-1} H_2O_2.

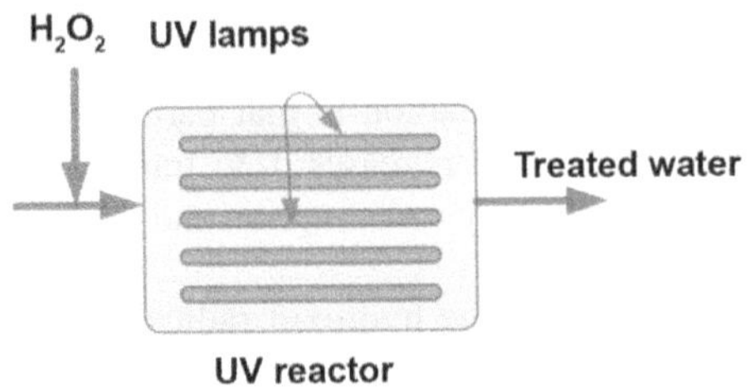

Figure 10.48. *Schematic representation of an AOP with H_2O_2–UV. For a color version of this figure, see www.iste.co.uk/gaid/watertreatment3.zip*

Pesticides such as metaldehyde do not absorb UV light and are not degraded, whereas atrazine shows high conversion, with a high molar absorption coefficient (3,683 m^{-1}·cm^{-1}) and high quantum yield.

Hormones are adequately degraded with a conventional UV dose (300 mJ·cm^{-2}) in the presence of H_2O_2. The most commonly used lamps are medium-pressure (MP) and Low-Pressure (LP) mercury vapor lamps. MP lamps emit a broad spectrum of light, between 200 and 800 nm, whose range between 235 and 300 nm is relevant for the UV–H_2O_2 coupling. This wavelength range matches the absorbance spectrum of many compounds and results in a high photolytic capacity, despite an overall lower efficiency and a higher potential for the formation of by-products.

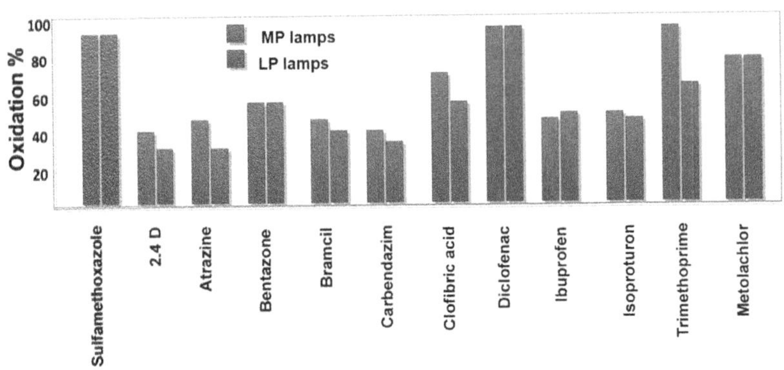

Figure 10.49. *Pesticides and drugs oxidation rate with UV–H_2O_2. For a color version of this figure, see www.iste.co.uk/gaid/watertreatment3.zip*

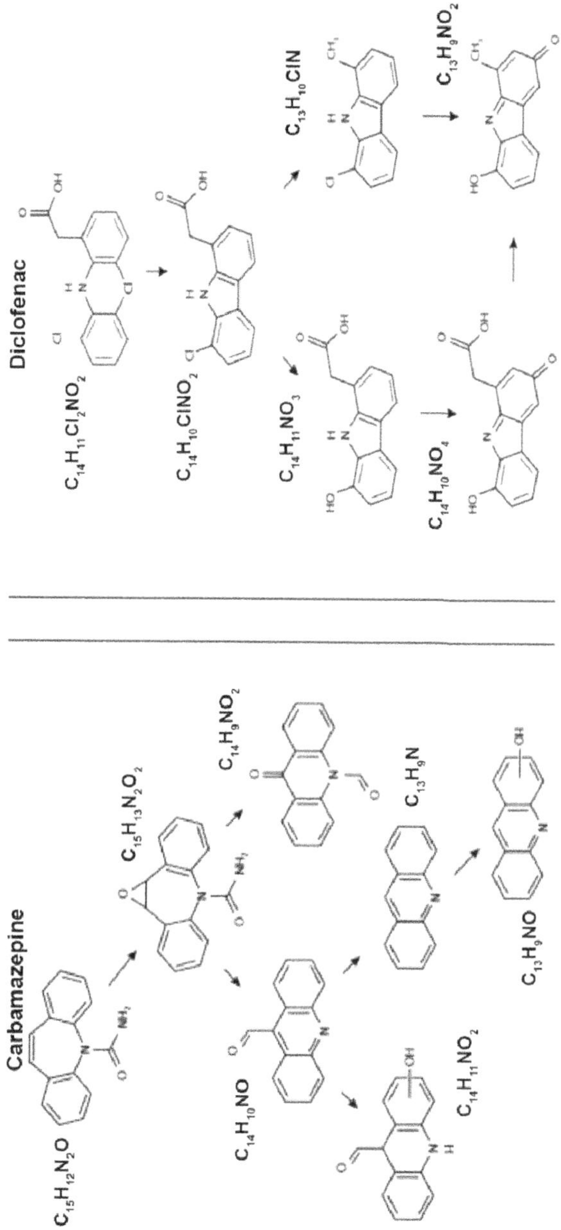

Figure 10.50. *By-products identified with UV–H_2O_2 for carbamazepine and diclofenac*

LP lamps emit UV light at a single wavelength (253.7 nm). Therefore, they can only photolyze compounds with an absorbance around 254 nm.

The oxidation results with an MP lamp (2,200 W) and a LP lamp (300 W) and a H_2O_2 dosage of 5 mg·L^{-1} are summarized in Figure 10.49 for pesticides and drugs. Water transmittance is 80%.

For some compounds, the UV–H_2O_2 treatment involving the use of a 300 mJ·cm^{-2} dose associated with a H_2O_2 5–10 mg·L^{-1} injection shows a difference between the efficiency of LP and MP lamps (such as 2.4 D and trimethoprim), whereas for other compounds (sulfamethoxazole, carbendazim, diclofenac, metolachlor, isoproturon), no difference is observed. As shown in Figure 10.50 for carbamazepine and diclofenac, the by-products from oxidation are varied.

10.4.5. *Nanofiltration and reverse osmosis membranes*

Due to their very low cut-off concentration, nanofiltration (NF), low-pressure reverse osmosis (LPRO) and reverse osmosis (RO) membranes are a well-known barrier for various ions and molecules. The efficiency of this technology has been widely proven in relation to natural organic substances, MPs organics and minerals. The information provided by the manufacturers is limited to the cut-off concentration expressed in molecular weight cut-off (MWCO) or by Daltons for a compound removed at 90%. Larger molecules are expected to have a retention over 90%, whereas small molecules should have a retention below 90%, depending on the membrane cutoff.

The separation mechanism and its efficiency depend on the type of membrane (cut-off concentration, morphology, surface electric charge, structure, etc.), the characteristics of ions and organic substances (solubility and diffusivity within the membrane) and isobaric equivalent temperature. Sorption–diffusion mechanisms contribute to retention by membranes and are explained by the hydrophobic interactions or hydrogen bonds between organic molecules and the membrane's surface.

NF enables the retention of high molecular weight molecules, greater than or equal to 250 Da, whereas LPRO favors the retention of molecules with a molecular weight greater than or equal to 150 Da. Depending on the membrane's cut-off concentration, performances may be reduced if the membrane is rather "open", such as the NF 270 membrane, for example, in comparison with the NF 90 membrane, which is tighter and enables higher reductions. But efficiency also depends on the electrostatic repulsion between the membrane and the MP.

A two- or three-stage membrane configuration shows that the second stage can remove compounds, which were not completely removed after passing through the first stage if, however, the conditions for its removal are possible (molecule's configuration, geometric structure, etc).

In all cases, it is necessary to identify the right membrane to be used depending on the family of MPs to be removed. Nevertheless, for a successful implementation of this technology, it is necessary to obtain information on the retention performances of a given membrane. This is traditionally done by pilot testing, either on-site or at the laboratory.

Veolia has tested several high-pressure membranes such as nanofiltration (NF), Low-Pressure Reverse Osmosis (LPRO) and reverse osmosis (RO) and these tests have made it possible to identify the most suitable membrane depending on the family of MPs to be removed.

10.4.5.1. *Correlation between pore diameter and rejection*

The parameters used for estimating or calculating membrane performances in relation to molecules are:

– molecular length (Lm) and molecular width (lm), used for calculating the steric partition factor;

– the n-octanol/water partition coefficient (log Kow) is used for estimating the rejection of hydrophobic compounds;

– the Stokes radius and the diffusion coefficient are estimated from the molecule's dimensions.

Two models are considered: the non-spherical model and the spherical model, leading to the calculation of the pore radius. The main property of a semi-permeable membrane is its molecular sieving function, for which the pore radius is an important parameter. Its evaluation is based on the rejection of hydrophilic organic compounds since these compounds have little or no interaction with the pore wall. The Stokes radius serves to represent the compound's effective size, assuming it has a spherical shape, even if – unlike inorganic ions – almost all of the uncharged organic compounds do not have that shape. With nanofiltration (NF), LPRO and reverse osmosis (RO) membranes, the rejection of hydrophilic organic compounds is correlated to the molecule's width. Rejection increases uniformly as the molecule's width increases too. Since this information is not always available, we refer to the molecular mass in rough estimate.

The molecule's shape can influence transport processes including diffusion and convection mechanisms. For hydrophilic organic compounds, it can be represented by a rectangular parallelepiped, which is calculated from the molecule's length and width.

For uncharged organic compounds, the molecular shape can be spherical or parallelepipedic.

In all cases, the membrane pore is considered cylindrical. In the pore transport model, it is assumed that the membrane's pores are uniform and cylindrical, and molecules are spherical. Thus, the ratio λ between the molecule radius (rm) and the pore radius (rp) is:

$$\lambda = \frac{rm}{rp}$$

When the molecule radius is larger than the pore radius, the molecule is rejected by the membrane. When the molecule radius is smaller than the pore radius, the molecule can enter the pore and a steric partition coefficient, Φ, is determined. For spherical molecules, Φ is calculated directly from λ, according to the following equation:

$$\Phi = (1 - \lambda)^2$$

For uncharged organic compounds which are not spherical, the value of parameter Φ is affected by the molecule's shape. For uncharged organic compounds with a parallelepiped shape, the parameter Φ is expressed as a function of molecular length (Lm) and molecular width (lm). The length of the molecule's projection on the pore surface is the effective molecular length (Lp) with an angle α, which is the angle between the molecule's axis L and the pore surface.

Lp is expressed by the equation:

$$Lp = Lm \cos\alpha + 2\ MW \sin\alpha$$

where MW is the molecular weight.

Length (Lm) is defined as the longest axis through the molecular structure, and width (lm) as the mean lateral length of a "rectangle" enveloping the molecule in a perpendicular direction to the length axis.

Figure 10.51 represents a theoretical curve between the rejection rate and the molecule radius to pore radius ratio. Actually, as pore distribution is not homogeneous in a membrane, differences are observed in the rejection rate. In addition, interactions between substance and membrane surface are also involved in this observed difference.

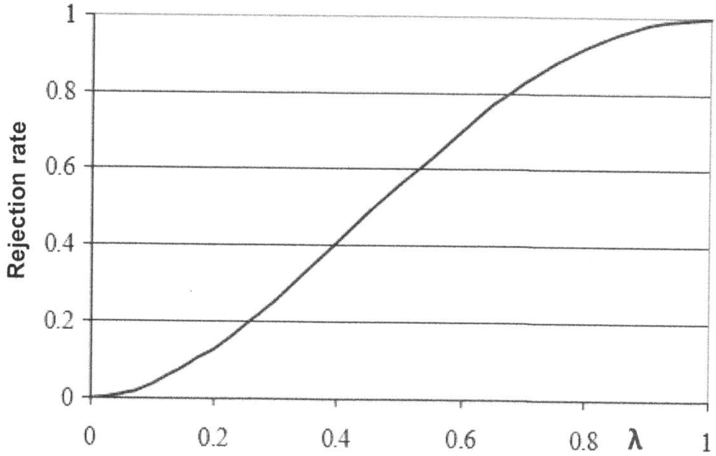

Figure 10.51. *Evolution of the rejection rate as a function of the ratio molecule radius/pore radius*

10.4.5.2. *The case of pesticides*

The vast majority of pesticides and their metabolites have molecular weights greater than 200 Da. Therefore, the use of membranes having a cut-off concentration greater than 200 Da is considered sufficient to obtain a good removal of these molecules. Membrane effectiveness (nanofiltration, LPRO or RO) against pesticides is based on key parameters such as molecular weight, molecular dimensions (length and width), hydrophobia/hydrophilicity (log Kow) and log D related to the acid dissociation constant (pKa).

10.4.5.2.1. Relationship between molecular weight and membrane pore diameter

In general, the size of pesticides in relation to the membrane's pore radius is the most important factor determining the pesticide rejection. The pore size of nanofiltration and LPRO membranes is often close to or smaller than the size of pesticides (<1 nm). It is evident that the larger the pesticide molecule, the greater the membrane's rejection effect.

As a result, the membrane's MWCO concentration is correlated with the rejection of pesticides having a variable molecular weight. MWCO is defined as the molecular weight at which a 90% rejection is achieved. Due to the ready availability of molecular weight, it is a parameter that can be quickly used as a performance indicator.

Although molecular weight is not a direct measure of a molecule's dimensions, it still reflects molecular size and is an easily accessible parameter. This curve illustrates the correlation between molecular weight and molecule diameter (Figure 10.52):

$$\text{Molecule diameter (nm)} = 0.065\ (\text{MW})^{0.438}$$

This correlation is valid for the molecular weight range for which nanofiltration is generally chosen (up to ± 600 Da).

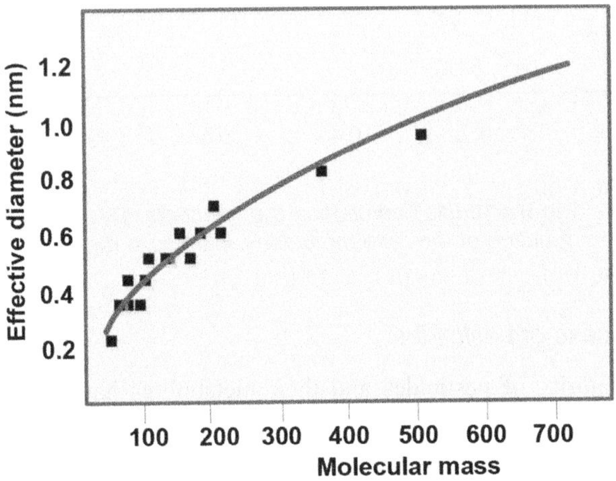

Figure 10.52. *Molecular mass and molecule diameter correlation. For a color version of this figure, see www.iste.co.uk/gaid/watertreatment3.zip*

The curve (Figure 10.52) can be a first indicator of the performances of nanofiltration or reverse osmosis membranes in relation to organic MPs. Depending on the membrane's cut-off concentration, it is possible to have an initial estimate of performances.

Table 10.18 and the curves in Figure 10.53 show the performance results of the NF 90 membrane against pesticides, assuming they have a spherical shape. As they have a molecular mass greater than 150 Da, methyl chloride (CH_3Cl) and dichloromethane (CH_2Cl_2) were also added.

Nevertheless, there may be deviations from this correlation, meaning that other parameters may be involved. In point of fact, the use of molecular weight does not take into account the spatial geometry of pesticides, which is important when the pesticide length and width are very different (Figure 10.54). The rejection of small pesticide molecules by membranes made of larger pores can be influenced, not only by design parameters such as the relative size of the pesticide and membrane pore, but also by the physicochemical interactions between these organic molecules and the membrane surface.

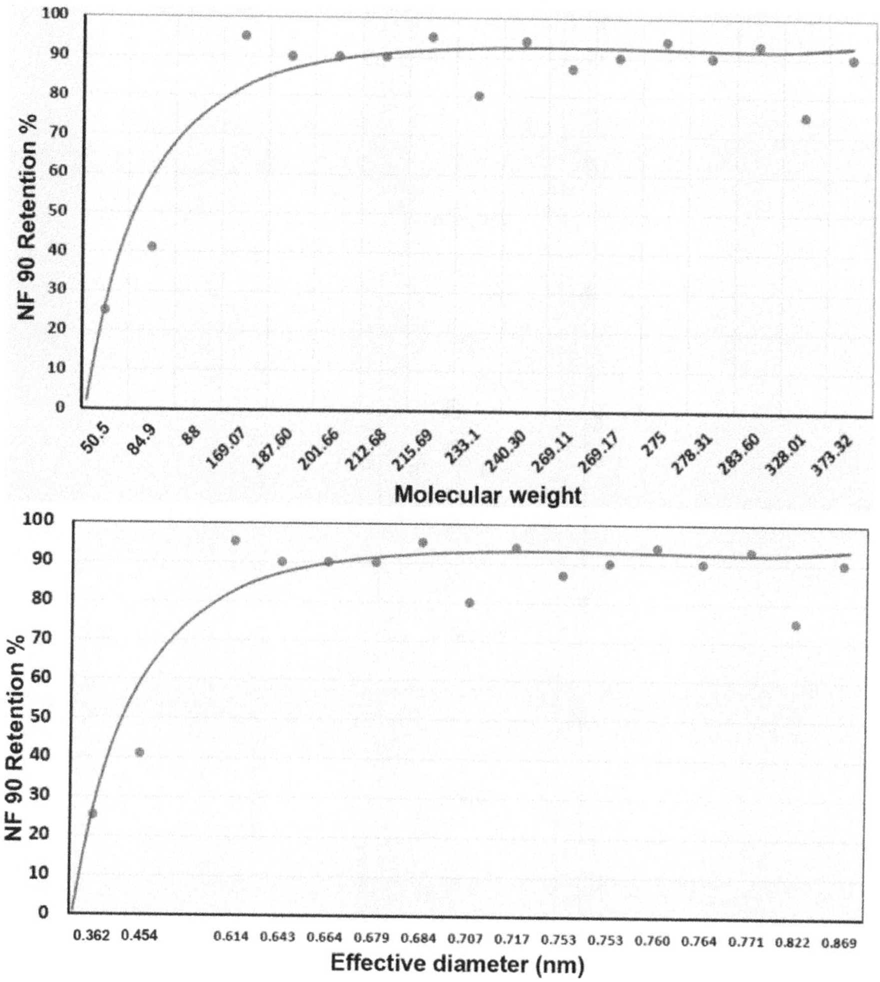

Figure 10.53. *Percent retention for various molecular weights and effective diameters for NF 90*

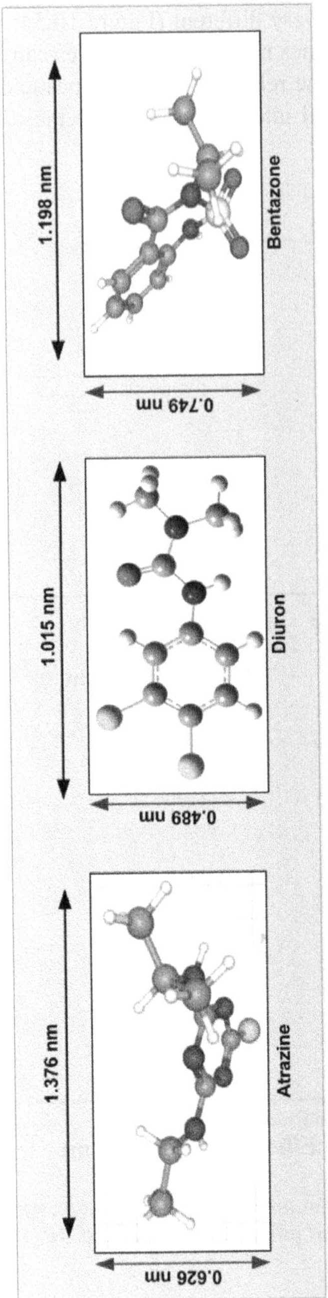

Figure 10.54. *Dimensions of atrazine, diuron and bentazone pesticides*

For example, the removal of diuron (MM 233.1 g·mole^{-1}, surface area 49.6 Å[exp]^{2}) with the NF 200 membrane (MWCO 300 Da) depends both on size and polarity, and leads to an average efficiency of 35–40%. A good correlation between pesticides dimensions and membrane pore diameter is a first indicator of the membrane's efficiency.

The membrane's porosity, the MP's molecular weight and their dimensions globally govern retention on the NF 270 and NF 90 membranes. This means that, by way of steric exclusion, a membrane with a 0.60 nm pore diameter (NF 90) can reject every molecule having a molecular diameter greater than that of the membrane. This is the case with most pesticides. It is valid if the molecule is considered to be spherical. It can penetrate the pores perpendicularly and flatly.

Pore	Membrane diameter (nm)	Cut-off limit or threshold (Daltons)
NF 70 (DOW/Filmtec)	0.68	200–250
NF 270 (DOW /Filmtec)	0.76–0.80	300
ESNA (Hydranautics)	–	250
NF 90 (DOW/Flimtec)	0.60	200
ESPA (Hydranautics)	0.62	200
HMA (Toray)	0.61	150
XLE (DOW/Fimtec)	0.60	150
BW30 (DOW/Filmtec)	–	100

Table 10.18. *Pore diameter of some nanofiltration and low-pressure reverse osmosis membranes*

When the molecule is not spherical, it can penetrate the pores perpendicularly due to its elongated shape.

Physical parameters can be included to predict membrane performances, such as the molecule's equivalent width, for example, defined by the root of the geometric mean between the molecule's height and width. MP retention increases with the equivalent width by steric exclusion.

10.4.5.3. *Mechanisms including size and load exclusion*

The two main mechanisms for pesticide retention are charge exclusion and size exclusion. Charge exclusion (Donan effect) is caused by the electrostatic interactions between a charged membrane and charged molecules. Size exclusion is based on the ratio between the pesticide and the membrane pore size, even if other mechanisms can be involved, such as the physicochemical interactions between the molecules and the membrane.

Ionizable molecules are acids or bases, characterized by one or more acidity constants (pKa). Depending on pesticide pKa and the water pH level, pesticides can be in a neutral, positively charged or negatively charged form. The presence of charge-bearing groups in a pesticide's molecular structure can strongly influence their rejection by the membrane. This explains that the rejection of neutral pesticides is done by steric exclusion, whereas the rejection of positively charged pesticides is done by steric exclusion and adsorption onto the membrane. The rejection of negatively charged pesticides tends to occur rather by steric exclusion, which outweighs the repulsion of negative charges from the membrane. This is because most NF membranes have a negative surface charge under the pH conditions of the water to be treated (pH 6–9). Due to the existence of acidic functional groups in the active layer, they reject negatively charged MPs not only by size exclusion but also by electrostatic repulsion.

Table 10.19 shows the percentage retention of NF 90 (DOW), Toray TMHA and XLE (DOW/Filmtec) as a function of molecular weight and effective diameter of some pesticides.

Negatively charged pesticides experience a higher rejection than expected and than the one based on steric considerations alone. The membrane's negative charge can also affect the rejection of neutral pesticides via their dipole moment. The pesticide's hydrophobicity (log Kow > 2) can also influence rejection because it determines the pesticide's degree of adsorption onto the membrane.

Table 10.20 presents the influence of the physical characteristics of some pesticides on their rejection by the NF 200 membrane.

The data presented (Table 10.20) show that the rejection rates for a large number of pesticides are quite high, with the exception of small molecules, which are less well retained due to their surface, which is smaller than that of the pores from the NF 200 membrane (0.474 nm^2). The rejection values for small molecules (dichlorvos, diuron, isoproturon, chlortoluron, monolinuron) are significantly lower than those for larger molecules such as, for example, sulcotrione and diazinon. The rejection of meco-prop in its neutral form is identical to that of diuron, which has an equivalent size, thus justifying the size exclusion.

Pesticides	Molecular weight g·mole^{-1}	Effective diameter nm	NF 90 retention % (tight)	LPRO Toray HMA retention %	Retention % LPRO DOW XL
AMPA	186.17	0.641	80	85	85
Atrazine	215.69	0.684	95	95	95
Acetochlor	269.11	0.753	87	92	92
Acetochlor ethanesulfonic acid (ESA)	315.11	0.807	87	95	95
Acetochlor oxanilic acid (OXA)	265.13	0.748	87	95	95
Alachlor	269.17	0.753	90	95	95
Alachlor ESA	270	0.754	95	95	95
Alachlor OXA	265	0.748	90	92	95
Bentazone	240.30	0.717	94	95	95
Chlortoluron	212.68	0.679	90	95	95
Desethylatrazine	187.60	0.643	90	95	95
Deisopropylatrazine	173.60	0.622	90	95	95
DEDIA	215.68	0.684	92	94	940
Diuron	233.1	0.707	90	95	95
Dimethenamid	275	0.760	95	95	95
Glyphosate	169.07	0.614	80	85	85
Heptachlor	373.32	0.869	90	95	95
Metaldehyde	176	0.625	80	80	90
Metolachlor	283.60	0.771	95	95	95
Metolachlor ESA	329.11	0.823	95	95	95
Metolachlor OXA	279.14	0.765	95	95	95
Oxadixyl	278.31	0.764	90	92	92
Simazine	201.66	0.664	90	95	95
Sulcotrione	328.01	0.822	75	90	90

Table 10.19. *Percentage retention of NF 90 (DOW), Toray TMHA and XLE (DOW/Filmtec) as a function of molecular weight and effective diameter for some pesticides*

Pesticides	Molecular weight g·mole^{-1}	Equivalent width nm	Effective diameter nm	Pesticide diameter/ Pore diameter	Surface nm^2	Rejection % NF200
Pirimicarb	238	0.690	0.706	0.905	0.586	75
Cyanazine	240	0.710	0.710	0.910	0.865	80
Metribuzin	214	0.520	0.713	0.914	0.964	35
Dianizon	304.35	0.640	0.724	0.948	0.856	80
Dichlorvos	221	0.504	0.786	1.008	0.448	40.7
Chlortoluron	212.68	0.610	0.788	1.010	0.323	85
Diuron	233.10	0.560	0.789	1.011	0.327	35.2
Atrazine	215.69	0.740	0.805	1.032	0.627	82
Monolinuron	214.60	0.690	0.817	1.047	0.416	85
Bentazone	240.3	0.760	0.822	1.053	0.749	90
Simazine	201.66	0.640	0.838	1.074	0.627	75
Isoproturon	206.30	0.660	0.911	1.167	0.323	87
Metamitrone	328.8	0.550	0.970	1.243	0.780	95

Table 10.20. *Influence of the physical characteristics of some pesticides on their rejection by the NF 200 membrane*

The presence of natural organic substances (humic substances) and various other pesticides and organic MPs may affect the efficiency of membrane rejection. With tight NF membranes (NF 70 or 90-type) or LPRO membranes, which have a relatively small pore diameter, the rejection may be lower than expected due to competitive adsorption onto the different pesticide molecules. The occupation of an adsorption site by a voluminous pesticide may restrict the adsorption of other pesticides. In the presence of humic substances, pore blocking by these large molecules reduces the pesticides' transport toward the adsorption pores. Therefore, only the micropores and part of the mesopores become active sites for retention.

A contrario, membrane fouling acts as an additional barrier for the passage of pesticides, and as a result, rejection may increase.

10.4.5.4. *The case of pharmaceutical residues*

The cut-off concentration of NF 90 (MWCO 200 Da) compared to that of NF 200 (MWCO 250–300 Da) promotes the rejection of MPs and achieves a higher efficiency, mainly due to size exclusion mechanisms. Table 10.21 presents the rejections of NF 270, NF 90, HMA and XLE membranes in relation to the medicinal products present in the water.

Medicinal products	Molecular mass (g·mole^{-1})	Log Kow	Log D	Molar volume (cm^3·mole^{-1})	Effective diameter (nm)	Equivalent width (nm)	LPRO membrane retention %	LPRO membrane retention %	Nanofiltration membrane retention %	
							DOW XLE	Toray HMA	NF270 (loose)	NF90 (tight)
Acetaminophen (Paracetamol)	151	0.46	0.23	120.90	0.704	0.53	80	80	22	75
Caffeine	194	–0.07	–0.45	133.30	0.668	0.70	92	92	70	90
Carbamazepine	236	2.45	2.58	186.50	0.785	0.73	95	95	80	95
17β-Estradiol	272	4.01	3.94	232.60	0.902	0.74	95	95	80	96
Bisphenol A	228	3.32	3.86	199.50	0.864	0.79	92	92	65	89
17α-Ethynilestradiol	296	3.67	1.48	225.60	0.785	0.85	95	95	80	95
Sulfamethoxazole	253	0.89	–0.45	173.10	0.850	0.64	95	95	84	96
Fenoprofen	242	–0.02	0.37	180.90	0.816	0.83	95	95	91	94
Ketoprofen	254	–0.52	–0.13	187.90	0.816	0.83	95	95	91	95
Naproxen	230	3.18	0.41	192.20	0.802	0.61	95	95	92	95
Ibuprofen	206	3.97	0.77	200.30	0.859	0.71	95	95	89	96
Gemfibrozil	250	4.77	1.58	221.90	0.997	0.78	95	95	90	94
Atenolol	266.3	0.15	–2.21	369.8	1.084	0.69	95	95		95
Bezafribrate	361.8	4.25	1.20	212.8	0.85	–	95	95	92	95
Ciproflaxacin	331.3	0.38	0.10	496.7	0.858	–	95	95	90	95
Clarithromycin	748	3.16	0.80	–	1.79	–	95	95	90	95
Diclofenac	296.1	4.75	0.48	182	0.851	0.900	95	95	80	92
DEET	191.2	2.40	1.96	191.5	0.648	0.81	95	95	91	95
Erythromycin	733.9	3.06	0.81	611	1.170	0.585	95	95	85	97
Estriol	288.4	2.45	2.53	–	0.776	–	95	95	85	95
Fluoxetine	309.3	4.05	1.41	412.4	0.801	0.85	95	95	88	95
Fenofibrate	360.8	5.20	1.34	644.2	0.857	–	95	95	88	95
Fluvoxamine	318.3	2.80	1.75	454.7	0.811	–	96	96	85	95
Furosemide	330.7	2.03	–0.82	206.6	0.824	–	92	90	75	90
Gabapentin	171.2	–1.10	–1.42	155.1	0.618	–	94	94	90	95
Iopamidol	777.1	–2.42	–0.66	–	1.19	–	95	95	95	95
Iohexol	821.1	–3.05	2.52	373.2	1.22	–	95	95	95	95
Iopromide	791.1	2.49	– 0.36	364	1.20	–	95	95	95	95
Metformin	129.1	–2.60	–3.24	184.4	0.540	–	85	85	85	85

Medicinal products	Molecular mass (g·mole^{-1})	Log Kow	Log D	Molar volume (cm^3·mole^{-1})	Effective diameter (nm)	Equivalent width (nm)	LPRO membrane retention %	LPRO membrane retention %	Nanofiltration membrane retention %	
							DOW XLE	Toray HMA	NF270 (loose)	NF90 (tight)
Metoprolol	267.3	1.88	–0.77	477.3	1.127	0.730	95	95	90	95
Nonylphenol	220.3	5.76	5.88	231.1	1.083	0.660	95	95	89	94
Omeprazole	345.4	2.23	1.80	246.7	0.840	–	98	98	85	98
Oxazepam	286.7	2.24	2.80	819.1	0.774	–	95	95	92	98
Ofloxacin	361.4	–0.39	–	240.9	0.857	–	95	95	90	95
Progesterone	272.4	4.01	3.83	232.8	0.757	–	95	95	85	95
Primidone	218.2	0.91	–0.84	364.2	0.786	0.765	95	95	87	95
Propanolol	259.3	3.48	0.21	235.6	0.741	–	95	95	80	95
Sotalol	272.3	0.24	–1.70	–	0.936	0.78	95	95	85	95
Tramadol	299.8	3.01	1.13	483.5	0.790	–	97	97	90	97
Triclocarban	315.6	4.34	5.80	–	0.808	–	95	95	89	95
Triclosan	289.5	4.76	5.17	186.7	0.778	0.64	–	–	30	55
Trimethoprim	290.3	0.91	0.20	–	0.780	0.88	95	95	75	98
Valsartan	495.5	1.49	–0.80	–	0.97	–	95	95	90	95
Vanlafaxine	277.4	3.20	0.70	616.4	0.763	–	95	95	85	96
Androstenedione	286.4	2.75	2.72	–	0.774	0.74	95	95	85–55	95
Meprobamate	218.2	0.70	0.70	178.8	0.687	0.86	98	98	82	98
Testosterone	288.4	3.32	3.00	262.2	0.776	0.63	95	95	90	98
Enalapril	376.4	0.07	–1.01	–	0.873	–	98	98	95	97
Androsterone	290.4	3.69	3.93	218.3	0.779	0.650	98	98	98	98
Salicylic acid	138.1	2.26	–1.20	635.3	0.562	0.260	85	85	60	85
Phenacetin	179	1.58	1.68	163	0.809	0.540	75	75	42	72
Metrodinazole	171	–0.02	–0.27	117.8	0.617	0.660	70	70	46	71
Phenazone	188	0.38	0.54	162.7	0.663	0.660	93	93	73	91
Estrone	270.4	2.60	3.80	225.3	0.908	0.76	78	78	70	75

Table 10.21. *Efficacy of nanofiltration and low pressure reverse osmosis membranes in relation to medicinal products*

Size exclusion is considered to be the main cause of rejection of neutral compounds by NF 200 membranes, even though there are some interactions between these compounds and membranes. Assuming that the molecules are spherical, a correlation is obtained between the NF 200 membrane's rejection % and the molecule diameter (dm)/pore diameter (dp) ratio, as represented by Figure 10.55 for substances such as androsterone, metronidazole, eprobamate, meprazole, paracetamol, phenacetin, phenazone, primidone, testosterone, triclosan and trimethoprim.

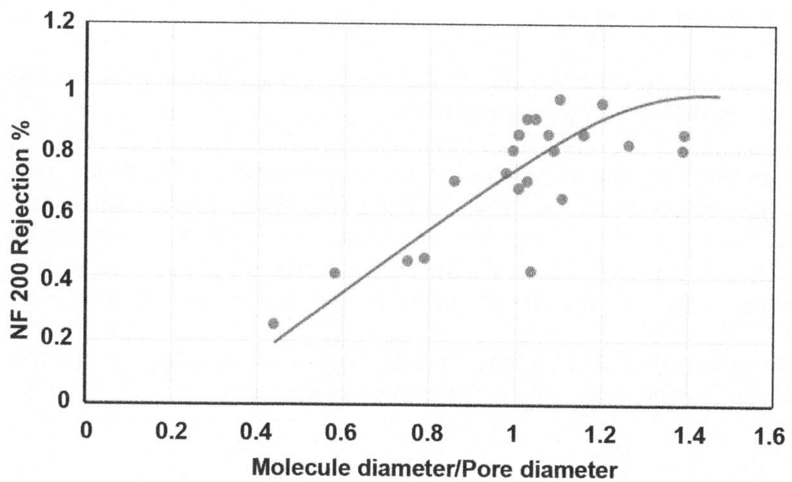

Figure 10.55. *Relationship between the rejection % of neutral substances and the dm/dp ratio with NF 270*

The rejections of positively charged compounds are higher than those of neutral compounds, due to the effects of electrostatic interactions between the membrane's surface negative charge and the compounds' positive charge. In the case of NF 90, a combined effect between electrostatic repulsion and size exclusion leads to higher rejections than those obtained with NF 270. NF 90 is also very effective for removing positively charged compounds (solalol, metoprolol, atenolol).

The charge effect seems to be much weaker for NF 70 (negatively charged membrane). The difference in retention between a positively charged compound and a negatively charged one is small, meaning that the charge effect is only of minor importance for membranes such as NF 70 or NF 90. Molecules are mainly retained due to size effects.

The rejections of negatively charged substances (sulfamethoxazole, naxoprene, ibuprofen, gemfibrozil, diclofenac, DEET, ketoprofen) are higher with NF 90 than with NF 270 due to a significant removal by size exclusion mechanisms.

Bisphenol A is moderately removed (45–50%) by NF 200, whereas caffeine (very hydrophilic) is fairly well removed (70%), despite a negative log Kow (very hydrophilic) and a smaller molar volume than that bisphenol A. A combination of the pore size of a membrane, such as NF 270, and the molecular size and hydrophobicity of a compound, such as bisphenol A, can cause the compound to be divided across the membrane.

MPs such as dexamethazone, febantel, ibuprofen and propranolol are removed over 90% by the three types of membranes.

In general, we can conclude that the charge effect is important when the molecules are much smaller than the pores. Membrane charge can increase or decrease retention due to electrostatic interactions. When the molecules are approximately the same size as the pores, charge effects are only of minor importance, since the molecules are mainly retained due to a sieving effect.

The following equation offers an indication of the membrane's efficiency in relation to the rejection of MPs such as medicinal products:

$$R(\%) = \frac{100}{1 + e(^{-B})}$$

where:

B = 11.372 deff + 6.795 leff + 107.305 RS – 0.399 log D – 116.496

deff is the effective diameter or molecule diameter (nm) given by the ratio:

$$deff = \frac{L}{2} + \frac{h}{3{,}14}$$

L is the molecule's length (nm) and h is its height (nm), whereas leff is the equivalent width (nm) given by:

$$leff = \sqrt{h.l}$$

RS is the membrane's salt rejection (%) and log D is a measure of the log Kow at a given pH for a compound with a certain pKa. This parameter therefore takes into account the concentration of ionized molecules.

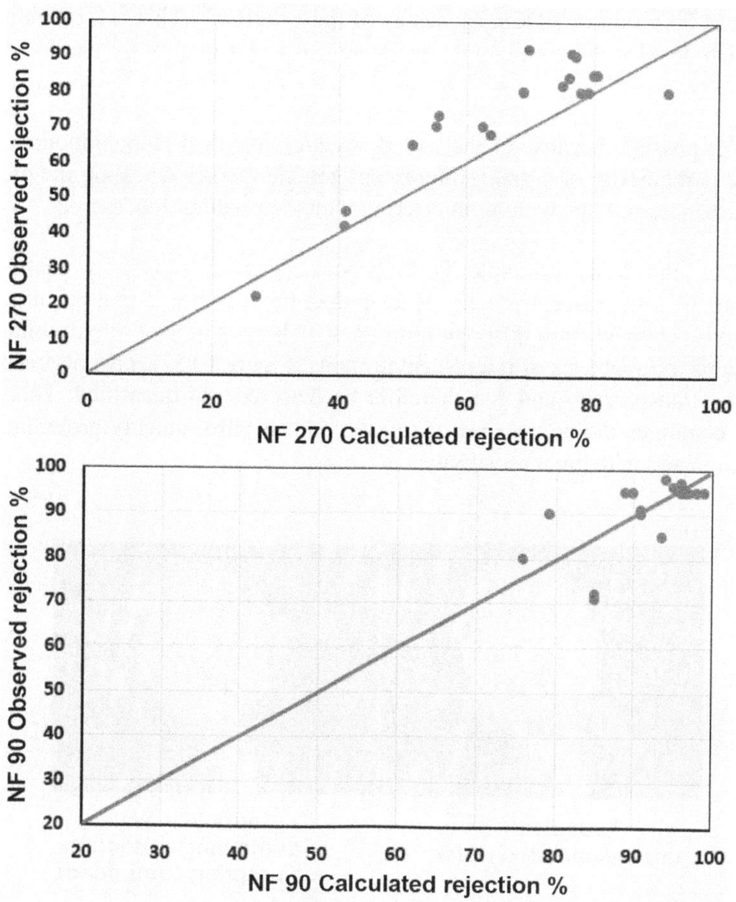

Figure 10.56. *Prediction of pharmaceutical residue rejection with NF 270 and NF 90*

Figure 10.56 illustrates the observed and calculated rejection percentages for various drugs with the NF 270 and NF 90 membranes. A fairly good similarity was observed between many compounds.

10.5. Frogbox®: an effective monitoring and control tool

Tools for the continuous and comprehensive assessment of water quality are important in determining ecological costs and risks. Nowadays, the problem of EDCs is increasingly singled out. Through its subsidiary EPAS-International and in partnership with WatchFrog, Veolia offers an innovative and original tool for the continuous detection of EDCs, by means of the FrogBox®. This is a biomonitoring tool which enables the continuous assessment of water quality by the detection of EDCs.

This is possible because of the use of aquatic biological tadpole models, which have the particularity of a green fluorescent genetic marker revealing the effects of the disturbance, and this without interfering with their natural functioning.

In fact, the basic principle is to produce genetic constructs enabling the expression of a fluorescent protein. If an endocrine disruptor is present, it activates the response element from different hormones such as estrogen or thyroid hormones and triggers the synthesis of fluorescent proteins (Figure 10.57). This fluorescence is visible by transparency and can therefore be detected and quantified. This testing method combines the advantages of in vivo and in vitro, quickly providing better information and with great sensitivity.

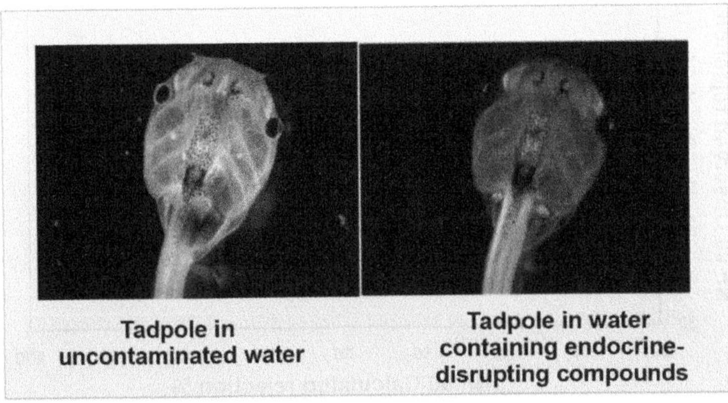

Figure 10.57. *Tadpole sensitivity to endocrine-disrupting compounds. For a color version of this figure, see www.iste.co.uk/gaid/watertreatment3.zip*

The sensitivity concentration to EDCs is in the order of nanograms per liter. The transgenic tadpoles, aged between 8 days and 3 weeks, are placed in contact with the samples in microplates. In the event of a physiological reaction, fluorescence appears after only a few minutes and within 3 days. Xenopus larvae have the advantage of being translucent, effortlessly observable, requiring little space and being easily reproducible. In addition, tadpoles enable polydetection. For example, the same larva can light up in red to signal an impact on the nervous system, in green when an effect on the muscles occurs, and in blue to reveal the presence of toxins. The level of fluorescence is evaluated by the tool enabled by a high-performance optical system and original software.

Frogbox® can be used on a large number of effluents such as wastewater treatment stations, drinking water plants and for industrial wastewater (pharmaceuticals, cosmetics, agriculture and chemical production). The results can be used in several ways: measuring the reduction rate of a water treatment system, controlling the quality of effluents, controlling drinking water quality, evaluating the impact of an effluent on the environment, monitoring water quality, as well as for identifying contamination sources.

Figure 10.58. *Frogbox®. For a color version of this figure, see www.iste.co.uk/gaid/watertreatment3.zip*

10.6. The evolution of micropollutants in drinking water plants

In terms of MPs, drinking water plants supplied with surface water are conditioned by the discharge from wastewater treatment stations, soil leaching and rainfall. The input concentration of MPs can be very variable, but in all cases contains one or more MP categories.

A drinking water plant including the coagulation–flocculation–settling–filtration–disinfection stages is nonetheless ineffective against MPs. When pesticides are present in the resource, a GAC filtration stage is always included in the water treatment station. This stage removes a large number of MPs. A regular monitoring of filter saturation level is necessary.

Since 2010, consulting firms have been recommending local authorities to instead install PAC reactors, because these make it possible to constantly monitor the quality of the water produced by adjusting the PAC dose when required.

When the drinking water plant contains an ozonation stage, it is generally considered that the injected ozone doses are sufficient to carry out a pre-disinfection, but may prove to be limited for a significant removal of MPs, risking the formation of by-products if such dosage is increased.

Disinfection (chlorine and its derivatives, UV) is ineffective for the removal of MPs with the doses selected for this treatment stage.

Nowadays, the trend is rather to work at the source, at the industrial manufacturers' level, in order to reduce the weight of these substances during their manufacturing process, to install a tertiary or quaternary treatment at the treatment stations with a view to removing these MPs before rejection. Many advanced technologies have shown to improve the removal of these MPs in wastewater treatment effluents: higher sludge age-specific biological secondary processes (such as membrane bioreactors) or tertiary treatments (such as membrane separation), adsorption and chemical oxidation using ozone and advanced oxidation processes. The widespread belief is that it is preferable to include a barrier to MPs at the wastewater treatment station level, by reinforcing the existing means, rather than at the drinking water production plant level.

PAC reactor-type treatments, or GAC-filter ozonation, make it possible to significantly improve water quality at the biological treatment outlet, and thus remove the majority of analyzed MPs below 0.1 $\mu g \cdot L^{-1}$.

Microfiltration and ultrafiltration membranes have no effect on MPs. High-pressure membranes such as nanofiltration and LPRO are effective against MPs.

Veolia now has a range of technologies designed to remove MPs which are present both in wastewater and drinking water. Anticipating future standards is the leitmotiv for preparing to cope with this new type of contamination.

It is clear that, in the minds of the public, EDCs and pharmaceutical residues are a constant concern. At present, prime contractors already include this parameter and the associated guarantees for water treatment stations because cities, civil associations and communities are demanding for these substances to also be treated at their drinking water plants, even if there are no current standards. This is the case for not only large cities like Nantes but also small settlements such as Fouet Banegan, le Craon and so on.

10.7. References

Abellan, M.N., Gebhardt, W., Schroder, H.F. (2008). Detection and identification of degradation products of sulfamethoxazole by means of LC/MS and -MS(n) after ozone treatment. *Water Sci. Technol.*, 58(9), 1803–1812.

Antoniou, M.G., Hey, G., Vega, S.R., Spiliotopoulou, A., Fick, J., Tysklind, M., Jansen, J., Andersen, H.R. (2013). Required ozone doses for removing pharmaceuticals from wastewater effluents. In *8th IWA Specialized Conference on "Assessment and Control of Micropollutants and Hazardous Substances in Water"*, Zurich.

Bellin, I. (2015). Eau potable : des traitements complémentaires pour éliminer les polluants spécifiques. *L'Eau, L'Industrie, Les Nuisances*, 384, 35–40.

Bellona, C.J., Drewes, J.E., Xu, P., Amy, G. (2004). Factors affecting the rejection of organic solutes during NF/RO treatment. *Water Res.*, 38(12), 2795–2809.

Bendeddouche, N., Ben Hocine, H., Kaoua, F., Gaid, K. (1994). Prédictions de l'adsorption des acides humiques sur charbon actif l'aide de la fonction logistique généralisée. *Annales Maghrébines de l'Ingénieur*, 8(1), 65–72.

Besnault, S., Ruel, M.S., Baig, S., Heiniger, B., Esperanza, M., Budzinski, H., Miège, C., Coquery, M. (2013). Occurrence of selected micropollutants in treated wastewater and removal with optimised tertiary treatments. In *8th IWA Specialized Conference on Assessment and Control of Micropollutants and Hazardous Substances in Water*, Zurich.

Boehler, M., Brocker, S., Kobler, S., Le Goaziou, Y., Moser, R., Rettby, R., Rigault, M. (2012). Elimination of micro pollutants out of municipal wastewater by means of powder activated carbon in the Actiflo® Carb process and by ozone at WWTP Schönau, Cham (GVRZ). Report, the Federal Office for the Environment/Alpha Umwelttechnik/Veolia/ GVRZ/Labosafe,Hunziker-Betatech/Eawag, Dübendorf.

Bouchonnet, S., Kinani, S., Souissi, Y., Bourcier, S., Sablier, M., Roche, P., Boireau, V., Ingrand, V. (2011). Investigation of the dissociation pathways of metolachlor, acetochlor and alachlor under electron ionization – Application to the identification of ozonation products. *Rapid Commun. Mass Spectrom.*, 25, 93–103.

Broseus, R., Vincent, S., Aboulfadl, K., Daneshvar, A., Sauvé, S., Barbeau, B., Prevost, M. (2009). Ozone oxidation of pharmaceuticals, endocrine disruptors and pesticides during water treatment. *Water Res.*, 43, 4707–4717.

Comerton, M.A., Andrews, C.R., Bagley, M.D., Hao, C. (2008). The rejection of endocrine disrupting and pharmaceutically active compounds by NF and RO membranes as a function of compound and water matrix properties. *J. Mem. Sci.*, 313, 323–335.

Corwin, C.J. and Summer, R.S. (2012). Controling trace organic contaminants with GAC adsorption. *J. Am. Water Works Assoc.*, 104, 36–47.

Cotton, G., Feliers, C., Kecili, K., Colon, A.F. (2019). Performance du traitement membranaire (NDF et OIBP) : rétention des micropolluants et enjeux énergétiques. *Techniques Sciences Méthodes*, 4(104), 29–35.

Cougnaud, A., Faur, C., Le Cloirec, P. (2005). Removal of pesticides from aqueous solution: Quantitative relationship between activated carbon characteristics and adsorption properties. *Environ. Technol.*, 26, 852–866.

Dantas, R., Canterino, M., Marotta, R., Sansa, C., Esplugas, S., Andreozzi, R. (2007). Bezafibrate removal by means of ozonation: Primary intermediates, kinetics and toxicity assesment. *Water Res.*, 41(12), 2525–2532.

Dolar, D., Vukovic, A., Asperger, D., Kosutic, K. (2012). Efficiency of RO/NF membranes at the removal of veterinary antibiotics. *Water Sci. Technol.*, 65(2), 317–323.

Drault-Pezard, C. (2012).Ozonation et procédés d'oxydation avancée : solutions d'avenir pour traiter les polluants émergents.*L'Eau, L'Industrie, Les Nuisances*, 351, 55–60.

El Najjar, N.H., Touffet, A., Deborde, M., Journel, R., Vel Leitner, N.K. (2013). Levofloxacin oxidation by ozone and hydroxyl radicals: Kinetic study, transformation products and toxicity. *Chemosphere*, 93(4), 604–611.

Gaid, K. (1989). Modélisations mathématiques de l'épuration biologique sur charbon actif. In *Prés. 2ème Colloque maghrébin sur les modèles numériques de l'ingénieur*. 23–24 November, Rabat.

Gaid, K. and Benmoussa, H. (1989). Contribution au mécanisme de fixation du substrat par les bactéries fixées. *Journal of Technology*. 7, 86–89.

Gaid, K. and Philipot, J.M. (2005). Le concept OPALINE : une solution à la sécurité sanitaire en eau potable. *L'Eau, L'Industrie, Les Nuisances*, 277, 73–78.

Gaid, K. and Sauvignet, P. (2012). Élimination des micropolluants en eaux usées avec réacteur charbon actif en poudre, (Actiflo® Carb) et suivi de la qualité du traitement. In *Journées Informations Eaux*, Poitiers.

Gaid, K. and Sauvignet, P. (2013). High rate PAC reactor combined with ozone oxidation for enhanced micro-pollutants removal from urban waste water. In *8th IWA Specialized Conference on "Assessment and Control of Micropollutants and Hazardous Substances in Water"*, Zurich.

Gaid, K. and Sauvignet, P. (2015). Substances organiques, perturbateurs endocriniens, micropol-luants : une solution simple le Filtraflo carb. *L'Eau, L'Industries & Les Nuisances*, 382, 89–93.

Gaid, K. and Sauvignet, P. (2019a). Les micropolluants organiques : présence dans les effluents urbains. *Techniques. Sciences. Methodes. TSM.* 7(8), 81–92.

Gaid, K. and Sauvignet, P. (2019b). Oxydation avancée in situ du charbon actif : traitement des micropolluants, pesticides et matières organiques. *L'Eau, L'Industrie, Les Nuisances*, 419, 1–7.

Gaid, K., Cavelier, C., Martin, G. (1979). Observations au microscope électronique des populations bactériennes sur charbon actif. *J. français d'hydrologie.* 10, fasc. 2, 29, 125–132.

Gaid, K., Cavelier, C., Martin, G., Lemarchand, D. (1980). Adsorption de micropolluants sur charbon actif. *Revue des sciences de l'eau*, 125, 85–92.

Gaid, K., Le Cloirec, P., Bernard, T., Martin, G. (1982). Evolution physico-chimique et biologique d'un substrat marqué (phénol) aprés son adsorption sur charbon actif. *Environ. Technology Letters*, 3, 329–339.

Gaid, K., Le Cloirec, P., Martin, G., Lemarchand, D. (1983). Etude de la rétention des substances humiques sur charbon actif. *Environ. Technology Letters*, 4, 297–305.

Gaid, K., Kaoua, F., Bendedouche, N., Benhocine, H. (1994). Modélisations de composés organiques sur charbon actif à l'aide de la fonction logistique. *AQUA -Water RT.* 43(1), 17–21.

Gaid, K., Sauvignet, P., Bourdon, F., Ingrand, V., Rigot, M. (2015). Les réacteurs CAP "Multiflo-Actiflo" la solution pour l'élimination des perturbateurs endocriniens et des substances dangereuses. *L'Eau, L'Industrie, Les Nuisances*, 353, 68–75.

Gaid, K., Poussade, Y., Sauvignet, P., Saudrais, I., Daines, C. (2017). Removal of contaminants of emerging concern and pathogens on an innovative hybrid PAC/membrane process. *Conference European Desalination Society*, 27 October, Montpellier.

Gaid, K., Sauvignet, P., Cotte, C., Decerle, D. (2019). Les métabolites du Métolachlore et de l'Alachlore Quels traitements et quelle efficacité ? *L'Eau, L'Industrie, Les Nuisances*, 423, 110–121.

Gur-Reznik, S., Koren-Menashe, I., Heller-Grossman, L., Rufel, O., Dosoretz, C.G. (2011). Influence of seasonal and operating conditions of the rejection of pharmaceutical active compounds by RO and NF membranes. *Desalination*, 277, 250–256.

Guyard, C. (2012). Dégradation de la ressource en eau : faire face aux polluants émergents. *L'Eau, L'Industrie, Les Nuisances*, 356, 45–50.

Hollender, G. (2009). Elimination of organic micropollutants in a municipal wastewater treatment station upgraded with a full-scale post ozonation followed by sand filtration. *Environ. Sci. Technol.*, 43(20), 7862–7869.

Huber, M.M., Canonica, S., Park, G.Y., Von Gunten, U. (2003). Oxidation of pharmaceuticals during ozonation and advanced oxidation processes. *Environ. Sci. Technol.*, 37(5), 1016–1024.

Kaoua, F. and Gaid, K. (1990). Adsorption de composés organiques présents dans l'eau. Cas du charbon actif et d'une argile minérale. *Annales de Chimie*. 4, 221–228.

Kimura, K., Amy, G., Drewes, J.E., Watanabe, Y. (2003). Adsorption of hydrophobic compounds onto NF/RO membranes: An artifact on overestimation of rejection. *J. Mem. Sci.*, 221, 89–101.

Kisoa, Y., Nishimuraa, Y., Kitaob, T., Nishimurab, K. (2000). Rejection properties of non-phenylic pesticides with nanofiltration membranes. *J. Membr. Sci.*, 171, 229–237.

Košutić, K. and Kunst, B. (2002). Removal of organics from aqueous solutions by commercial RO and NF membranes of characterized porosities. *Desalination*, 142, 47–56.

Košutić, K., Dolar, D., Ašperger, D., Kunst, B. (2007). Removal of antibiotics from a model wastewater by RO/NF membranes. *Sep. Purif. Technol.*, 53, 244–249.

Larcher, S., Delbès, G., Robaire, B., Yargeau, V. (2012). Degradation of 17α-ethinylestradiol by ozonation – Identification of the by-products and assessment of their estrogenicity and toxicity. *Environ. Int.*, 39, 66–72.

Löwenberg, J., Zenker, A., Baggenstos, M., Koch, G., Kazner, C., Wintgens, T. (2013). Comparison of two PAC/UF processes for the removal of micropollutants from wastewater treatment station effluent: Process performance and removal efficiency. In *8th IWA Specialized Conference on Assessment and Control of Micropollutants and Hazardous Substances in Water*, Zurich.

Luster-Teasley, S.L., Yao, J.J., Herner, H.H., Trosko, J.E., Masten, S.J. (2002). Ozonation of chrysene: Evaluation of by-product mixtures and identification of toxic constituent. *Environ. Sci. Technol.*, 36, 869–876.

Margot, J., Magnet, A., Thonney, D., Chèvre, N., de Alencastro, F., Rossi, L. (2011). Traitement des micropolluants dans les eaux usées : rapport final sur les essais pilotes à la STEP de Vidy, Lausanne. Report, STEP, Lausanne.

Margot, J., Kienle, C., Magnet, A., Weil, M., Rossi, L., Felippe de Alencastro, L., Abegglen, C., Thonney, D., Chèvre, N., Schärer, M., Barry, D.A. (2013). Treatment of micropollutants in municipal wastewater: Ozone or powdered activated carbon? *Sci. Total Environ.*, 461–462, 480–498.

McDowell, D.C., Huber, M.M., Wagner, M., Von Gunten, U., Ternes, T.A. (2005). Ozonation of carbamazepine in drinking water: Identification and kinetic study of major oxidation products. *Environ. Sci. Technol.*, 39(20), 8014–8022.

Mestankova, H., Bramaz, N., Schirmer, K., Canonica, S., Linden, K., Von Gunten, U. (2013). Advanced oxidation processes: Toxicity of transformation products. In *8th IWA Specialized Conference on Assessment and Control of Micropollutants and Hazardous Substances in Water*, Zurich.

Nghiem, L.D. (2005). Removal of emerging trace of organic contaminants by naofiltration and reverse osmosis. PhD Thesis, University of Wallongong.

Nghiem, L.D., Schäfer, A.I., Elimelech, M. (2005). Pharmaceutical retention mechanisms by nanofiltration membranes. *Environ. Sci. Technol.*, 39, 7698–7705.

Rattier, M., Reungoat, J., Joss, A., Gernjak, W., Keller, J. (2013). Evaluation of the role of adsorption and biodegradation of micropollutants during biological activated carbon filtration. In *8th IWA Specialized Conference on Assessment and Control of Micropollutants and Hazardous Substances in Water*, Zurich.

Rosal, R., Rodríguez, A., Perdigón-Melón, J.A., Petre, A., García-Calvo, E., Gómez, M.J., Agüera, A., Fernandez-Alba, A.R. (2010). Occurrence of emerging pollutants in urban wastewater and their removal through biological treatment followed by ozonation. *Water Res.*, 44, 578–588.

Ruhl, A.S., Altmann, J., Meinel, F., Zietzschmann, F., Sperlich, A., Jekel, M. (2013). Enhancing the efficiency of powdered activated carbon for micro pollutant removal by immobilization in a fixed bed. In *8th IWA Specialized Conference on Assessment and Control of Micropollutants and Hazardous Substances in Water*, Zurich.

Sauvignet, P., Gaid, K., Marlin, P. (2018). Utilisation du charbon actif en poudre pour éliminer le métaldéhyde par clarification Actiflo carb. In *48ème Congrès du Groupe Français des pesticides*. University of Limoges, 30 May–1 June.

Schaar, P. (2010). Micropollutant removal during biological wastewater treatment and a subsequent ozonation step. *Environ. Pollut.*, 158(5), 1399–1404.

Sein, M.M., Zedda, M., Tuerk, J., Schmidt, T.C., Golloch, A., Von Sonntag, C. (2008). Oxidation of diclofenac with ozone in aqueous solution. *Environ. Sci. Technol.*, 42, 6656–6662.

Seung-Woo, N., Dae-Jin, C., Seung-Kyu, K., Namguk, H., Kyung-Duk, Z. (2014). Adsorption characteristics of selective hydrophilic & hydrophobic micropollutants in water using activated carbon. *J. Hazard. Mater.*, 270, 144–152.

Snyder, S.A., Adham, S., Redding, A. (2007a). Role of membranes and activated carbon in the removal of endocrine disruptors and pharmaceuticals. *Desalination*, 202(1–3), 156–181.

Snyder, S.A., Adham, S., Redding, A., Cannon, F., DeCarolis, J., Oppeenheimer, J., Yoon, Y. (2007b). Role of membranes and activated carbon in the removal of endocrine disruptors and pharmaceuticals. *Desalination*, 202(1–3), 156–181.

Ternes, H. (2003). Ozonation: A tool for removal of pharmaceuticals, contrast media and musk fragrances from wastewater. *Water Res.*, 37(8), 1976–1982.

Tootchi, L., Seth, R., Tabe, S., Yang, P. (2013).Transformation products of pharmaceutically active compounds during drinking water ozonatrion. *Water Sci. Technol., Water Supply*, 13(6), 1576–1582.

Van der Bruggen, B. and Vandecasteele, C. (2002). Modelling of the retention of uncharged molecules with nanofiltration. *Water Res.*, 36, 1360–1368.

Van der Bruggen, B., Schaep, J., Wilms, D., Vandecasteele, C. (1999). Influence of molecular size, polarity and charge on the retention of organic molecules by nanofiltration. *J. Membr. Sci.*, 156, 29–41.

Verliefde, A.R.D., Heijman, S.G.J., Cornelissen, E.R., Amy, G., Van der Bruggen, B., Van Dijk, J.C. (2007). Influence of electrostatic interactions on the rejection with NF and assessment of the removal efficiency during NF/GAC treatment of pharmaceutically active compounds in surface water. *Water Res.*, 41(15), 3227–3240.

Von Gunten, U. (2003). Ozonation of drinking water: Part I. Oxidation kinetics and product formation. *Water Res.*, 37, 1443–1467.

Wert, E.C., Rosario-Ortiz, F.L., Drury, D.D., Snyder, S.A. (2007). Formation of oxidation by-products from ozonation of wastewater. *Water Res.*, 41, 1481–1490.

Wert, E.C., Rosario-Ortiz, F.L., Snyder, S.A. (2009). Effect of ozone exposure on the oxidation of trace organic contaminants in wastewater. *Water Res.*, 43, 1005–1014.

Yangali-Quintanilla, V., Verliefde, A., Kim, T.U., Sadmani, A., Kennedy, M., Amy, G. (2009). Artificial neural network models based on QSAR for predicting rejection of neutral organic compounds by polyamide nanofiltration and reverse osmosis membranes. *J. Membr. Sci.*, 342, 251–262.

Yargeau, V., Segura, P.A., Kaplan, P. (2013). Nature and estrogenic activity of ozonation transformation products of estrone. In *8th IWA Specialized Conference on Assessment and Control of Micropollutants and Hazardous Substances in Water*, Zurich.

Yoon, Y., Westerhoff, P., Snyder, A., Wert, E.C. (2006). Nanofiltration and ultrafiltration of endocrine disrupting compounds, pharmaceuticals and personal care products. *J. Membr. Sci.*, 270, 88–100.

Yu, Z.S. and Peldszus, P.M. (2009). Huck: Adsorption of selected pharmaceuticals and an endocrine disrupting compound by granular activated carbon. Adsorption capacity and kinetics. *Environ. Sci. Technol.*, 43(5), 1467–1473.

Yu, J., Lv, L., Lan, P., Zhanga, S., Pana, B., Zhang, W. (2012). Effect of effluent organic matter on the adsorption of perfluorinated compounds onto activated carbon. *J. Hazard. Mater.*, 225–226, 99–106.

11

Removal of Perfluorinated Compounds

Perfluorinated compounds (PFCs) belong to a large family of chemical molecules known as fluorocarbons. These are gaseous halogen compounds endowed with high thermal, chemical and biological stability. They display unique properties: high surface activity, thermal and acid resistance, and are both hydrophobic and lipophobic. Many of these substances have been developed and manufactured for industrial, agricultural, military and household chemical applications: fire-fighting foams, Teflon, aerosols, non-stick food containers, non-stick cookware, stain-resistant textiles and other similar non-stick or waterproof materials.

PFAS is a general acronym for per- and polyfluoroalkyl substances. These chemicals belong to a class containing carbon-fluorine bonds with a charged functional group (usually a carboxylic acid or a sulfonic acid), which is extremely resistant to degradation, with half-life of several years in the environment or in human blood (3 to 8.5 years estimated half-life in human blood).

> The presence of PFAS in the environment is of anthropogenic origin. There are not only direct but also indirect sources of PFAS emissions into the environment. Direct emission sources are related to the production and use of these compounds, primarily due to their presence in industrial discharges. As for indirect sources, these exist as impurities or degradation products.

It is believed that over 3,500 PFAS may exist and may be found in the global environment. PFAS compounds are very persistent (the life span of perfluorooctanoic acid [PFOA] in water could reach 41 years), and are resistant to biological degradation. For the aforementioned reasons, PFAS compounds tend to accumulate in the human body (and in animals) as a result of multiple exposures. Bioaccumulation may reach a toxic level for people exposed to PFAS. The health effects of PFOA exposure include kidney cancer, testicular cancer, high cholesterol and thyroid effects.

Perfluorooctane sulfonic acid (PFOS) and PFOA are the two most closely monitored molecules at the resource level, as well as at the production unit level.

11.1. Physicochemical properties

PFOS and PFOA are part of the PFC family and belong to the PFAS class in particular. The difference between the members of the PFC family lies in the number of carbon atoms, which determines the perfluorinated chain. PFOS and PFOA are compounds with eight carbons and are therefore called C8.

Table 11.1 summarizes the proposed classification for perfluoroalkyl compounds.

<table>
<tr><th colspan="2">Classifications and chemical structures</th><th>CnF2n+1R, with R=</th></tr>
<tr><td rowspan="7">Perfluorinated alkyl acids</td><td>Perfluoroalkyl carboxylic acids (PFCA)</td><td>-COOH</td></tr>
<tr><td>Perfluoroalkyl carboxylates (PFCA)</td><td>-COO-</td></tr>
<tr><td>Perfluoroalkane sulfonic acids (PFSA)</td><td>$-SO_3H$</td></tr>
<tr><td>Perfluoroalkane sulfonates (PFSA)</td><td>$-SO_3$-</td></tr>
<tr><td>Perfluoroalkyl sulfinic acids (PFSiA)</td><td>$-SO_2H$</td></tr>
<tr><td>Perfluorophosphonic acids (PFPA)</td><td>$-P(=O)(OH)_2$</td></tr>
<tr><td>Perfluorophosphinic acids (PFPiA)</td><td>$-P(=O)(OH)(C_mF_{2m+1})$</td></tr>
<tr><td colspan="2">Perfluoroalkane sulfonyl fluorides (PASF)</td><td>$-SO_2F$</td></tr>
<tr><td colspan="2">Perfluoroalkane sulfonamides (FASA)</td><td>$-SO_2NH_2$</td></tr>
<tr><td colspan="2">Perfluoroalkanoyl fluorides (PAF)</td><td>-COF</td></tr>
<tr><td colspan="2">Perfluoroalkyl iodides (PFAI)</td><td>-I</td></tr>
<tr><td colspan="2">Perfluorinated aldehydes (PFAL), perfluorinated aldehyde hydrates (PFAL, H_2O)</td><td>-CHO and $-CH(OH)_2$</td></tr>
</table>

Table 11.1. *Classification and chemical structure of perfluoroalkylated molecules*

The molecules the most often analyzed and mentioned in the literature are PFOA and PFOS. Due to the surfactant properties of perfluoroalkane sulfonic acids (PFSAs) and perfluoroalkyl carboxylic acids (PFCAs), it is difficult to determine some of their properties such as the octanol-water partition coefficient (Kow), despite the fact that this parameter is very often used to predict the adsorption of compounds onto activated carbon.

PFOA is a fully fluorinated organic acid, whereas PFOS appears as a fluorinated anion commonly used in the salt form (sodium, potassium and ammonium), or incorporated into polymers (Table 11.2). The structure of PFOS (C8) includes a functional group R which can be OH, NH_2, etc

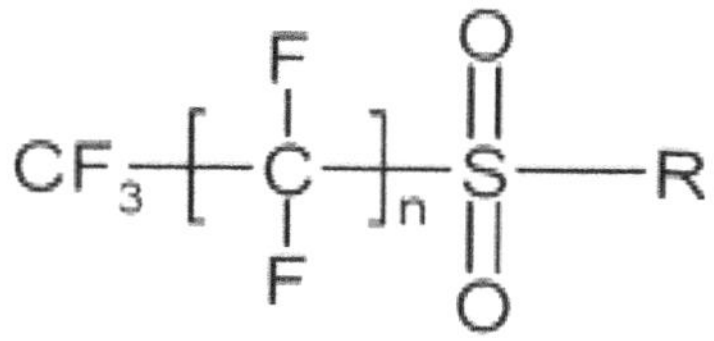

Figure 11.1. *PFOS structure*

Name	Perfluorooctane sulfonic acid (PFOS)	Perfluorooctanoic acid (PFOA)
Raw formula	$C_8F_{17}SO_3H$	$C_8F_{15}O_2H$
Molecular structure	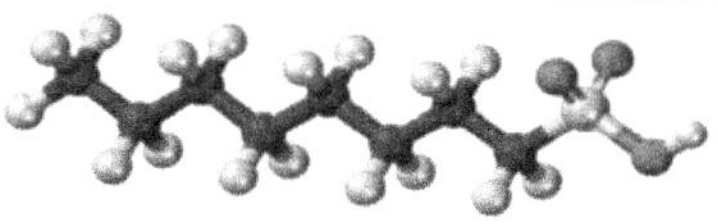	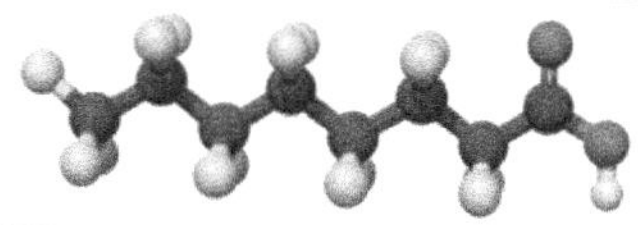
pKa	pKa = 0.14 (-SO3H) Anionic at pH 7	pKa = 2.8 (-COOH) Anionic at pH 7
Molar mass	500 g/mole	414 g/mole
Solubility in water	570 mg/L at 25 °C (1)	4340 mg/L at 25 °C (2)
Henry's constant	3.05, 10^{-9} atm·m^3 $mole^{-1}$	–
Log Kow	2.10 – 2, 57	1, 47 – 2, 06

Table 11.2. *Physicochemical characteristics of PFOS and PFOA. For a color version of this table, see www.iste.co.uk/gaid/watertreatment3.zip*

11.2. Presence in the water

PFAS are water soluble and can be present in groundwater, runoff water, discharges from sewage treatment stations and industrial discharges.

Some reports indicate that the presence of PFAS in wastewater (where they are not removed) could be the main source of contamination in surface waters. Other sources argue that they may be partly concentrated in sludge, and later discharged in the environment through sludge spreading.

Therefore, their presence in drinking water can create removal difficulties for many conventional treatment systems.

11.3. Drinking water regulations

The draft European directive on the quality of potable water proposes values of 0.1 $\mu g \cdot L^{-1}$ for each PFAS separately and 0.5 $\mu g \cdot L^{-1}$ for total PFAS.

By PFAS, the EU refers to each particular per- and polyfluorinated alkylated substance (chemical formula: $C_nF_{2n+1}{}^{-}R$), according to the OECD nomenclature.

The seven PFAS substances analyzed in potable water are the following: perfluorooctane sulfonate (PFOS), perfluorohexane sulfonate (PFHxS), perfluorobutane sulfonate (PFBS), perfluorooctanoic acid (PFOA), perfluoroheptanoic acid (PFHpA), perfluorohexanoic acid (PFHxA) and perfluoropentanoic acid (PFPeA). The quantification limit for PFOA and PFOS is 4 $ng \cdot L^{-1}$.

11.4. Treatments

11.4.1. *Coagulation–flocculation–settling (or flotation)*

Clarification stages are not effective. The results obtained by Veolia, during the study of the efficiency of treatment stations in South West Queensland in Australia, showed that conventional coagulation with aluminum salts (e.g. 10–60 $mg \cdot L^{-1}$ aluminum sulfate at pH 6.5–7) is ineffective (<10%) for the treatment of PFOS and PFOA. Other jar-tests have confirmed that a less than 15% removal of PFOS and PFOA is obtained with ferric chloride as a coagulant.

11.4.2. *Chemical oxidation*

Fluorine is the most electronegative element and as such resists oxidation in order to retain its electrons. PFAS do not contain aromatic bonds or phenolic structures. Ozone doses and contact times reaching 0.87–2 $mg \cdot L^{-1}$ and 120 min are efficient to oxidize PFOA and PFOS. The presence of the strong C-F bond as well as the electron-withdrawing functional groups -COOH and -SO_3H in the structures of PFCAs and PFASs explains why these compounds are resistant to oxidation, even by molecular ozone and hydroxyl radicals. PFOA does not easily react with hydroxyl radicals OH•, and the second-order reaction rate constant is very low.

Given the low reactivity of PFAS, as much with ozone as with advanced oxidation and due to their lower redox potentials (E0 = 1.36–1.50 V), chlorine-based oxidation processes do not oxidize PFAS under the operating conditions of drinking water plants.

Therefore, oxidants (chlorine, chlorine dioxide, ozone, $KMnO_4$) have no impact on the concentrations of these two molecules due to the high dissociation energy required to break C-F bonds.

11.4.3. *UV oxidation*

The tests carried out at 254 nm and 365 nm, for a 60 min contact time in the reaction zone, have shown that the absorbance values of PFOA increase sharply after irradiation, going as far as doubling the initial absorbance. This can be explained by the decomposition of PFOA into by-products due to absorbance titration and which interfere in the evaluation of the final concentrations.

11.4.4. *Activated carbon*

The adsorption of organic molecules onto the surface of activated carbon is based on mechanisms dependent on physical processes, such as electrostatic attractions (dipole–dipole interactions, van der Waals type-interactions), or chemisorption (Figure 11.2). Indeed, nonpolar molecules are attracted to one another via dipole–dipole interactions called London forces (hydrophobic bonds). They are called dipole–dipole because the interaction mechanism is based on the attractiveness of the positive charge at the end of polar molecules with the negative charge of non-polar molecules. Hydrogen bonds, which also belong to this type of interactions, are formed with molecules containing fluorine, oxygen or nitrogen atoms whose negative charge participates in this type of connection.

The parameters obtained using the adsorption isotherms (Langmuir and Freundlich) provide information about the adsorption mechanism, the surface properties as well as adsorbate–adsorbent affinity.

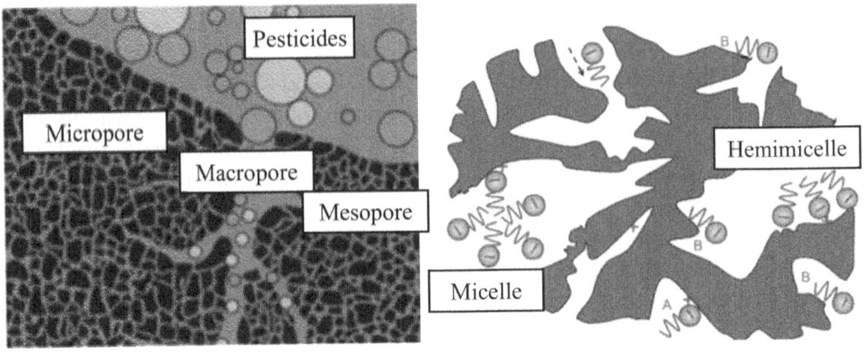

Figure 11.2. *Adsorption onto activated carbon. For a color version of this figure, see www.iste.co.uk/gaid/watertreatment3.zip*

11.4.4.1. *Granular activated carbon*

The literature presents numerous results of PFAS removal using different types of granular activated carbon (GAC). The results obtained show that the efficiency in the treatment of PFOS and PFOA reaches 95%, rapidly decreasing with time. For other PFAS substances, efficiency depends on the length of the carbon chain and its structure. During GAC filtration, the contents of PFAS with the longest carbon chains, such as PFNA (perfluorononanoic acid) and PFOS, are well removed, with output concentrations lower than the quantification limits (4 $ng \cdot L^{-1}$).

PFAS with short carbon chains, like perfluorobutanoic acid (PFBA) and perfluorobutane sulfonate (PFBS), are well removed but induce shorter saturation cycles. There are two categories of PFAS: short-chain (≤6 carbon atoms) and long-chain (≥8 carbon atoms). For long-chain molecules, sulfonated compounds are better removed than carboxylated compounds. Better efficiency is observed for PFOS compared to PFOA.

For a pH of 7.2 on contaminated groundwater, the removal capacity with Filtrasorb 400 GAC is around 85 $mg \cdot g^{-1}$ for PFOS and 45 $mg \cdot g^{-1}$ for PFOA. On surface waters comprising a variable dissolved organic carbon (DOC) between 3 and 4 $mg \cdot L^{-1}$, the removal capacity with GAC (wood base) is 48 $mg \cdot g^{-1}$ for PFOS and 32 $mg \cdot g^{-1}$ for PFOA.

The phenomenon of size exclusion is included in the adsorption mechanism: microporosity, which refers to the presence of high sorption energy sites in the micropores (<20 Å), facilitates the retention of molecules whose size is smaller than this diameter. The length and molecular diameter of a PFOA molecule are estimated at 8 Å and 3.55 Å, respectively. For PFOS, they are 13 Å and 4.0 Å, respectively. For this type of molecules, for example, it is coconut-based activated carbons, which are preferred, since they have a wide micropore distribution (Table 11.3) and are well suited to PFAS. In addition, the low molecular weight of PFAS (less than 550 g per mole) facilitates their transport to the micropores.

Bituminous carbons and wood-based carbons generally have quite developed porous structures, and should therefore lead to a good reduction of PFOS and PFOA. They adsorb organic matter satisfyingly and are well suited for surface waters. This is why laboratory tests or pilot tests are the best way to select the most suitable activated carbon for the water matrix to be treated.

A contact time of 10–15 min is recommended based on the composition of the water to be treated. The treatment of PFOS and PFOA in groundwater with a low concentration of DOC is easier and more effective than treating surface water with a high DOC concentration.

In fact, the presence of DOC accentuates the phenomenon of competition and decreases the lifetime of GAC. DOC leakage is faster and requires activated carbon renewal, although there may still be pore availability for PFOS and PFOA.

Figure 11.3 shows the evolution of PFAS concentrations in groundwater (DOC < 1 $mg \cdot L^{-1}$, in [PFAS] = 0.22 $\mu g \cdot L^{-1}$) processed on two types of GAC (Table 11.3). Figure 11.4 shows the evolution of PFAS concentrations in surface water (TOC = 7.4 $mg \cdot L^{-1}$, [PFAS] = 0.22 $\mu g \cdot L^{-1}$). In both cases, the contact times were 10 min.

Parameters	Coconut base	Coal base	Wood base
Micropore volume	High	Low to medium	Low to medium
Mesopore volume	Medium	Medium to high	Medium to high
Macropore volume	Low	Low	High
Mhos hardness	High	High	Medium
Ashes	<5%	3–8%	3–8%
Apparent density	0.35–0.50 $g \cdot cm^{-3}$	0.32–0.60 $g \cdot cm^{-3}$	0.24–0.39 $g \cdot cm^{-3}$
Iodine number	1,100 $m \cdot g^{-1}$	1,000 $m \cdot g^{-1}$	1,100 $m \cdot g^{-1}$
Specific surface	1,480 $m^2 \cdot g^{-1}$	1,400 $m^2 \cdot g^{-1}$	1,480 $m^2 \cdot g^{-1}$

Table 11.3. *Characteristics of some granular activated carbons tested*

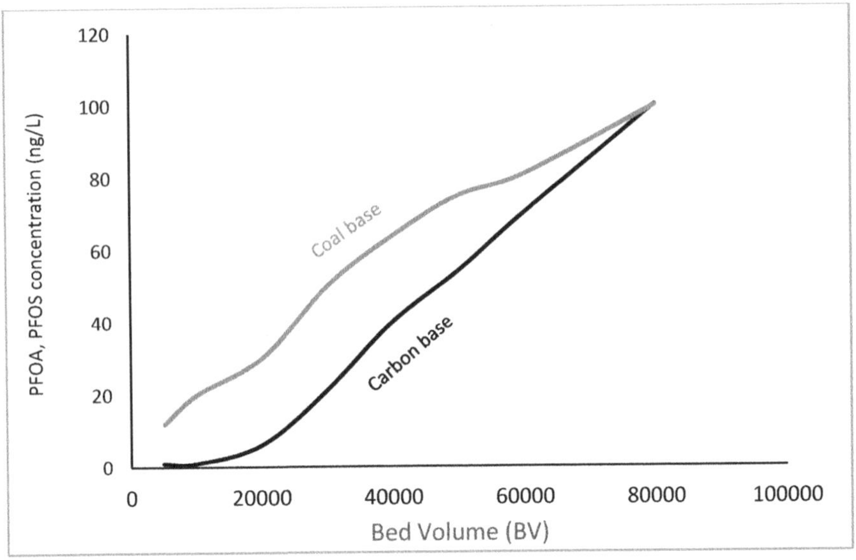

Figure 11.3. *PFOA and PFOS treatment on groundwater using GAC*

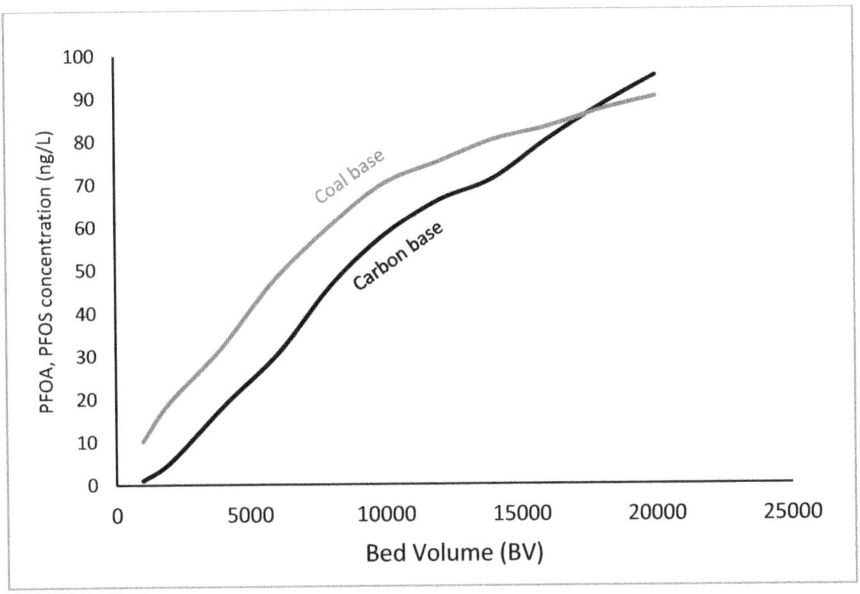

Figure 11.4. *PFOA and PFOS treatment on surface water using GAC*

The results show that a BV of 50,000 is obtained in groundwater, with a PFAS leak of 0.1 $\mu g \cdot L^{-1}$. On surface water (Figure 11.4), the BV is only 11,000 with a rapid increase in DOC and a PFAS outlet of 0.1 $\mu g \cdot L^{-1}$, which illustrates the effect of natural organic matter (NOM) on the competition for adsorption. It is therefore important for the treatment of DOC to be optimized (DOC inlet CAG < 2 $mg \cdot L^{-1}$) before GAC filtration to increase its life span.

11.4.4.2. *Powdered and µgrain activated carbon*

The parameters given for GAC are also valid for powdered activated carbon (PAC) or µgrain activated carbon (µGAC). However, other factors are at play regarding PAC, such as pore diffusion, which increases with the reduction of particle diameter (15–25 µm). The fine particle size associated with the large specific surface increases the exchange surface, and therefore, the number of sites available for the diffusion of PFOA and PFOS molecules.

The activated carbon's surface is negatively charged when the pH is higher than the PZC pH, whereas it is positively charged when the pH is lower than the PZC pH, in which case the activated carbon will have a great affinity with anionic compounds. The electrostatic repulsion with the surface of negatively charged activated carbon (PZC < pH solution) leads to lower PFOA and PFOS adsorptions.

The evaluation of adsorption capacities according to the Langmuir and Freundlich models leads to very interesting values. The carbon used is Aquasorb MP 25 (Jacobi).

Molecules	**Langmuir parameters**			**Freundlich parameters**		
	Qmax $mg \cdot g^{-1}$	**b $L \cdot mg^{-1}$**	**qe $mg \cdot g^{-1}$**	**K $mg \cdot g^{-1} \cdot L^{-1/n} \cdot mg^{-1/n}$**	**1/n**	**X/M $mg \cdot g^{-1}$**
PFOS	520	2,750	112	635	0.18	121
PFOA	278	1,950	54	343	0.2	54

Table 11.4. *Langmuir and Freundlich parameters for PFOA and PFOS adsorption onto powdered activated carbon (Veolia tests)*

The tests carried out on highly DOC-concentrated leachates and containing PFCs (PFOS and PFOA) are summarized in Table 11.5.

As DOC is high, it is important to remove organic matter as much as possible in order to reduce the competition of PFCs against adsorption mechanisms onto activated carbon. The DOC in leachates is made of molecules that are difficult to remove by conventional $FeCl_3$ doses. There is no comparison with the treatment of humic substances. A 500 mg·L^{-1} dose of commercial $FeCl_3$ (41%) was required to obtain a 31% reduction in DOC and a 1,200 mg·L^{-1} dose at pH 5.5–6 was necessary to attain a 63% DOC removal.

Parameters	Inlet leachate value	After Actiflo®	After PAC addition
pH	7.46	5.6	5.6
Turbidity (NTU)	80.7	0.37	<0.3
UV 254 (m^{-1})	57.6	21.4	<7
Alkalinity (mg$CaCO_3$·L^{-1})	492.9	72	–
Hardness (mg$CaCO_3$·L^{-1})	1,490	–	–
TOC (mg·L^{-1})	15.94	5.8	<2
PFOS (ng·L^{-1})	200	190	5.1
PFOA (ng·L^{-1})	75	75	1.9

Table 11.5. *Physicochemical composition of leachate contaminated with fluorinated compounds (Veolia tests)*

The addition of PAC (Norit SAE) downstream of this coagulation stage (DOC = 5.8 mg·L^{-1}) involves a PAC dosage of 150 g·m^{-3} to attain a PFOS concentration of 0.1 µg·L^{-1}, as well as a PAC dose of 300 mg·L^{-1} to reach around 6 ng·L^{-1} (97%) in PFOS and less than 2 ng·L^{-1} in PFOA.

A wood-based (thermally activated) PAC only removed a maximum of 55% of PFDA (C10) and PFOS in the presence of 4.5 mg·L^{-1} of TOC. The significant removal of PFBA, PFPeA, PFHxA, PFHpA and PFBS from drinking water may not be obtained with conventional PAC doses. Unlike the Freundlich model, the Langmuir model can illustrate maximum sorption. The results of the Langmuir model adjustment indicate that the sorption capacity of PFOS (72.2 ± 19.0 mg·g^{-1}) is greater than that of PFOA (52.8 ± 17.1 mg·g^{-1}) and PFHpA (41.3 ± 9.2 mg·g^{-1}).

The high doses of PAC are justified by the fact that the competition of DOC is high compared to the low concentrations of PFCs. Identical results were obtained with µGAC (Microsorb 400 R).

The DOC at the output of these tests was between 1 and 2.5 mg·L^{-1} for an input of 15.9 mg·L^{-1} DOC, indicating a global 85.5–93.7% removal of the organic matter present. PFCs are removed over 97%.

As these compounds are not removed by coagulants based on metallic salts, it is necessary to reduce the competitiveness of DOC. A two-stage removal is recommended: the first stage of TOC removal using $FeCl_3$ in an acid medium, followed by a PAC reactor for the removal of PFCs.

A treatment with a PAC reactor (Actiflo® Carb or Multiflo® Carb) or a μGAC reactor (Filtraflo® Carb or Opacarb® FL) would consume 30 times less material due to the carbon mass present in the reactor and which is continuously renewed. As a result, the performance of PAC is better than that of GAC. Veolia technologies make it possible to respond to the removal of these compounds in addition to removing organic carbon, pesticides and organic micropollutants. Within these reactors, PAC adsorbs PFAS faster than GAC due to the particle size of activated carbon, shorter internal diffusion distances, and the surface functional groups available. At present, the municipal water market is emphasizing guarantees on DOC (THM) and micropollutants. PAC reactor technology satisfyingly responds to this problem by also removing PFOA and PFOS.

11.4.5. *High-pressure membranes: nanofiltration and reverse osmosis*

The increasing contamination of surface water and groundwater by various organic molecules (humic substances, endocrine-disrupting compounds, pharmaceutical residues, industrial products of various kinds) encourages the increasingly frequent use of nanofiltration (NF) and low-pressure reverse osmosis (LPRO) membrane processes. This recommendation is justified by advances in energy efficiency, operational efficiency, cost reduction and the ability of membranes to treat a wide range of water contaminants.

The cut-off concentration of these membranes is much lower than the molecular weight of PFAS (maximum 500 Da).

Veolia tests are performed with a NF membrane (200 Da cut-off limit/threshold) as well as a LPRO membrane using a test bench (Figure 11.6) and a pilot ($500\ m^3 \cdot d^{-1}$), as shown in Figure 11.5. The composition of the feed water is summarized in Table 11.6.

Conversion rates have been adjusted between 75 and 90%. The flow varied between 18 and 20 lmh. The results obtained with NF and LPRO membranes are summarized in Tables 11.7 and 11.8, showing the inlet concentrations and those of the permeate.

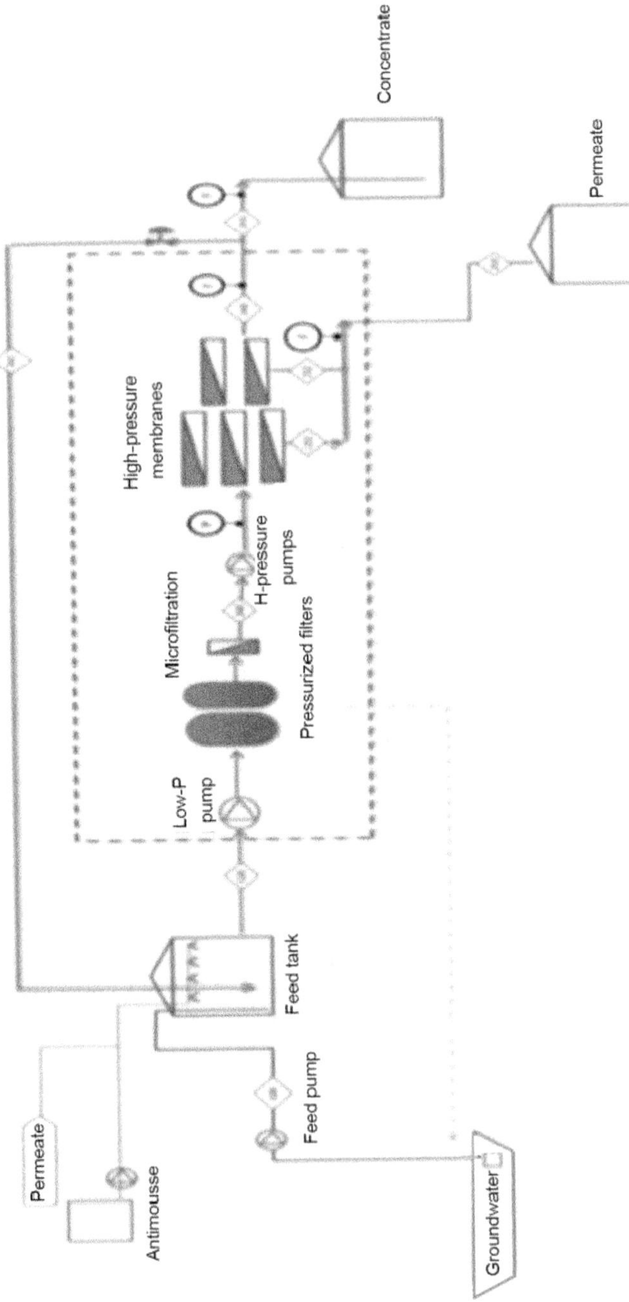

Figure 11.5. *Diagram of the test pilot with high-pressure membranes. For a color version of this figure, see www.iste.co.uk/gaid/watertreatment3.zip*

Parameters	Values	Parameters	Values
pH	7.64	Fe mg·L^{-1}	1.39
Conductivity	1,400	Mg mg·L^{-1}	94.2
Turbidity NTU	9.62	Mn mg·L^{-1}	0.78
Total dissolved solids mg·L^{-1}	944	Hg mg·L^{-1}	0.00
Alkalinity (mg $CaCO_3$·L^{-1})	701.1	NO_3 mg·L^{-1}	5.6
Hardness (mg $CaCO_3$·L^{-1})	615	NO_2 mg·L^{-1}	0.09
Cl mg·L^{-1}	189	P mg·L^{-1}	0.09
F mg·L^{-1}	0.8	K mg·L^{-1}	10.9
I mg·L^{-1}	109	$Si0_2$ mg·L^{-1}	10.9
Al mg·L^{-1}	0.8	Na mg·L^{-1}	120
NH_4 mg·L^{-1}	0.04	SO_4 mg·L^{-1}	72.5
Ba mg·L^{-1}	0.26	Zn mg·L^{-1}	0.02
Ca mg·L^{-1}	192	B mg·L^{-1}	0.12

Table 11.6. *Feed water chemical composition*

The results reveal that reverse osmosis grants a reduction over 95%. NF enables a variable rejection percentage between 84 and 98%, depending on the 11 PFCs (Figure 11.7), their molecular weight and geometric structure. The complete removal (99%) of PFOA and PFOS by an LPRO membrane has also been reported. Data collected from these studies suggest that high-pressure membranes are capable of removing fluorinated compounds with high efficiencies.

Although size is probably the dominant factor, solute-membrane interactions which depend on factors such as charge, hydrophobicity and dipole moment must be significant if the molecular weight of the solute is near or below the membrane's MWCO. But this is not the case for the PFCs tested, which all have a molecular weight greater than 200 Da. The membrane pore size cut-off to molecular weight cut-off (MWCO) ratio probably plays the most important role in relation to the rejection of PFAS by NF/LPRO membranes.

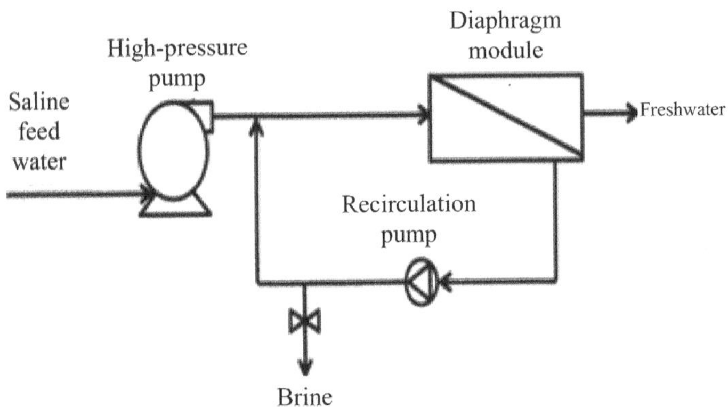

Figure 11.6. *Diagram of the test pilot with high-pressure membranes*

Low-pressure reverse osmosis	Inlet C ($mg \cdot L^{-1}$)	Permeate concentration at different conversion rates ($mg \cdot L^{-1}$)				
Compounds		75	80	85	90	95
1H.1H.2H.2H-perfluorodecane sulfonic acid (8: 2 FTS)	0.019	0.000	0.000	0.000	0.000	0.000
1H.1H.2H.2H-perfluorooctane sulfonic acid (6: 2 FTS)	0.41	0.027	0.027	0.027	0.028	0.028
Perfluorobutane sulfonic acid (PFBS)	0.24	0.025	0.025	0.026	0.026	0.026
Perfluorobutanoic acid (PFBA)	0.34	0.038	0.039	0.039	0.040	0.040
Perfluoroheptanoic acid (PFHpA)	0.24	0.023	0.023	0.024	0.024	0.024
Perfluorohexane sulfonic acid (PFHxS)	1.50	0.095	0.096	0.098	0.099	0.101
Perfluorohexanoic acid (PFHxA)	2.30	0.239	0.242	0.245	0.249	0.253
Perfluorononanoic acid (PFNA)	0.014	0.000	0.000	0.000	0.000	0.000
Perfluorooctane sulfonic acid (PFOS)	2.30	0.029	0.029	0.030	0.030	0.031
Perfluorooctanoic acid (PFOA)	0.60	0.031	0.032	0.032	0.033	0.033
Perfluoropentanoic acid (PFPeA)	0.52	0.055	0.056	0.056	0.057	0.058

Table 11.7. *Efficacy of LPRO on perfluorinated substances (Veolia tests)*

Nanofiltration	Inlet C (mg·L^{-1})	Permeate concentration at different conversion rates (mg·L^{-1})				
Compounds		75	80	85	90	95
1H.1H.2H.2H-perfluorodecane sulfonic acid (8: 2 FTS)	0.01	0.000	0.000	0.000	0.000	0.000
1H.1H.2H.2H-perfluorooctane sulfonic acid (6: 2 FTS)	0.37	0.000	0.000	0.000	0.000	0.000
Perfluorobutane sulfonic acid (PFBS)	0.21	0.031	0.032	0.032	0.032	0.033
Perfluorobutanoic acid (PFBA)	0.33	0.034	0.035	0.035	0.036	0.036
Perfluoroheptanoic acid (PFHpA)	0.20	0.008	0.008	0.008	0.008	0.008
Perfluorohexane sulfonic acid (PFHxS)	1.30	0.057	0.058	0.059	0.060	0.061
Perfluorohexanoic acid (PFHxA)	2.20	0.108	0.109	0.111	0.113	0.115
Perfluorononanoic acid (PFNA)	0.008	0.000	0.000	0.000	0.000	0.000
Perfluorooctane sulfonic acid (PFOS)	1.60	0.006	0.006	0.006	0.006	0.006
Perfluorooctanoic acid (PFOA)	0.54	0.008	0.008	0.008	0.008	0.008
Perfluoropentanoic acid (PFPeA)	0.45	0.033	0.033	0.034	0.034	0.035

Table 11.8. *Efficiency of NF on perfluorinated substances (Veolia tests)*

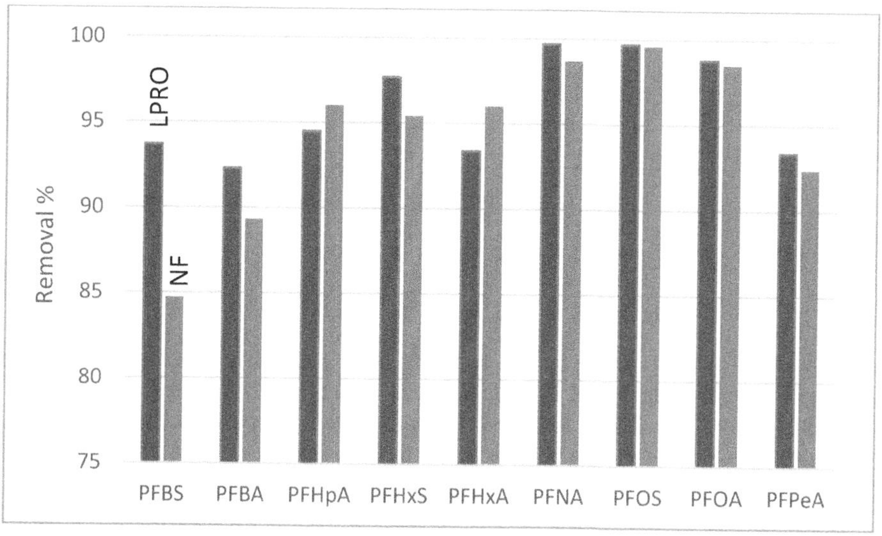

Figure 11.7. *PFAS removal on LPRO and NF membranes, conversion efficiency: 85%, T: 25°C, flow: 19 lmh (Veolia tests). For a color version of this figure, see www.iste.co.uk/gaid/watertreatment3.zip*

11.5. Conclusion

It appears that coagulation, flocculation and sedimentation (or flotation) only lead to an efficiency of <10%. The free residual chlorine commonly used for disinfection or maintenance of distribution systems is ineffective for removing PFAS. Oxidation and advanced oxidation processes (AOP), under the operating conditions applied to drinking water plants, do not oxidize most PFAS. To a certain extent, the oxidation of FTOHs (fluorotelomer alcohols) and FASAs (FASA) may be possible. However, these could simply be oxidized into other PFAS. The UV irradiation used, both at conventional disinfection doses (400 $mJ \cdot cm^{-2}$) and at much higher doses, has also proven to be ineffective.

Compounds	MM	Flotation/ Filtration	Coag/Floc/Set/ Filtration	Resins	GAC	NF	RO	Oxidation
PFOA	414							
PFOS	500							
PFBA	214							
PFPeA	264							
PFHxA	314							
PFHpA	364							
PFNA	464							
PFDA	514							
PFBS	300							
PFHxS	400							

Table 11.9. *Effectiveness of various treatments: <10% shown in yellow; 10–60% shown in green; >90% shown in purple. For a color version of this table, see www.iste.co.uk/gaid/watertreatment3.zip*

Granular, micrograin and PAC are effective and reach reductions over 90%. Long-chain PFASs are better removed than short-chain compounds. However, short chain PFAS such as PFBA and PFBS quickly saturate the GAC and these compounds should be monitored regularly when removal is required. The efficiency of GAC may be compromised in the presence of NOM, because the presence of DOC accentuates the competition phenomenon. The GAC lifetime is guided much more by DOC saturation than by PFOS and PFOA saturation. DOC leakage is faster and requires activated carbon renewal, although there may still be pore availability for PFOS and PFOA. PAC and µGAC are effective when used in reactors where the concentration of activated carbon is high and enables the frequent renewal of fresh carbon, which is key for obtaining constant efficiency in PFAS removal. The biodegradation of most PFAS in GAC filters is inefficient. NF/LPRO membranes are very effective for most PFAS. However, low molecular weight PFAS (such as PFBA and PFPeA) may be removed less successfully, depending on the type of NF membrane used. Table 11.9 summarizes the efficiency of the different treatments.

11.6. References

Anses (2014). Note d'appui scientifique et technique de l'Agence nationale de sécurité sanitaire de l'alimentation, de l'environnement et du travail relative aux composés perfluorés présents dans les eaux destinées à la consommation humaine. Report, Saisine, Maison Alfort.

Appleman, T.D., Dickensona, E.R.V., Bellonac, C., Higgins, C.P. (2013). Nanofiltration and granular activated carbon treatment of perfluoroalkyl acids. *J. Hazard. Mater.*, 260, 740–746.

Arcadis (2016). Environmental fate and effects of poly- and perfluoroalkyl substances (PFAS). Report, Concawe/NICOLE, Brussels.

Boiteux, V., Dauchy, X., Bach, C., Colin, A., Hemard, J., Sagres, V., Rosin, C., Munoz, J.F. (2017). Concentrations and patterns of perfluoroalkyl and polyfluoroalkyl substances in a river and three drinking water treatment stations near and far from a major production source. *Sci. Total Environ.*, 583(1), 393–400.

Buck, R.C., Franklin, J., Berger, U., Conder, J.M., Cousins, I.T., Voogt, P.D., Jensen, A.A., Kannan, K., Mabury, S.A., Van Leeuwen, S.P.J. (2011). Perfluoroalkyl and polyfluoroalkyl substances in the environment: Terminology, classification, and origins. *Integr. Environ. Assess. Manag.*, 7(4), 513–541.

Carter, K.E. and Farrell, J. (2010). Removal of perfluorooctane and perfluorobutane sulfonate from water via carbon adsorption and ion exchange. *Sep. Sci. Technol.*, 45, 762–767.

Corwin, C.Y. and Summers, R.S. (2010). Scaling trace organic contaminant adsorption capacity by granular activated carbon. *Environ. Sci. Technol.*, 44, 5403–5408.

Dickenson, E. and Drewes, J. (2010). Quantitative structure property relationships for the adsorption of pharmaceuticals. *Water Sci. Technol.*, 62, 2270–2276.

Eric, R.V. (2016). Treatment mitigation strategies for poly- and perfluoroalkyl substances. Working document, Water Research Foundation, Henderson.

Eschauzier, C., de Voogt, P., Brauch, H.J., Lange, F.T. (2012). Polyfluorinated chemicals in European surface waters, ground and drinking waters. In *Polyfluorinated Chemicals and Transformation Products, Handbook of Environmental Chemistry*, Knepper, T., Lange, F.T. (eds). Springer, Berlin.

Faria, P.C., Orfao, J.J., Pereira, M.F. (2004). Adsorption of anionic and cationic dyes on activated carbons with different surface chemistries. *Water Res.*, 38(8), 2043–2052.

Flores, C., Ventura, F., Martin-Alonso, J., Caixach, J. (2013). Occurrence of perfluorooctane sulfo-nate (PFOS) and perfluorooctanoate (PFOA) in N.E: Spanish surface water and their removal in a drinking water treatment station that combines conventional and advanced treatment in parallel lines. *Sci. Total Environ.*, 461–462, 618–626.

Gnagy, M. (2019). Rapport interne Veolia sur les technologies d'élimination des PFOA et PFOS. Report, Veolia, Aubervilliers.

Hansen, M.C., Borresen, M.H., Schlabach, M., Cornelissen, G. (2010). Sorption of perfluorinated compounds from contaminated water to activated carbon. *J. Soils Sedim.*, 10(2), 179–185.

Kaiser, M.A., Barton, C.A., Botelho, M., Buck, R.C., Buxton, L.W., Gannon, J., Kao, C.C., Larsen, B.S., Russel, M.H., Wang, N., Waterland, R.L. (2006). Understanding the transport of anthropogenic fluorinated compounds in the environment. *Organohalogen Compound*, 68, 675–678.

Kucharzyk, K.H., Darlington, R., Benotti, M., Deeb, R., Hawley, E. (2017). Novel treatment technologies for PFAS compounds: A critical review. *J. Environ. Manage.*, 204, 757–764.

Kudo, N. and Kawashima, Y. (2003). Toxicity and toxicokinetics of perfluorooctonoic acid in humans and animals. *J. Toxicol. Sci.*, 28(2), 49–57.

Lau, C. (2012). Perfluorinated compounds. *EXS*, 101, 47–86.

Lipp, P., Sacher, F., Baldauf, G. (2010). Removal of organic micro-pollutants during drinking water treatment by nanofiltration and reverse osmosis. *Desalination Water Treat.*, 13(1–3), 226–237.

Loi-Brügger, A., Panglisch, S., Hoffmann, G., Buchta, P., Gimbel, R., Nacke, C. (2008). Removal of trace organic substances from river bank filtrate – Performance study of RO and NF membranes. *Water Sci. Technol. Water Supply*, 8(1), 85–92.

Matsui, Y., Knappe, D.R., Takagi, R. (2002). Pesticide adsorption by granular activated carbon adsorbers: Effect of natural organic matter preloading on removal rates and model simplification. *Environ. Sci. Technol.*, 36, 3426–3431.

Matsui, Y., Fukuda, Y., Inoue, T., Matsushita, T. (2003). Effect of natural organic matter on powdered activated carbon adsorption of trace contaminants: Characteristics and mechanism of competitive adsorption. *Water Res.*, 37, 4413–4424.

McCleaf, P., Englund, S., Östlund, A., Lindegren, K., Wiberg, K., Ahrens, L. (2017). Removal efficiency of multiple poly and perfluoroalkyl substances (PFASs) in drinking water using granular activated carbon (GAC) and anion exchange (AE) column. *Water Res.*, 120, 77–87.

McNamara, J.D., Franco, R., Mimna, R., Zappa, L. (2018). Comparison of activated carbons for removal of perfluorinated compounds from drinking water. *J. Am. Water Works Assoc.*, 110, 1–14.

Ochoa-Herrera, V. and Sierra-Alvarez, R. (2008). Removal of perfluorinated surfactants by sorption onto granular activated carbon, zeolite and sludge. *Chemosphere*, 72, 1588–1593.

OECD (2013). Synthesis paper on per and polyfluorinated chemicals. Working document, OECD/UNEP, Paris.

Post, G.B., Cohn, P.D., Cooper, K.R. (2012). Perfluorooctanoic acid (PFOA), an emerging drinking water contaminant: A critical review of recent literature. *Environ. Res.*, 116, 93–117.

Qu, Y., Zhang, C., Li, F., Bo, X., Liu, G., Zhou, Q. (2009). Equilibrium and kinetics study on the adsorption of perfluorooctanoic acid from aqueous solution onto powdered activated carbon. *J. Hazard. Mater.*, 169(1–3), 146–152.

Rahman, M.F., Peldszus, S., Anderson, W.B. (2014). Behaviour and fate of perfluoroalkyl and polyfluoroalkyl substances (PFASs) in drinking water treatment: A review. *Water Res.*, 50, 318–340.

Shivakoti, B.R., Fujii, S., Nozoe, M., Tanaka, S., Kunacheva, C. (2010). Perfluorinated chemicals (PFCs) in water purification plants (WPPs) with advanced treatment processes. *Water Sci. Technol. Water Supply*, 10(1), 87–95.

Steenland, K., Fletcher, T., Savitz, D.A. (2010). Epidemiologic evidence on the health effects of perfluorooctanoic acid (PFOA). *Environ. Health Perspect.*, 118(8), 1100–1108.

Steinle-Darling, E. and Reinhard, M. (2008). Nanofiltration for trace organic contaminant removal: Structure, solution, and membrane fouling effects on the rejection of perfluorochemicals. *Environ. Sci. Technol.*, 42(14), 5292–5297.

Stock, N.L., Muir, D.C.G., Mabury, S. (2009). Perfluoroalkyl compounds. In *Persistent Organic Pollutants*, Harrad, S. (ed.). Wiley, New York.

Takagi, S., Adachi, F., Miyano, K., Koizumi, Y., Tanaka, H., Mimura, M., Watanabe, I., Tanabe, S., Kannan, K. (2008). Perfluorooctanesulfonate and perfluorooctanoate in raw and treated tap water from Osaka, Japan. *Chemosphere*, 72(10), 1409–1412.

Takagi, S., Adachi, F., Miyano, K., Koizumi, Y., Tanaka, H., Watanabe, I., Tanabe, S., Kannan, K. (2011). Fate of perfluorooctanesulfonate and perfluorooctanoate in drinking water treatment processes. *Water Res.*, 45(13), 3925–3932.

Tang, C.Y.Y., Fu, Q.S., Robertson, A.P., Criddle, C.S., Leckie, J.O. (2006). Use of reverse osmosis membranes to remove perfluorooctane sulfonate (PFOS) from semiconductor wastewater. *Environ. Sci. Technol.*, 40, 7343–7349.

Thompson, J., Eaglesham, G., Reungoat, J., Poussade, Y., Bartkow, M., Lawrence, M., Mueller, J.F. (2011). A removal of PFOS, PFOA and other perfluoroalkyl acids at water reclamation plants in South East Queensland Australia. *Chemosphere*, 82, 9–17.

Vecitis, C.D., Park, H., Cheng, J., Mader, B.T., Hoffmann, M.R. (2009). Treatment technologies for aqueous perfluorooctanesulfonate (PFOS) and perfluorooctanoate (PFOA). *Front. Environ. Sci. Eng.*, 3(2), 129–151.

Veolia (2019). Direction technique et Performance, élimination de PFOS et PFOA sur des lixiviats. Report, Veolia, Aubervilliers.

Xiao, F., Simcik, M.F., Gulliver, J.S. (2013). Mechanisms for removal of perfluorooctane sulfonate (PFOS) and perfluorooctanoate (PFOA) from drinking water by conventional and enhanced coagulation. *Water Res.*, 47, 49–56.

Yu, Z., Peldszus, S., Huck, P.M. (2009). Adsorption of selected pharmaceuticals and an endocrine disrupting compound by granular activated carbon. Adsorption capacity and kinetics. *Environ. Sci. Technol.*, 43(5), 1467–1473.

Yu, Q., Zhanga, R., Denga, S., Huanga, J., Yua, G. (2009). Sorption of perfluorooctane sulfonate and perfluorooctanoate on activated carbons and resin: Kinetic and isotherm study. *Water Res.*, 43, 1150–1158.

12

Biological Removal of Ammonia

In groundwater, the presence of ammonia does not constitute a pejorative criterion, but its occurrence should be monitored to make sure that there is no contamination by wastewater infiltration.

In surface waters, ammonia originates from a phase in the degradation process of organic matter, be it a wastewater discharge (whose nitrogen has not been sufficiently treated), or the spreading of nitrogenous fertilizers, discharged by soil leaching.

In the natural environment, like all nitrogen compounds, ammonia promotes the growth of algae, thus accelerating the eutrophication of water collections. The presence of ammonia in distributed water can promote the development of microbial flora, entailing the appearance of unpleasant tastes. It also has the disadvantage of requiring an increase in the amount of chlorine used for disinfection, a part of which is transformed into chloramines

At present, the two most widely used techniques are chlorination beyond the "break point" and biological nitrification. Breakpoint chlorination is extensively detailed in Chapter 21.

12.1. The principle of biological nitrification

The oxidation of ammoniacal nitrogen to nitric nitrogen is done biologically. Nitrification is caused by autotroph bacteria and strict aerobes, which are always present in water. They are also necessarily chemolithotrophs, using the oxidation of mineral substrates as the sole source of energy, and carbon dioxide as the exclusive source of carbon.

– The phenomenon takes place in two stages:

a) Oxidation of the ammonium ion to nitrite:

$$NH_4^+ + 3/2\ O_2 \rightarrow NO_2^- + H_2O + 2\ H^+ + 351\ kJ$$

This phase, called nitritation, is caused by *Nitrosomonas*, *Nitrobacter*, *Nitrosococcus*, and *Nitrospira* bacteria.

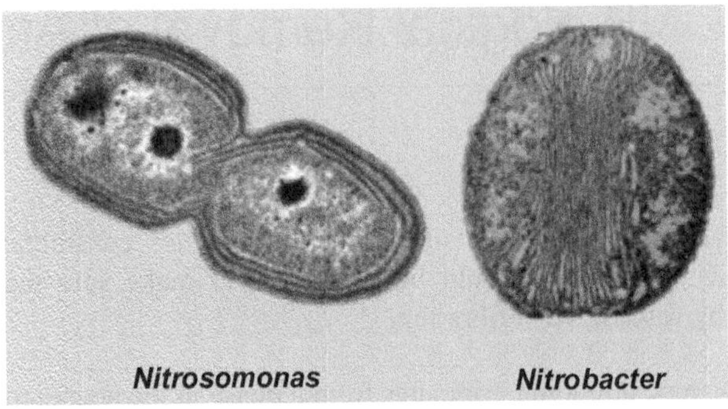

Figure 12.1. *Nitrosomonas and Nitrobacter bacteria. For a color version of this figure, see www.iste.co.uk/gaid/watertreatment3.zip*

b) An oxidation of nitrites into nitrates, for which bacteria of the genus *Nitrobacter* are responsible:

$$NO_2^- + 1/2\ O_2 \rightarrow NO_3^- + H_2O + 2\ H^+ + 74\ kJ$$

In broad terms, the oxidation equation is written as:

$$NH_4^+ + 2\ O_2 \rightarrow NO_3^- + H_2O + 2\ H^+ + 425\ kJ$$

As we can see, the process takes place without the external addition of a reagent, apart from oxygen at the theoretical rate of 4.57 g O_2 per gram of ammonia nitrogen (or 3.55 g O_2 per g NH_4^+). The general nitrification equation also acknowledges the use of ammonia for the synthesis of bacteria. The global equation is the following:

$$NH_4^+ + 1.83\ O_2 + 1.98\ HCO_3^- + 0.747\ H^+ \rightarrow 0.021C_5H_7NO_2 + 3.29\ H_2O + 1.88\ CO_2 + 0.98\ NO_3^-$$

$C_5H_7NO_2$ represents the biomass produced.

The nitrification of one gram of NH_4^+ leads to the production of 0.16 g of bacteria, which have to be partially extracted from the reactor. This process produces little sludge. This reaction is also accompanied by a consumption of alkalinity and pH, that is, a consumption of 7.07 mg $CaCO_3$ per mg of N-NH_4^+.

The biological process requires 0.2 mg·L^{-1} of phosphates (in P). The optimum pH is between 7.2 and 8.2. Biological activity is greatly slowed down below 10°C and comes to a full stop below 4°C.

Optimal temperatures for the growth of nitrifying bacteria are between 25 and 30°C.

In addition to its main separation and clarification functions, filtration makes it possible for biological treatments to be carried out. In this sense, the filtering mass not only acts as a physical barrier against suspended solids but behaves as a support enabling the attachment and development of an active biomass. The combination of physical and biological effects in the same filter is generally called biofiltration or biological filtration. This biological reactor and its design are based on specific parameters.

This denomination applies to every drinking water biological filtration process involving the intervention of bacteria (denitrification, iron removal, manganese removal, slow filtration, DOC-BDOC removal). In the presence of ammonia, filters are called nitrifying filters.

12.2. Design parameters

It is the physicochemical characteristics of raw water which are evaluated when choosing the most suitable treatment to comply with drinking water quality standards. The main difference between groundwater and surface water is that the former is often anoxic, whereas the latter is aerobic. Anoxic conditions favor the presence of compounds in their reduced form, such as NH_4^+, Fe^{2+}, Mn^{2+}, CH_4 and H_2S. Unlike surface waters, the organic carbon concentration is normally low.

12.2.1. *Dissolved oxygen*

The amount of ammonia nitrogen N-NH_4^+ that is biologically nitrifiable depends on the concentration of dissolved oxygen available. The saturation concentration of O_2 in water as a function of temperature (T °C) and at atmospheric pressure is given by:

$$Cs\ O_2\ (mg{\cdot}L^{-1}) = 14.161 - 0.3943\ T + 0.007714\ T^2 - 0.0000646\ T^3$$

$Cs\ O_2$ is equal to 8.84 $mg{\cdot}L^{-1}$ at 20°C.

Knowing that 3.55 mg of oxygen is required to nitrify 1 $mg{\cdot}L^{-1}$ of NH_4^+, Table 12.1 summarizes the theoretical amount of NH_4^+ removed and the theoretical quantity of nitrates formed, as a function of the temperature for water saturated with O_2.

T°C	**9**	**10**	**11**	**12**	**13**	**14**	**15**	**16**	**17**	**18**	**19**	**20**
$Cs\ O_2$ $mg{\cdot}L^{-1}$	11.19	10.92	10.67	10.43	10.20	9.98	9.76	9.56	9.37	9.19	9.01	8.84
NH_4^+ $mg{\cdot}L^{-1}$	3.44	3.36	3.28	3.21	3.14	3.07	3.00	2.94	2.88	2.83	2.77	2.72
NO_3 formed $mg{\cdot}L^{-1}$	11.84	11.56	11.29	11.04	10.79	10.56	10.33	10.12	9.92	9.72	9.54	9.36

Table 12.1. *Theoretical quantity of biologically nitrifiable ammonia ($mg{\cdot}L^{-1}$)*

Since water is never at saturation but often around 5 mg $O_2{\cdot}L^{-1}$ for surface water, the amount of nitrifiable ammonia is 1.53 $mg{\cdot}L^{-1}$ at 20°C under optimum conditions for biological activity. And as water temperature varies throughout the year, the conditions for nitrification at 10°C are reduced, even if the concentration of dissolved oxygen is higher.

The same applies for groundwater, even if the temperature of this water is fairly constant (12–13°C).

As the real efficiency of biological oxidation does not exceed 90%, it is assumed that biological nitrification on a sand filter is 1.3 mg $NH_4^+{\cdot}L^{-1}$ at the most favorable temperature (20°C). The biological removal of NH_4^+ drops to 0.4 mg $NH_4^+{\cdot}L^{-1}$ at 10°C.

Biological nitrification performance significantly improves, depending on the support material. Thus, a porous support such as activated carbon increases the removal of NH_4^+ to 1.5 $mg{\cdot}L^{-1}$ at 20°C, and to 0.6–0.7 $mg{\cdot}L^{-1}$ at 10°C.

In the presence of an NH_4^+ concentration greater than 1.3 $mg{\cdot}L^{-1}$, water aeration must be forced so that dissolved oxygen is no longer a limiting parameter. In that case, continuous aerated filters are used. These technologies are described in Section 12.5.

12.2.2. *Filtration rate*

Filtration rates are between 5 and 10 $m{\cdot}h^{-1}$. The filtration rate is related to the contact time and the material's height, which are the two key parameters.

12.2.3. *NH_4^+ concentration removed as a function of temperature (°C)*

Figure 12.2 illustrates the influence of temperature on the biological oxidation of NH_4^+. At 20°C, with a sand filter 1.4 m height, a velocity of 5 $m{\cdot}h^{-1}$ and an effective size of the material support of 1 mm, a removal of 1.3 $mg{\cdot}L^{-1}$ is obtained. The contact time is 16.8 min. Removal decreases to 0.7 $mg{\cdot}L^{-1}$ at 10°C. A removal of 1.6 mg $NH_4^+{\cdot}L^{-1}$ at 28°C on water saturated with dissolved oxygen has been observed on sand filters.

A granular activated carbon (GAC) filter placed on the second stage and preceded by inter-ozonation which saturates water with O_2, leads to higher removal efficiencies. This is explained by the material's internal porosity, something which promotes the development of biomass in addition to a high concentration of O_2 available over the entire filter's height. Thus, at 15°C, about 2.2 mg $NH_4^+{\cdot}L^{-1}$ is removed on this material.

12.2.4. *Applicable volume load*

The loading mass applicable to nitrifying filters is approximately 0.2–0.4 $kg{\cdot}d^{-1}$ per m^3 of material, depending on the type of the material, its effective size, its height and for a 20°C temperature.

For example, water containing 1.3 mg $NH_4^+{\cdot}L^{-1}$ can be treated on a filter with a sand height of 1.0 m at a 6 $m{\cdot}h^{-1}$ velocity, if the necessary amount of oxygen is well provided, and that the temperature is optimal. In practical terms, as the temperature changes throughout the year, it is recommended to operate at velocities below 10 $m{\cdot}h^{-1}$ and to implement higher layer height in order to maintain an acceptable removal when the water is at a low temperature.

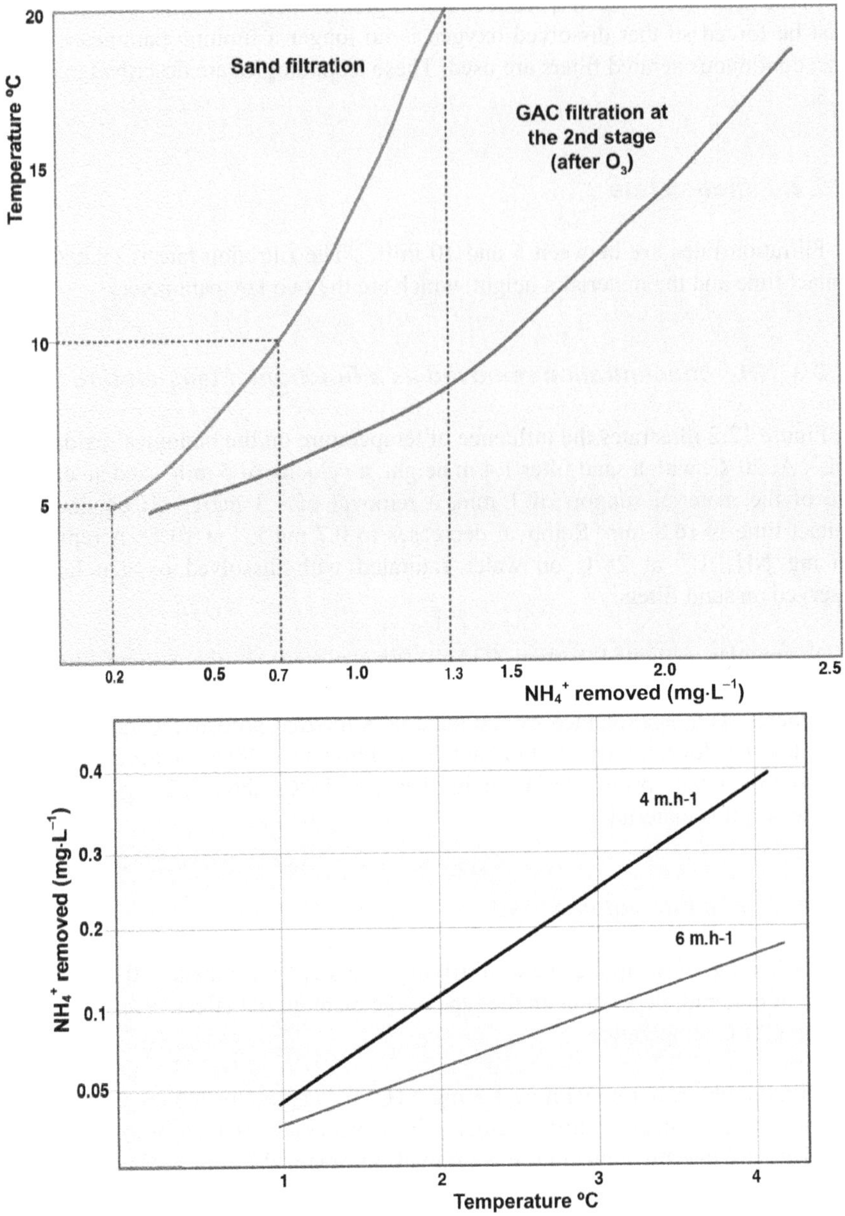

Figure 12.2. *NH_4^+removed on biological filters as a function of temperature. For a color version of this figure, see www.iste.co.uk/gaid/watertreatment3.zip*

12.2.5. *Contact time*

Contact time is considered one of the key parameters in the design and operation of biological filters, regardless of the material present in the filter. As the real contact time is frequently unknown, it is replaced by the empty bed contact time:

$$tc\ (\text{min}) = \frac{V \times 60}{Q} = \frac{H \times 60}{v},$$

where:

– V: media volume (m^3);

– Q: water flow ($m^3 \cdot h^{-1}$);

– H: material's height (m);

– v: filtration velocity ($m \cdot h^{-1}$).

The contact times required for the biological oxidation of NH_4^+ in sand filters are 30, 15 and 12 min at 5, 10 and 20°C, respectively. This is applicable for NH_4^+ concentrations lower than 1.5 $mg \cdot L^{-1}$.

12.2.6. *Material height*

Sand's height varies between 1 and 1.4 m, depending on contact time and filtration velocity.

12.3. Factors limiting oxygen

Given the above equation, the removable ammoniacal nitrogen in a water saturated with O_2 will be around 1.5–2 $mg \cdot L^{-1}$ because the reaction kinetics decrease rapidly as the oxygen concentration decreases. If the ammonia content of the water to be treated is higher than this value, and/or if the water to be treated presents an oxygen deficiency (a frequent case in deep waters), an external supply of oxygen will be necessary.

12.3.1. *Mineral carbon*

The reaction is accompanied by a consumption of alkalinity and in pH, translating a consumption of 7.07 mg $CaCO_3$ per mg of $N\text{-}NH_4^+$. In general, the mineralization of the majority of waters is sufficient.

12.3.2. *pH*

The maximum efficiency of the reaction is obtained at pH 8.5. However, even in pH values close to neutrality, the removal efficiencies are still very high; in this case a pH correction is not necessary.

12.3.3. *Temperature*

The nitrification reaction follows the laws of thermodynamics. For low temperature ranges, kinetics decrease linearly following the Q_{10}rule (nitrification rate reduced by half for a temperature drop of 10°C).

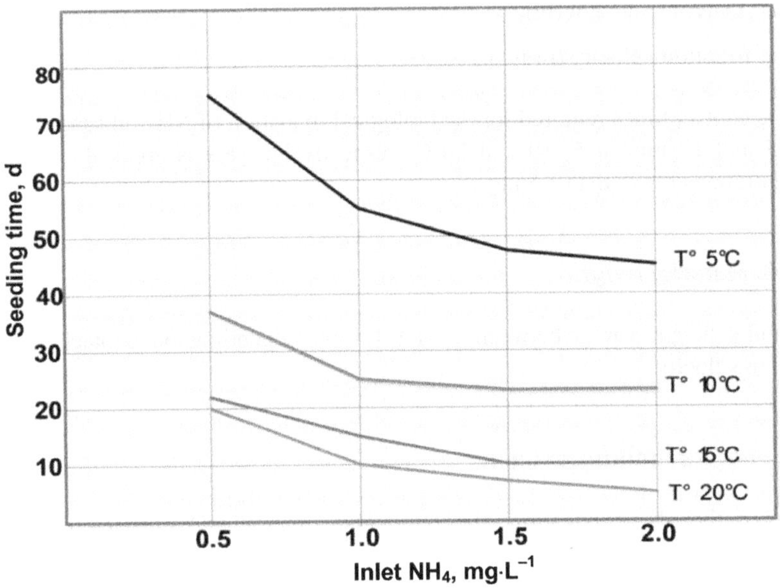

Figure 12.3. *Seeding times for nitrifying sand filters as a function of temperature. For a color version of this figure, see www.iste.co.uk/gaid/watertreatment3.zip*

The growth rate of bacteria increases with temperature from 10 to 22°C, remains constant between 30 and 35°C and decreases from 35°C. The optimum temperature for the growth rate of nitrifying bacteria is therefore around 25–30°C.

The evolution of the bacterial growth rate (μmax) as a function of temperature is described by an Arrhenius-type law for temperatures below 20°C. The equation is written out:

$$\mu max\ (T) = \mu max\ (20°C) \cdot 1.071^{T-20}$$

Experience shows that nitrification takes place with very low oxidation efficiencies at a temperature of 4°C. The design of the filters takes into consideration the lowest water temperature registered. The seeding time of a nitrifying sand filter takes between a few days to several weeks for a nitrifying biomass to be developed. Figure 12.3 presents the seeding times observed at several temperatures and for several NH_4^+ concentrations in raw water.

12.3.4. *Other elements*

The trace elements essential for the growth of nitrifying bacteria are generally present in sufficient amounts in the water to be treated. It may nonetheless be necessary to add a small amount of phosphorus when the water presents a particularly low amount of this element (content less than 0.1 $mg \cdot L^{-1}$). The main inhibitors are heavy metals and certain organic sulfur or aromatic compounds.

However, the tolerance limits are above the concentrations that would make the water unfit for consumption. As bacterial metabolites only inhibit nitrification at high concentrations (in the range of 0.5 $g \cdot L^{-1}$), they do not affect drinking water applications.

12.3.5. *Biological filter washing*

The washing frequency of filters varies between once every 3 days to once every 15 days. In general, they are more spaced in cold water. This is done to maintain a minimum concentration of nitrifying bacteria which will make it possible to launch an NH_4^+ removal once again after a backwash. These are more frequent in warm water, due to a higher biological activity.

Biofilter washing is similar to that of conventional filters. It is performed in an up flow condition (backwashing), only using water or water in combination with air, either consecutively or simultaneously. When washing nitrifying sand filters, the rinsing water flow rates are 25 $m^3 \cdot m^{-2} \cdot h^{-1}$, whereas for Biocarbon®-type filters, flow rates are 50 $m^3 \cdot m^{-2} \cdot h^{-1}$.

Microorganisms and abiotic (non-living) particles tend to accumulate in the filter bed. Microorganisms are better retained than abiotic particles, suggesting that optimal backwash conditions for the removal of abiotic particles do not lead to an excessive loss of biomass.

The washing method chosen influences the performance of biofiltration. Rinsing water must not contain chlorine, which would risk killing part of the biomass. This is especially true considering that the growth of the nitrifying flora is slow and NH_4^+ removal performances can be hindered while waiting for the flora to develop once again.

Washing may induce attrition for porous media like Biodagen®, pumice, anthracite or GAC (reduction into fine particles under the mechanical abrasive effect), in particular during the air and air/water phases. The loss of material caused by successive washes (either by entrainment and/or by attrition) is approximately 2% per year.

12.4. Implementation

Two types of processes are offered by OTV-Veolia, each presenting their specific options.

12.4.1. *Sand filtration*

The filters used are F or N filters, described in detail in Chapter 8. Rapid sand filters can be made of a single sand layer, or a double anthracite/sand or pumice/sand layer. Monolayer filters are often used in cases where the turbidity of settled water is low (<2 NTU). Dual media filters are used when the turbidity of settled water is higher than 3–5 NTU, or when the required turbidity of filtered water is very low (<0.2 NTU). The typical depth of the sand bed varies between 1 and 2.5 m, with a gravel layer of 0.1 m. Table 12.2 summarizes the main filter design characteristics.

Parameters	**Rapid sand** filtration
Filtration velocity ($m \cdot h^{-1}$)	6–15
Material grain size (d_{10} mm)	0.85–1.20
Material's height (m)	1–2.5
Contact time (min)	8–30
Washing frequency (days)	3–10
Washing method	Counter-current (air, air/water, water)

Table 12.2. *Characteristics of biological sand filters*

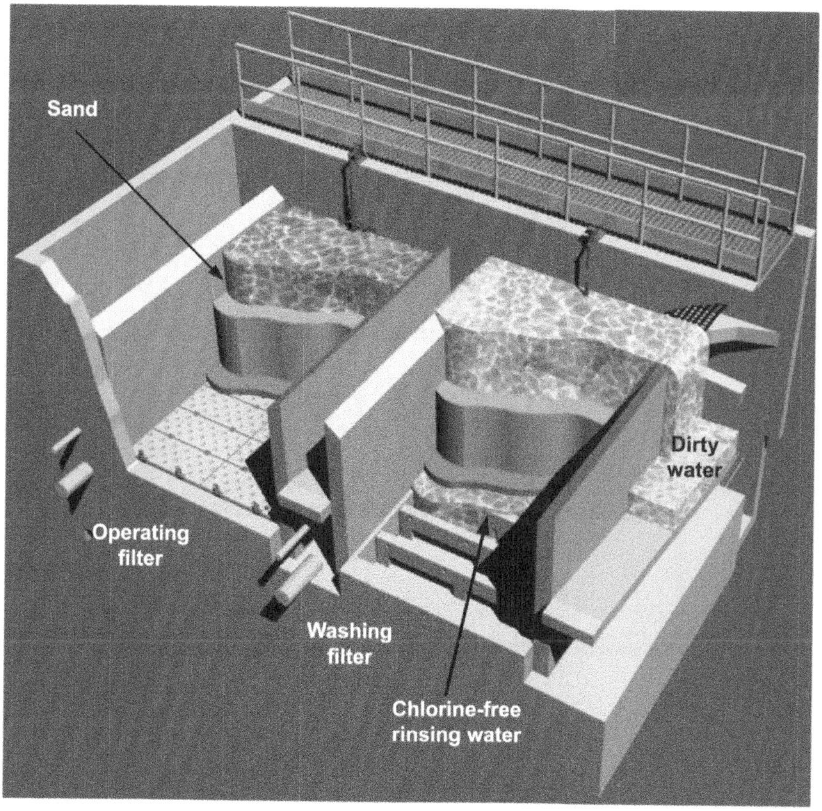

Figure 12.4. *Sand filter used for biological nitrification. For a color version of this figure, see www.iste.co.uk/gaid/watertreatment3.zip*

The water to be treated by nitrification must have the following characteristics (for optimum performance):

– temperature: ≥10°C;

– free oxidants (Cl_2 or O_3): <0.1 mg·L^{-1};

– pH 7–7.5;

– ammoniacal nitrogen (NH_4^+) ≤ 1.5 mg·L^{-1};

– absence of toxic products;

– dissolved oxygen: ≥5 mg·L^{-1} (3.25 mg O_2 per NH_4^+ to be nitrified);

– alkalinity 8.54 mg HCO_3^- per mg N-NH_4^+.

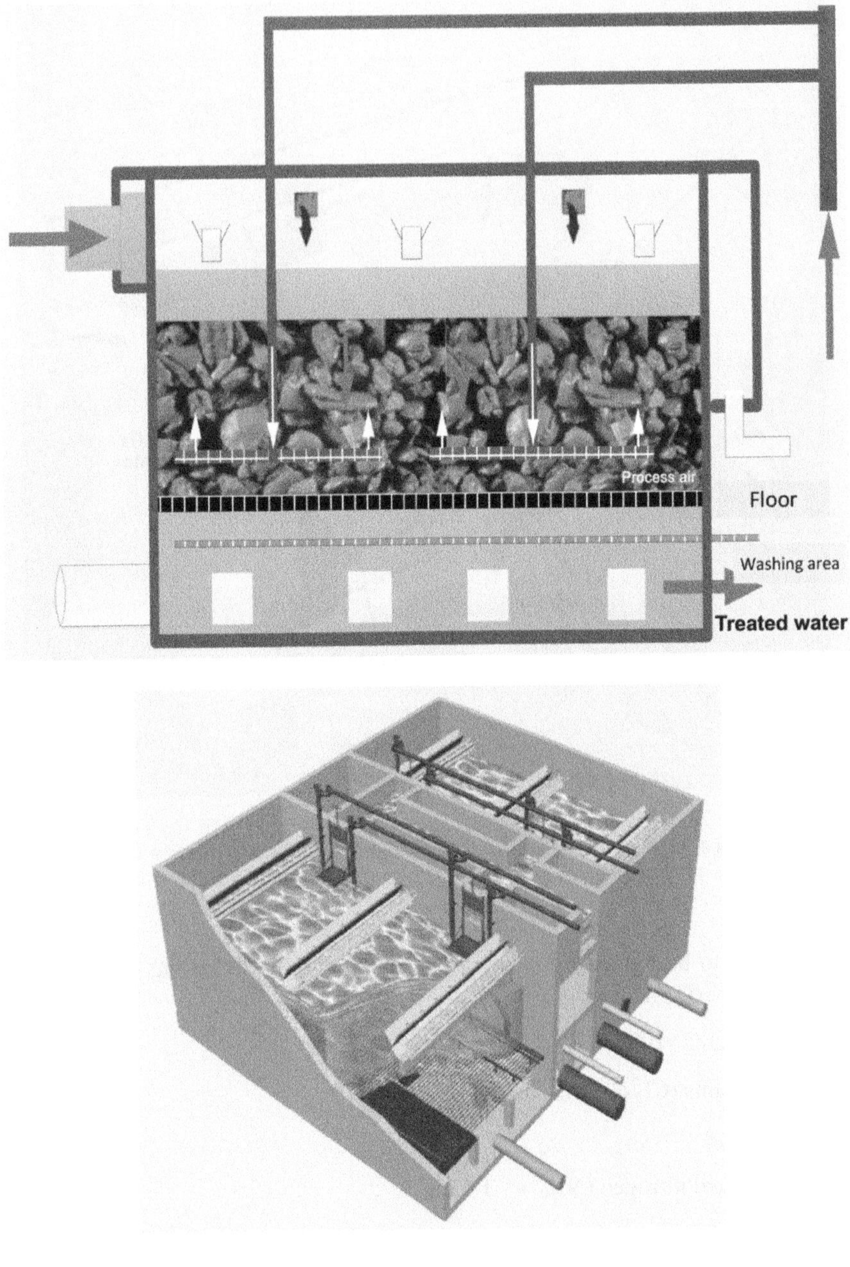

Figure 12.5. *Biocarbon® process with Biodagen®. For a color version of this figure, see www.iste.co.uk/gaid/watertreatment3.zip*

12.5. Biofilters (Biocarbon® process)

The principle of the Biocarbon® process comes down to percolating the water to be treated from top to bottom through the bed (downward flow) and through a layer of Biodagen® material (effective size between 1.5 and 3.0 mm). It works following a downward countercurrent filtration with a flow of continuously injected air at the bottom of the biofilter from a floor made up of specific strainers (Figure 12.5).

The volumetric air/water ratio is between 0.3 and 1.0. Biodagen® height depends on the contact time and filtration rate. Table 12.3 summarizes the operating conditions for biological nitrification on Biocarbon®.

Parameters	Characteristics
NH_4^+ concentration, $mg{\cdot}L^{-1}$	>1.5–8
Height, m	2–2.5
Contact time, min	15 min at 20°C, 20 min at 10°C
Filtration velocity, $m{\cdot}h^{-1}$	6 (N)–8 (in N^{-1})
Volume load, kg $NH_4^+{\cdot}m^{-3}{\cdot}d^{-1}$	0.36–1.00
Air/water ratio	0.8–2.0
Frequency between two washes, days	3–10
Washing conditions	
Air ($Nm^3{\cdot}m^{-2}{\cdot}h^{-1}$) for 1 min	70
Air + water ($m^3{\cdot}m^{-2}{\cdot}h^{-1}$) for 8 min	12 (water) and 70 (air)
Water only ($m^3{\cdot}m^{-2}{\cdot}h^{-1}$) for 12–15 min	50

Table 12.3. *Biocarbon® design parameters*

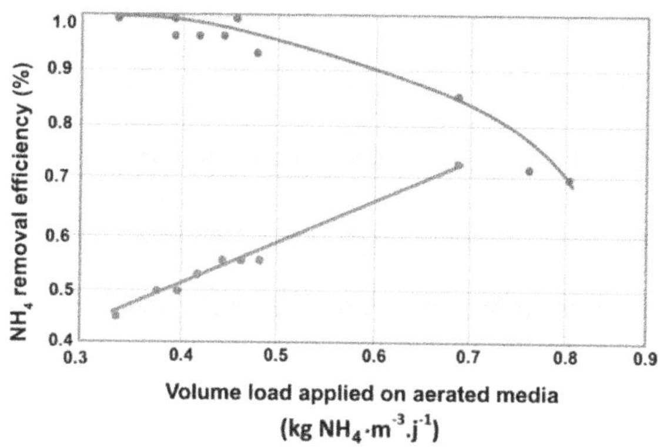

Figure 12.6. *Biocarbon® performances on NH_4^+ depending on the load applied. For a color version of this figure, see www.iste.co.uk/gaid/watertreatment3.zip*

The height of the material depends on the contact time, the filtration rate and the NH_4^+ concentration. A water containing 3.0 mg $NH_4^+ \cdot L^{-1}$ requires a height of about 2.4 m for a flow rate of 6 $m \cdot h^{-1}$, for variable temperatures between 5 and 20°C.

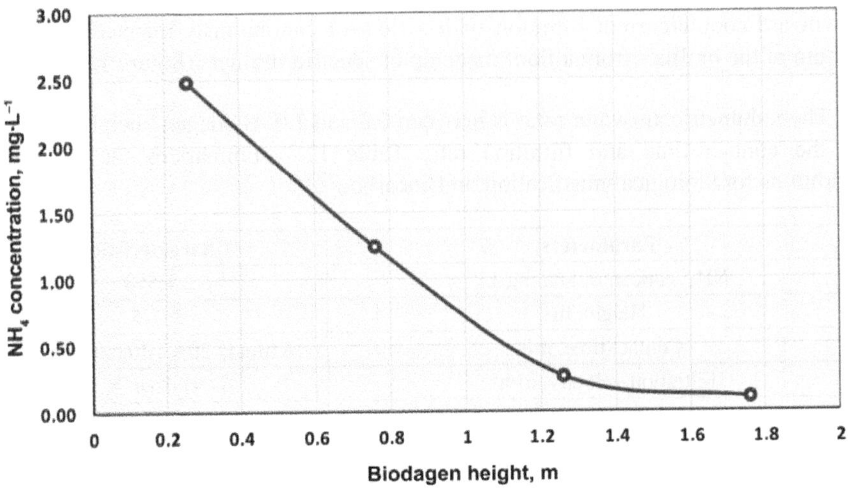

Figure 12.7. *NH_4^+ removal profile depending on Biodagen® height*

Figure 12.7 shows an NH_4^+ removal profile on Biocarbon®. Ammonia is well removed with 1.3 m. A height of 1.8 m is foreseen because the NH_4^+ concentration may increase to more than 3 $mg \cdot L^{-1}$. The final NH_4^+ concentration obtained is less than 0.2 $mg \cdot L^{-1}$. The temperature varied between 16 and 22°C. Biocarbon® is washed by applying air, then air and water simultaneously, and finally rinsing with water alone. The total duration of a washing cycle varies between 23 and 25 min.

The rinsing water should be chlorine free. The advantage of a downdraft filter like Biocarbon® is that it accepts turbidities up to 15–20 NTU and can be directly installed on raw water, unlike updraft biofilters which can only be used on very slightly turbid water, such as groundwater.

Biocarbon® is followed by sand-refining or dual media filtration (anthracite or pumice/sand) and disinfection.

One option with the Biocarbon® process is to replace Biodagen® with a dual media GAC-sand filter. Aeration is then blown through an intermediate ramp between the activated carbon and the sand (Figure 12.8). This option removes ammonia, adsorbs micropollutants and oxidizes part of the organic matter. Sand is responsible for refining.

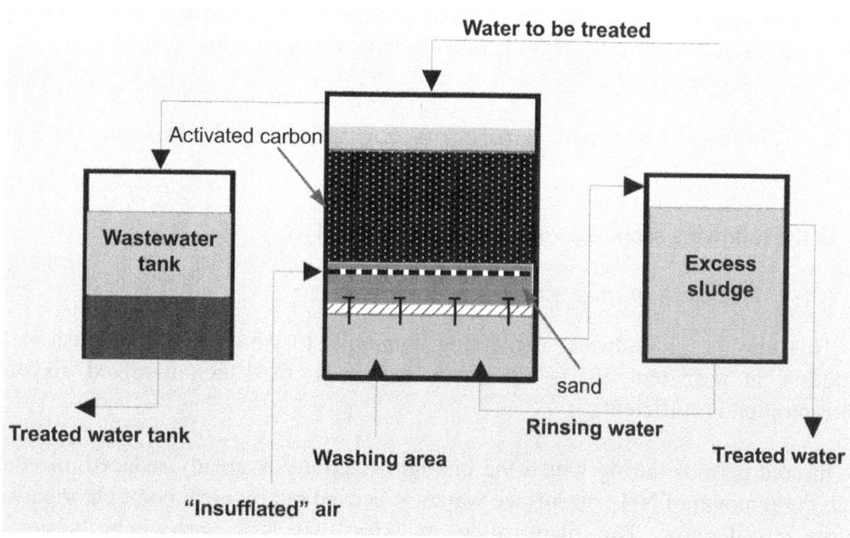

Figure 12.8. *Principle of the Biocarbon® process with activated carbon and sand. For a color version of this figure, see www.iste.co.uk/gaid/watertreatment3.zip*

The advantage of GAC is related to the multiple capabilities offered by this material: large specific surface area, macro and microporosity, good biological affinity (very good bacterial adhesion). Activated carbon has the following advantages:

– GAC is a particularly useful support for enhancing bacteria adhesion;

– preservation of aerobic conditions by blowing air into the bed;

– development of biological activity in almost the entire bed thanks to the agitation caused by air insufflation;

– bed's lower layer held at rest (below the air insufflation system) to guarantee a sufficient mechanical filtration effect.

Among all the possible applications of the biological treatment in the field of drinking water, ammonia nitrification is the most widespread because this process undoubtedly satisfies a current need. In particular, it solves the difficulties presented by the breakpoint chlorination process. It is effective and easy to implement.

12.6. Water treatment stations

The insertion of an ammonia biological treatment at a water treatment unit can be done in various ways. The design mainly depends on the characteristics of raw water, its variations as well as all the other treatment stages at the plant.

12.6.1. *Treatment stations with conventional sand, dual media or GAC filtration*

In the following sections, several cases are treated.

12.6.1.1. *The case of micropollutant-free water*

This may be groundwater containing ammonia, in which case biological sand filtration is sufficient. It is advisable to check that the dissolved oxygen concentration is sufficient.

In cold periods during which the biological activity is greatly reduced, or even null, the removal of NH_4^+ in surface waters is carried out by breakpoint chlorination before disinfection. The plant often includes two NH_4^+ processing systems: biological filtration, which operates at temperatures where the biomass has sufficient activity, and a chemical treatment with chlorine during the cold periods. An ammoniometer is installed at the outlet of the filter and the chlorine dosage is then associated with the residual ammonia. An alternative for monitoring ammonia is the increased demand for chlorine, which signals its presence. Chlorine is subjected to this demand until residual free chlorine is obtained.

In case of the presence of Fe^{2+}, Mn^{2+} and NH_4^+, the redox potential is the parameter which determines the oxidation sequence for these elements. Iron is oxidized at low redox values (<200 mV), ammonia at higher values (200–400 mV), whereas manganese requires even higher redox values (>400 mV) at neutral pH.

In the presence of iron and manganese, a prior chemical oxidation is required for these two elements, and their retention is first carried out on a sand filter. The removal of NH_4^+ is then carried out on a second filter, set at the second stage. Monitoring the residual oxidant and its neutralization should be done upstream of the biological filter.

The chemical oxidation of iron and Mn followed by nitrification can take place on a single filter, but it is necessary to control (or monitor) the residual of the oxidant, to provide greater media height in order to retain the oxidized particles in the first layers of the material, and to promote nitrification in the following layers.

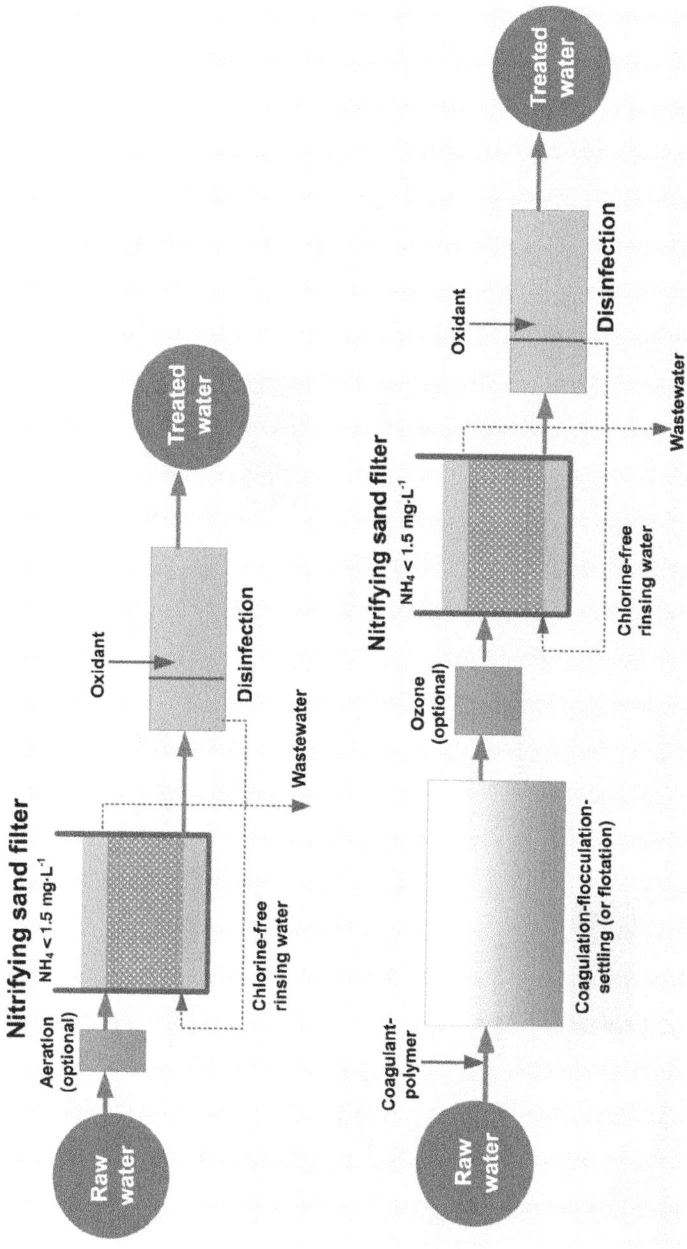

Figure 12.9. *Nitrifying sand filter on groundwater or on surface water. For a color version of this figure, see www.iste.co.uk/gaid/watertreatment3.zip*

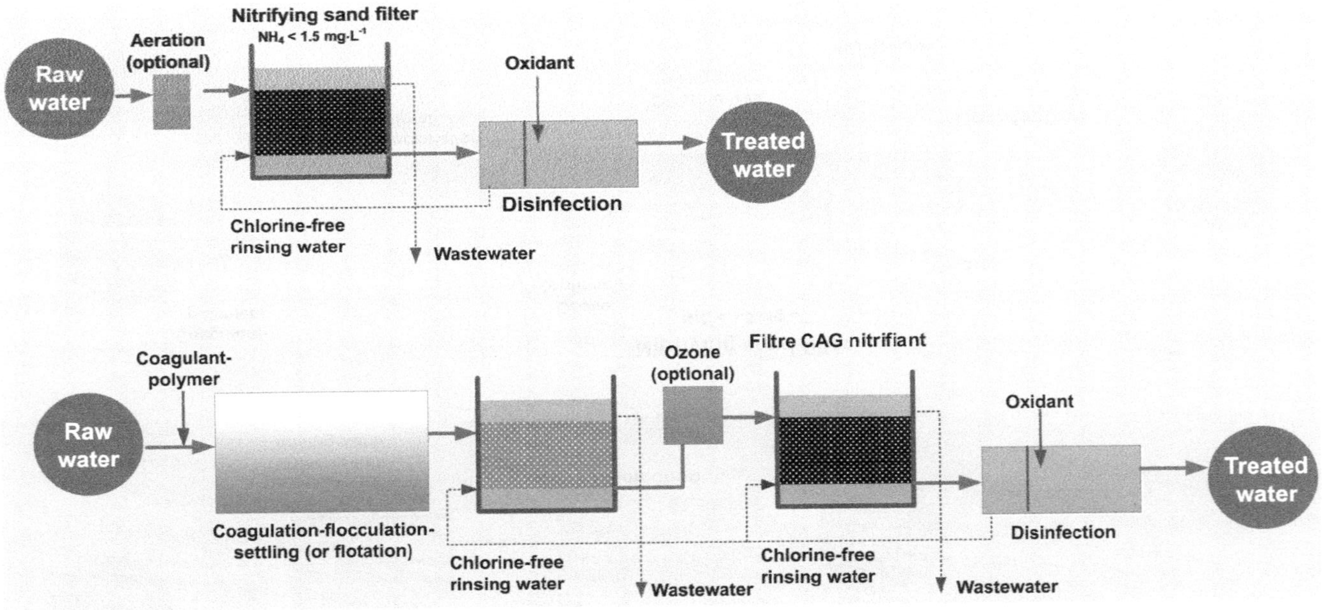

Figure 12.10. *Nitrification with a GAC filter placed at the first or second stage. For a color version of this figure, see www.iste.co.uk/gaid/watertreatment3.zip*

In the presence of heavily loaded surface water with turbidity or organic matter, clarification by coagulation–flocculation–settling is carried out upstream of the sand filter. A control of dissolved oxygen ensures that its concentration is sufficient for ammonia oxidation. If this concentration is not sufficient in relation to the amount of NH_4^+ to be oxidized, intermediary aeration is required.

Inter-ozonation is interesting if the presence of residual color, toxins, metals or certain micropollutants so forth, is reported. However, a neutralization of residual ozone is required.

Ozonation can be integrated into the treatment process upstream of the sand filters. Actually, ozonation does not interfere with nitrification. Numerous trials have shown that it is rather favorable to the development of bacterial activity. It also contributes to water oxygenation. The fact of ozonation water before filtration (inter-ozonation) promotes the development of biomass in the filter, thus increasing the biological activity within the filter. Residual O_3 must be controlled. In all cases, the bromate parameter should be estimated before opting for ozonation.

12.6.1.2. *The case of water containing micropollutants*

In the presence of micropollutants and NH_4^+ in groundwater, GAC filtration at the first stage can be interesting (Figure 12.10).

In the case of surface water containing micropollutants, the treatment process includes clarification by coagulation–flocculation–settling, a first stage of sand filtration thanks to which a first removal of NH_4^+ is obtained. The second stage, involving the use of a GAC filter, removes micropollutants and NH_4^+. Inter-ozonation ensures the removal of certain substances and enriches the water with dissolved oxygen.

This strategy (ozonation upstream of GAC filtration) promotes the growth of biomass in the filter thanks to the activated carbon's porous structure, which promotes bacterial development. GAC is considered a granular material promoting biodegradation processes faster and more efficiently than a sand filter or an anthracite-sand dual media filter.

12.6.2. *Treatment stations with Biocarbon® filters*

In the presence of an NH_4^+ concentration greater than 1.5 $mg \cdot L^{-1}$, and if this concentration changes throughout the year showing higher values, a Biocarbon® filter is recommended (Figure 12.11).

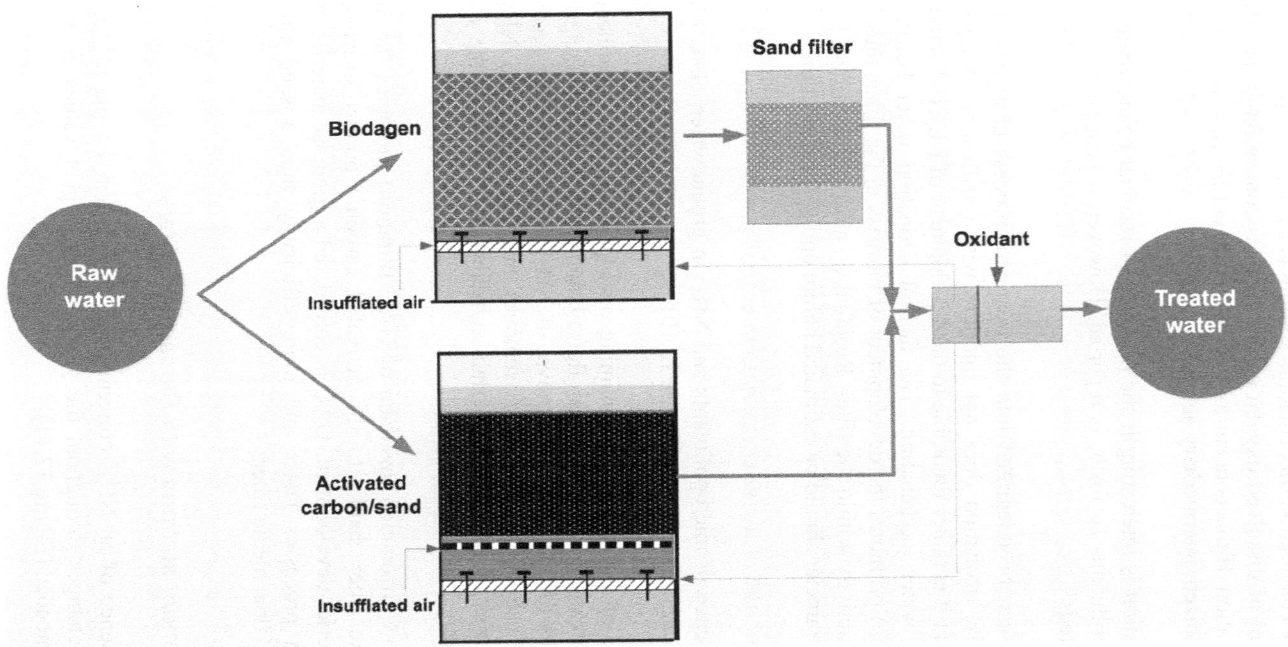

Figure 12.11. *Biocarbon® process with Biodagen material and activated carbon. For a color version of this figure, see www.iste.co.uk/gaid/watertreatment3.zip*

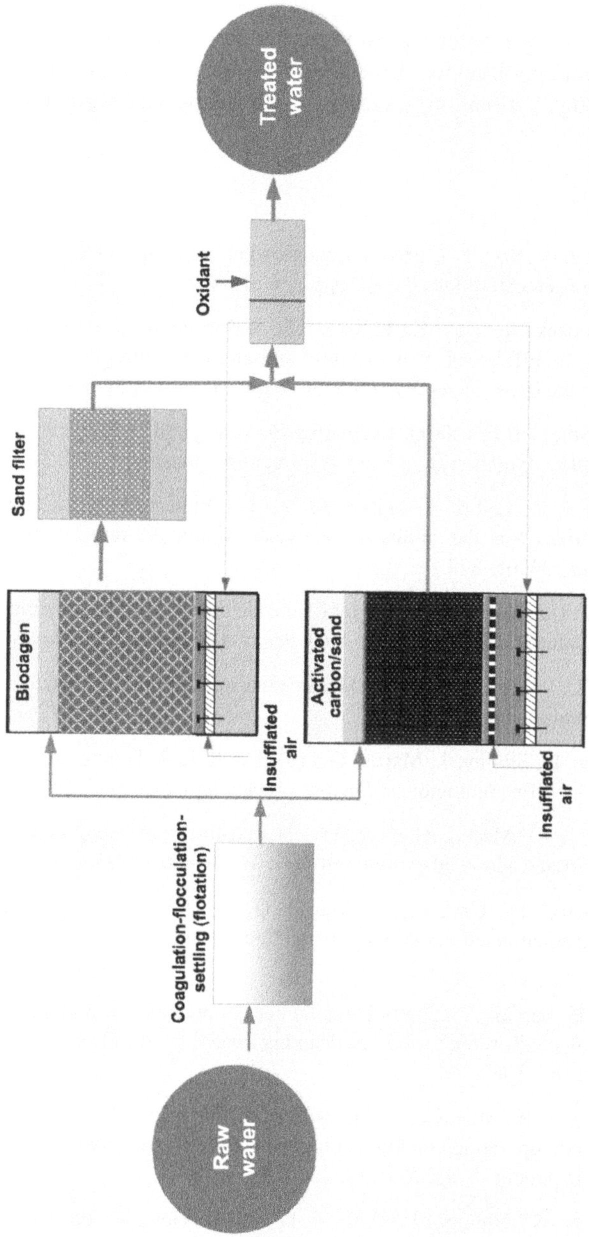

Figure 12.12. *Biocarbon® process on surface water. For a color version of this figure, see www.iste.co.uk/gaid/watertreatment3.zip*

With Biodagen®, sand filtration must be installed downstream. With activated carbon, sand filtration can be included in the biofilter.

When treating surface water loaded with turbidity, algae and containing organic matter or even micropollutants, Biocarbon® is installed after the coagulation–flocculation–settling (or flotation) stage. It is followed by sand filtration.

12.7. References

Bourdon, F., Jestin, J.M., Roy, F. (1988). La nitrification biologique dans les filtres à sable : avantage du report de chloration. *Water Supply*, 6(3), 77–87.

Bray, R. and Olanczuk-Neyman, K. (2001). The influence of changes in groundwater composition on the efficiency of manganese and ammonia nitrogen removal on mature quartz sand filtering beds. *Water Sci. Technol.*: *Water Supply*, 1, 91–98.

Chandran, K. and Smets, B.F. (2001). Estimating biomass yield coefficients for autotrophic ammonia and nitrite oxidation from batch respirograms. *Water Res.*, 35, 3153–3156.

Falas, P., Andersen, H.R., Ledin, A., Jansen, J.L. (2012). Impact of solid retention time and nitrification capacity on the ability of activated sludge to remove pharmaceuticals. *Environ. Technol.*, 33, 65–872.

Gaid, K., Martin, G., Lemaréchal, A. (1981). Le procédé biocarbone, un traitement biologique appliqué à la production d'eau potable. *Techniques & Sciences Municipales*, 4, 158–166.

Gaid, K., Cavelier, C., Martin, G. (1982a). Mécanismes de l'épuration biologique sur charbon actif. *Water Research*, 16(1), 7–13.

Gaid, K., Le Cloirec, P., Sibony, J., Martin, G. (1982b). Rôle de la porosité du charbon actif lors du lavage des filtres biologiques. *Environ, Technology Letters*, 3, 257–265.

Gaid, K., Leroux, A.Y., Martin, G. (1982c). A non-linear problem occurring in water purification in aerated activated carbon biofilters. *Mathematical Modelling*, 3, 38–47.

Gaid, K., LeMarchand, D., Cavelier, C., Martin, G. (1983). Improving filter media, by oxidative pretreatment with ozone and persulfate. *Ozone, Science and Engineering*, 5, 113–118.

Kihn, A., Laurent, P., Servais, P. (2000). Measurement of potential activity of fixed nitrifying bacteria in biological filters used in drinking water production. *J. Ind. Microbiol. Biotechnol.*, 24, 161–166.

Lee, C.O., Boe-Hansen, R., Musovic, S., Smets, B.F., Albrechtsen, H.J., Binning, P.J. (2014). Effects of dynamic operating conditions on nitrification in biological rapid sand filters for drinking water treatment. *Water Res.*, 64, 226–236.

Lopato, L., Rottgers, N., Binning, P.J., Arvin, E. (2013). Heterogeneous nitrification in a full-scale rapid sand filter treating groundwater. *J. Environ. Eng. ASCE*, 139, 375–384.

Lytle, D.A., Sorg, T.J., Wang, L.L., Muhlen, C., Rahrig, M., French, K. (2007). Biological nitrification in a full-scale and pilot-scale iron removal drinking water treatment station. *J. Water Supply: Res. Technol. -Aqua*, 56, 125–136.

Madoni, P., Davoli, D., Fontani, N., Cucchi, A., Rossi, F. (2001). Spatial distribution of microorganisms and measurements of oxygen uptake rate and ammonia uptake rate activity in a drinking water biofilter. *Environ. Technol.*, 22, 455–462.

Queinnec, I., Ochoa, J.C., Wouwer, A.V., Paul, E. (2006). Development and calibration of a nitrification PDE model based on experimental data issued from biofilter treating drinking water. *Biotechnol. Bioeng.*, 94, 209–222.

Raguenes, I. (1994). L'azote. Document, Ecole de l'eau d'OTV, Paris, France.

Stembal, T., Markic, M., Ribicic, N., Briski, F., Sipos, L. (2005). Removal of ammonia, iron and manganese from groundwaters of northern Croatia – Pilot plant studies. *Process Biochem.*, 40, 327–335.

Tatari, K. (2014). Nitrification biokinetics in rapid sand filters for drinking water treatment. PhD Thesis, Technical University of Denmark, Lyngby-Taarbæk.

Tatari, K., Smets, B.F., Albrechtsen, H.J. (2013). A novel bench-scale column assay to investigate site-specific nitrification biokinetics in biological rapid sand filters. *Water Res.*, 47(16), 6380–6387.

Tekerlekopoulou, A.G. and Vayenas, D.V. (2007). Ammonia, iron and manganese removal from potable water using trickling filters. *Desalination*, 210, 225–235.

Tekerlekopoulou, A.G., Papazafiris, P.G., Vayenas, D.V. (2012). Effect of environmental and operating conditions on a full-scale trickling filter for well water treatment. *Desalination Water Treat.*, 39, 228–234.

Uhl, W. and Gimbel, R. (2000). Dynamic modeling of ammonia removal at low temperatures in drinking water rapid filters. *Water Sci. Technol.*, 41, 199–206.

Wolthoorn, A., Temminghoff, E.J.M., Van Riemsdijk, W.H. (2004). Effect of synthetic iron colloids on the microbiological NH_4 removal process during groundwater purification. *Water Res.*, 38, 1884–1892.

Zhang, Y., Love, N., Edwards, M. (2009). Nitrification in drinking water systems. *Crit. Rev. Environ. Sci. Technol.*, 39, 153–208.

Zhu, I.X., Getting, T., Bruce, D. (2010). Review of biologically active filters in drinking water applications. *J. Am. Water Works Assoc.*, 102, 67–77.

13

Nitrate Removal

The increase in nitrate levels in water to be rendered drinkable (groundwater and surface water) is primarily due to the massive use of nitrogen fertilizers in agriculture, as well as profuse urban, industrial or agricultural discharges.

Figure 13.1. *Use of fertilizers in agriculture and algae development in receiving environments. For a color version of this figure, see www.iste.co.uk/gaid/watertreatment3.zip*

The main consequences of the presence of high nitrate concentrations in water are not only eutrophication phenomena and their transfer to potable water but also the toxicity they generate.

As a matter of fact, in terms of toxicity, the nitrites resulting from the reduction of nitrates oxidize hemoglobin's ferrous ions into ferric ions. Hemoglobin then turns into methemoglobin, which is unable to transfer oxygen to tissues. Beyond a certain rate of methemoglobin, various symptoms appear, ranging from a light cyanosis to the appearance of consciousness disorders (and these even with a life-threatening potential, due to cellular anoxia). Infants are more sensitive. Their weakly acidic stomach contains a bacterial flora capable of transforming nitrates into nitrites, and they present a deficiency in the enzymes responsible for transforming methemoglobin into hemoglobin.

Conditions predisposing to methemoglobinemia or cyanosis in infants seem to prevail for infants whose mother regularly consumed water with high nitrate content. It appears that the majority of methemoglobinemia cases occir when the nitrate levels in drinking water exceed 100 $mg \cdot L^{-1}$.

The French Government Resolution of January 11, 2007 relating to the quality limits and reference values for raw water and potable water set the authorized quality limit for this element at 50 mg $NO_3 \cdot L^{-1}$ in potable water.

In various regions, the nitrate concentration in raw water reaches or exceeds this limit. In order to meet regulatory and public health requirements, technical solutions are implemented for their removal from potable water.

The various industrially applicable treatment techniques call upon biological treatment and physicochemical techniques (high pressure membranes, electrodialysis, ion exchange).

13.1. Biological treatment

The biological treatment of nitrates comes down to developing a biomass under anoxic conditions (absence of dissolved oxygen) and in the presence of a source of carbon (addition of ethanol). It involves the use of nitrates as a source of oxygen and rejects nitrogen gas. In drinking water, these reactions take place within a biological filter whose excess biomass is extracted by daily backwashing.

Denitrification is the biological process whereby nitrates ($N0_3^-$) are reduced to nitrites ($N0_2^-$), and then to nitrogen gas (N_2). Biological denitrification can be performed by autotrophic or heterotrophic microorganisms. Several inorganic compounds, such as sulfur and hydrogen, can act as electron donors for autotroph bacteria acting as an organic-type electron donor required for heterotrophic bacteria. The type of electron donor thus determines the main difference between heterotrophic and autotrophic biological denitrification.

For heterotrophic denitrification, the required carbon and energy are provided by the oxidation of an easily biodegradable organic substrate. This process ensures the transformation of nitrates (a non-acceptable form of nitrogen in environmental terms), into an acceptable form: gaseous nitrogen (an inert gas). The advantage of the biological treatment is its ecological aspect insofar as nitrates are transformed into gaseous nitrogen instead of being directly released into the natural environment in their concentrated form.

13.1.1. *Biochemical reactions*

In an anoxic environment, bacteria have the power to reduce nitrates to gaseous nitrogen thanks to oxidation–reduction reactions engaging an organic carbonaceous substrate (heterotrophic denitrification).

Denitrification reactions unfold as a chain of biological reactions during which specific enzymes participate at each nitrous nitrogen reduction stage, following a generic scheme (according to Matsubara):

$$NO_3^- \rightarrow NO_2^- \rightarrow NO \rightarrow N_2O \rightarrow N_2$$

The process basically involves two stages:

– stage 1: reduction of nitrates to nitrites;

– stage 2: reduction of nitrites to molecular nitrogen.

Since bacteria consume oxygen in the various nitrogen forms, we simply state that they "breathe in" nitrates. In the field of drinking water, it is the "strictly denitrifying reduction" route that is used. This concept reveals two reaction pathways: the energy pathway and the cell synthesis pathway. The availability of the electron donor is an important parameter for the smooth progress of denitrification, and its dosage is related to the stoichiometric ratio (carbonaceous substrate to nitrate).

The assimilable organic carbon contained in the water to be rendered drinkable is generally insufficient because it is in the groundwater that nitrates are most often present. For this reason, a carbonaceous substrate needs to be added, such as ethanol (CH_3CH_2OH) or acetic acid (CH_3COOH).

The biochemical reactions are as follows.

With *ethanol*, the reactions are:

– energy pathway equation:

$$5\ CH_3CH_2OH + 12\ NO_3^- \rightarrow 10\ HCO_3^- + 6\ N_2 + 9\ H_2O + 2\ OH^-$$

– synthetic pathway equation:

$$50\ CH_3CH_2OH + 97\ NO_3^- \rightarrow 5\ C_5H_7NO_2 + 75\ CO_2 + 46\ N_2 + 84\ H_2O + 97\ OH^-$$

The quantity of cells is theoretically expressed by the chemical formula $C_5H_7NO_2$. Ethanol consumption is 0.691 mg of ethanol per mg of nitrates. Ethanol is preferred over other carbonaceous substrates due to the high efficiency it provides for lower biomass production.

In the case of Biofilters (bed fixed media), in practice and depending on the dissolved oxygen concentration in raw water, a consumption of ethanol occurs in the first layers of the material, following the reaction:

$$3\ CH_3CH_2OH + 2O_2 + NO_3^- + H^+ \rightarrow C_5H_7NO_2 + CO_2 + 6\ H_2O$$

In the presence of O_2, experience shows that the overall ethanol consumption is:

$$C\ \text{ethanol}\ (mg \cdot L^{-1}) = 0.691\ [\Delta NO_3\ \text{removed}] + 0.55\ [O_2]$$

It is accepted that all O_2 is consumed. In addition, 1 mg of ethanol corresponds to 0.52 mg of total organic carbon.

With *acetic acid*, the overall reaction is as follows:

$$88\ CH_3COOH + 100\ NO_3^- \rightarrow 8.8\ C_5H_7NO_2 + 42\ CO_2 + 46\ N_2 + 107\ H_2O + 88\ HCO_3^-$$

or:

$$0.14386\ NO_3^- + 0.01886\ H^+ + 0.125\ CH_3COOH \rightarrow 0.0640\ CO_2 + 0.1250\ HCO_3^- + 0.0122\ C_5H_7NO_2 + 0.06583\ N_2 + 0.1542\ H_2O$$

The quantity of cells is theoretically expressed by the chemical formula $C_5H_7NO_2$. Cell synthesis requires 0.85 g of acetic acid per gram of NO_3^-.

Taking into account the consumption of acetic acid in the presence of oxygen:

$$0.16457\ O_2 + 0.0122\ NO_3^- + 0.125\ CH_3COOH \rightarrow 0.0640\ CO_2 + 0.1250\ HCO_3^- + 0.1128\ H^+ + 0.0122\ C_5H_7NO_2 + 0.0884\ H_2O$$

That is 1.42 mg of acetic acid per mg of oxygen, meaning that the consumption of acetic acid is:

$$C\ \text{acetic acid}\ (mg \cdot L^{-1}) = 0.85\ [\Delta\ NO_3^-\ \text{removed}] + 1.42\ [O_2]$$

It is accepted that all O_2 is consumed during the denitrification reaction. One milligram of acetic acid corresponds to 0.40 mg of total organic carbon.

Cell production is shown in Table 13.2.

Pure product name	Theoretical consumption $g \cdot g^{-1} NO_3^-$	Consumption due to O_2 $g \cdot g^{-1} O_2$	Consumption observed $g \cdot g^{-1} NO_3^-$
(100 %)	**Overall reaction**	**Reaction**	**At the treatment station**
Ethanol	0.69	0.55	0.70–0.90
Acetic acid	0.85	1.42	0.9–1.2

Table 13.1. *Theoretical consumptions and those observed at the treatment station for Biofilters*

Pure product name	Production of cells $g \cdot g^{-1} NO_3^-$	Production observed $g \cdot g^{-1} NO_3^-$
(100 %)	**Theory**	**At the treatment station**
Ethanol	0.10	0.10
Acetic acid	0.15	0.53

Table 13.2. *Theoretical cell production and actual production observed at the treatment station for Biofilters*

Since the O_2 concentration is often lower than 5 $mg \cdot L^{-1}$ (which is quickly consumed from the first layers of the material in the Biofilters), the production of biomass associated with it is generally not taken into consideration in terms of the total production, due to nitrates.

The synthesis reaction leads to an increase in alkalinity of 0.8 mg $CaCO_3$ per mg of NO_3^- removed with ethanol, and 0.71 mg $CaCO_3$ per mg of NO_3^- removed with acetic acid.

For the smooth progress of denitrification reactions, a minimum of phosphorus should also be granted. In the long run (1–3 weeks), a lack of phosphorus could lead to a malfunction in the denitrification process: a decrease in the reduction efficiency of NO_3^- nitrates and the appearance of NO_2^- nitrites, or even of H_2S (hydrogen sulfide).

13.1.2. *Nitrite formation*

A nitrite concentration is often observed at the outlet of denitrification Biofilters. It depends on various mechanisms occurring in the biomass:

– the nature of the carbonaceous substrate: sugars tend to form many more nitrites than alcohols;

– a deficit of ethanol or of acetic acid in the substrate;

– short contact times.

13.1.3. *The bacteriological aspect*

There are about 20 species of strictly denitrifying bacteria. Generally, these bacteria are non-spore-forming, Gram-negative and facultative anaerobes mainly belonging to the genera *Bacillus*, *Paracoccus* and *Pseudomonas* (*Paracoccus denitrificans*, *Pseudomonas Stutzeri*, *Bacillus licheniformis*). These bacteria can oxidize a large number of carbonaceous substrates. The carbonaceous substrate provides both the energy and the carbon required by the metabolism and cellular synthesis. For an equal number of carbon atoms, the more the substrate is reduced, the greater the number of electrons released. The shorter the chain, the faster its biological degradation. In drinking water denitrification, the substrate must be nontoxic.

13.1.4. *Biofilter description (Biodenit® process)*

Denitrification via the Biodenit® process implements the fixation of bacteria onto a specific mineral support (alkaline schist) whose OTV brand name is "Biodagen®".

Filtration is performed downward, which significantly limits the risk of clogging strainers, which will be only in contact with treated water. An injection of carbonaceous substrate and monosodium phosphate enables the operator to control the development of the nitrifying biomass (Figure 13.2).

The Biodenit® OTV-Veolia filter is available in a rectangular casing built in reinforced concrete, or as a closed metal filter. An "open" filter is enveloped by a removable cover. The access to the filter is therefore easy for maintenance purposes, and most importantly, the risks of water contamination are avoided.

The essential features of Biodenit® filter are as follows:

– a Biodagen® layer resting on a perforated draining floor;

– ample water level above the Biodagen® layer;

– air backwashing, followed by an air/water phase and then rinsing with water.

Water is let into the filter by a channel and is distributed over the filter by a flooded spillway. The equal distribution is homogeneous along the entire filter. The spillway is accessed from the filter cover.

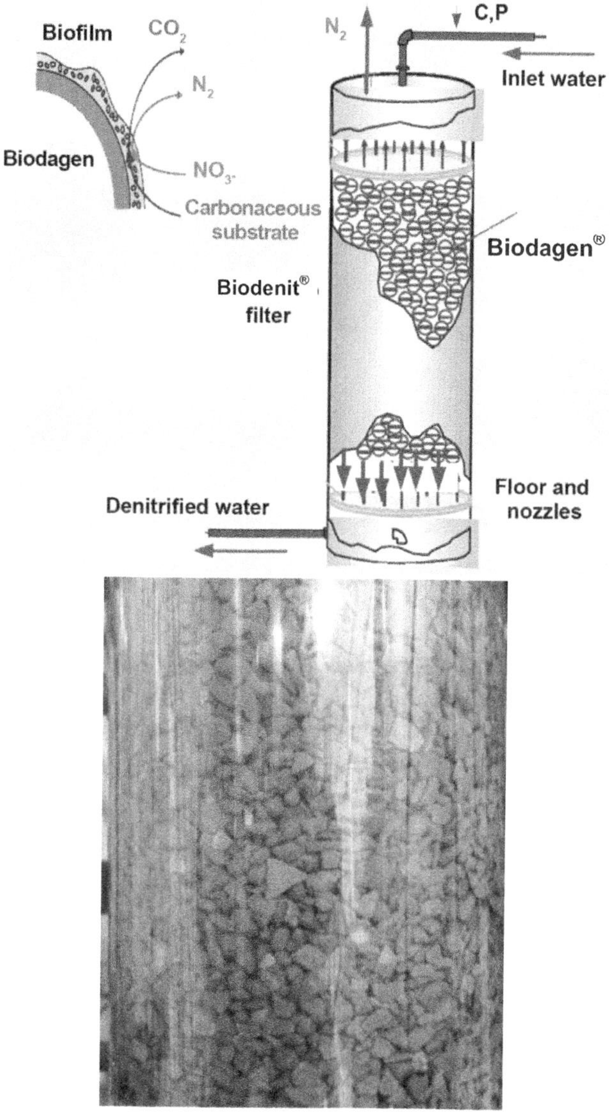

Figure 13.2. *Biodenit® process and Biodagen® material. For a color version of this figure, see www.iste.co.uk/gaid/watertreatment3.zip*

A regulation valve and a flow meter control the flow upstream of the Biodenit® filter.

The development of the biomass on the support results in grain magnification, reducing the size of the interstitial spaces and the passage offered to water. Clogging increases the head loss due to bacterial growth within the filter (Figure 13.3) and results in a gradual decrease in its flow rate, in the absence of devices to compensate for this head loss.

Figure 13.3. *Denitrifying biomass in the Biodenit® filter*

With the filter working at a given flow rate, the head loss becomes evident due to the difference in level between the water on the filter and the water in the filtered water basin, at the filter's outlet. The level of water in this filtered water basin is below the level of the false support floor of the filtering mass, in order to avoid its depression.

To increase the available head, the water level above the filter bed is adjusted. A regulator is responsible for maintaining a constant water level on the filter. The movements of the head loss compensator are regulated from the level of water on the filter.

Downstream regulation is designed so that:

– the filter automatically absorbs the entire inlet flow;

– the filtration velocity is stable;

– the filter medium is not uncovered when the water supply into the filter is interrupted.

The purpose of regulating the filtered water flow is to maintain the level of water at a constant level above the filter mass, regardless of the flow rate supplied and the variation in head loss caused by the progressive clogging of the filtering material. The filter is equipped with a modulating head loss compensating valve. Its opening/closing is controlled by two piezometric sensors placed at the filter supply channel, and in the filtered water basin. The draining floor on which Biodagen® rests has been especially designed by OTV-Veolia to ensure the filter's proper functioning and that it is washed correctly (Figure 13.4).

Figure 13.4. *Floor, slab and strainer. For a color version of this figure, see www.iste.co.uk/gaid/watertreatment3.zip*

It is the floor which has to ensure the perfect homogeneity of the fluid's distribution within the filtering mass during washing operations (water and air).

The floor is made of OTV plates, equipped with 50/m² strainers made of high-impact polystyrene, ensuring not only the flow of filtered water but most importantly, the distribution of injected backwashing water and air. The strainer heads are pierced with vertical slots, 0.4 mm wide, preventing any passage of filtering Biodagen® under the floor and securing a rigorous distribution of leaks, both during normal operation and during washes. During the washing operations, the water and the washing air are introduced under the floor through superimposed orifices (water at the bottom, air at the top), arranged between two support beams, in a sheath placed under a launder which receives the dirty washwater from the biofilter.

13.1.4.1. *Filter washing protocols*

Clogging cannot exceed a certain predefined value, at the risk of degrading the quality of filtered water, or even worse, the maximum value of the available head loss. When this value is reached, it is necessary to "unclog" the filter and to clean the filtering mass by removing the floc deposited there. As a general rule, such cleaning is performed by means of a water countercurrent initially associated with air insufflation, and continued after insufflation stops. The false support floor of the filtering mass is designed to ensure the passage of filtered water from top to bottom – without the material being entrained – for it to be distributed as uniformly as possible. With this same concern for distribution, the simultaneous passage of washwater and air in an ascending direction should be considered.

Washing is triggered by a clock, by a head loss concentration, or by a simple manual impulse (operator's decision). All washing operations are automated.

The washing equipment is designed for maximum flow rate values:

– low-flow washwater (8 min): surface flow: 10 $m^3 \cdot m^{-2} \cdot h^{-1}$;

– high-flow washwater (12–15 min): surface flow: 60 $m^3 \cdot m^{-2} \cdot h^{-1}$;

– washing air: surface flow: 70 $m^3 \cdot m^{-2} \cdot h^{-1}$.

13.1.4.2. *Biodagen® unloading and extraction*

Biodagen® is delivered by tank truck. It is unloaded using a hydro-ejector supplied with industrial water at a minimum of 5 bar obtained by tapping into the service water pipe. Filling in the filter is done through the removable cover, with a flexible tube.

13.1.4.3. *Reoxygenation*

Downstream of denitrification, it is necessary to reoxygenate the filtered water from the biofilter. Aeration can be implemented with a column of ozonated air (post-ozonation vents and/or fresh ozone), an ejector associated with a static mixer (in case of pressurized filters only), fine-bubble aeration (oxytube-type diffusers), or a cascade.

When compared to simple aeration, the advantage of ozonation is that it ensures oxygenation, the oxidation of any nitrites and a pre-disinfection.

13.1.4.4. *Refining by filtration*

The purpose of sand filtration downstream of the reoxygenation or ozonation stage is to remove any residual ethanol and to retain suspended solids (polishing treatment), before ending with disinfection.

13.1.5. *Water treatment stations including biological denitrification*

The choice of the water treatment depends on the physicochemical composition of raw water and various compounds.

13.1.5.1. *The case of low turbid water containing nitrates and micropollutants*

This is the most common case met for groundwater treatment. The treatment station includes a Biodenit® filter followed by an aeration stage (cascade or sprinkling or tank bubbling), or ozonation. A complementary stage with a monolayer GAC filter is used for the removal of micropollutants. Disinfection is implemented before distribution (Figure 13.5).

The GAC filter can be replaced by a dual media GAC-sand filter in order to polish and improve the particle size in the final filtered water.

13.1.5.2. *The case of low turbid water containing nitrates, hardness and micropollutants*

The treatment station includes a Biodenit® filter followed by an aeration stage (cascade, spraying or tank bubbling), followed by softening with a catalytic process (soda or lime) to reduce the hardness, sand or dual media filtration (anthracite-sand or pumice-sand) to remove the fine particle of remaining carbonates, and a filtration stage onto granular activated carbon to remove the micropollutants. Disinfection is implemented before distribution (Figure 13.6).

Acidification downstream of softening dissolves $CaCO_3$ particles at the softened water and produces good-quality water to feed filters (turbidity <3 NTU).

13.1.5.3. *The case of low turbid water containing nitrates and ammonia*

The system includes a Biodenit® filter followed by an ammonia removal step with a biological filter (see Chapter 12). Ozonation can be proposed for the final disinfection even if chlorine is also injected at the reservoir's outlet to protect the water in the network (Figure 13.7). The proportion of biologically removable ammonia with a polishing biofilter depends on the oxygen content obtained after the aeration step which must be installed after the Biodenit® filter (biological nitrification requires 3.55 mg of O_2·mg-1/nitrified NH4). This O_2 content depends on the temperature and the chosen aeration system. Nitrified ammonia into nitrates is counted among the final outlet nitrates ($\Delta NO_3 = 62/18 \times$ nitrified NH_4). Excess ammonia, which is not biologically nitrifiable, would be removed by breakpoint chlorination.

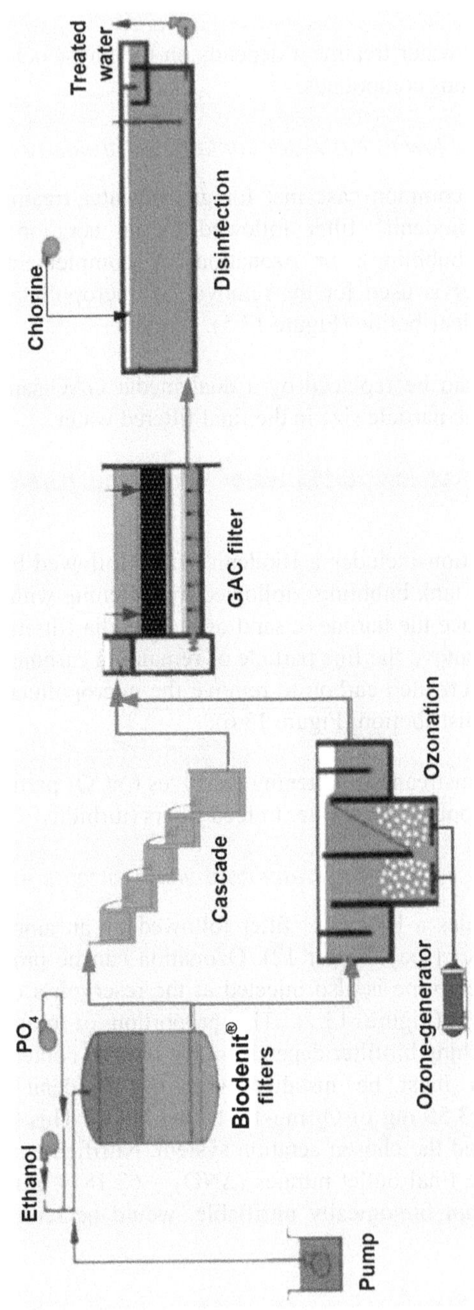

Figure 13.5. *Biological removal of nitrates, followed by activated carbon filtration. For a color version of this figure, see www.iste.co.uk/gaid/watertreatment3.zip*

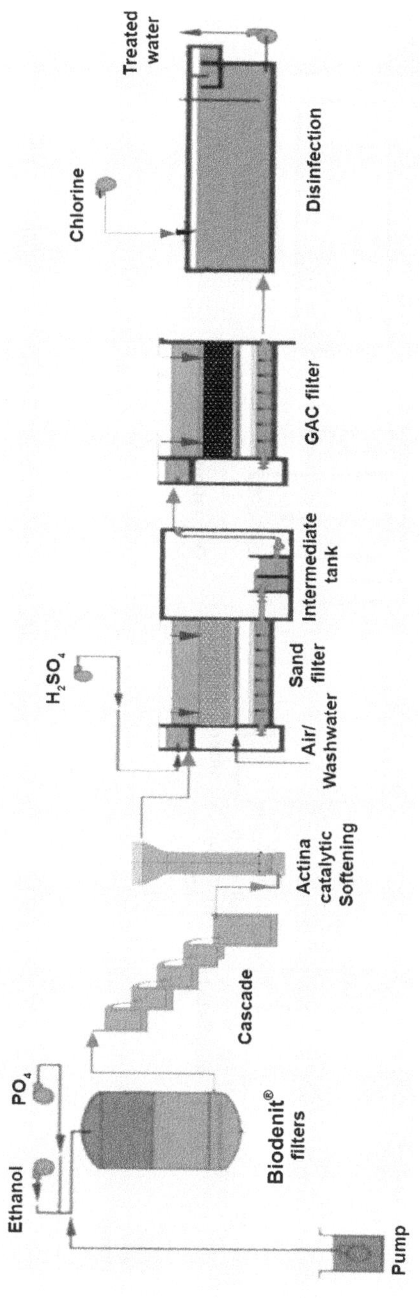

Figure 13.6. *Treatment station (biologically) removing nitrates, hardness, heavy metals and micropollutants. For a color version of this figure, see www.iste.co.uk/gaid/watertreatment3.zip*

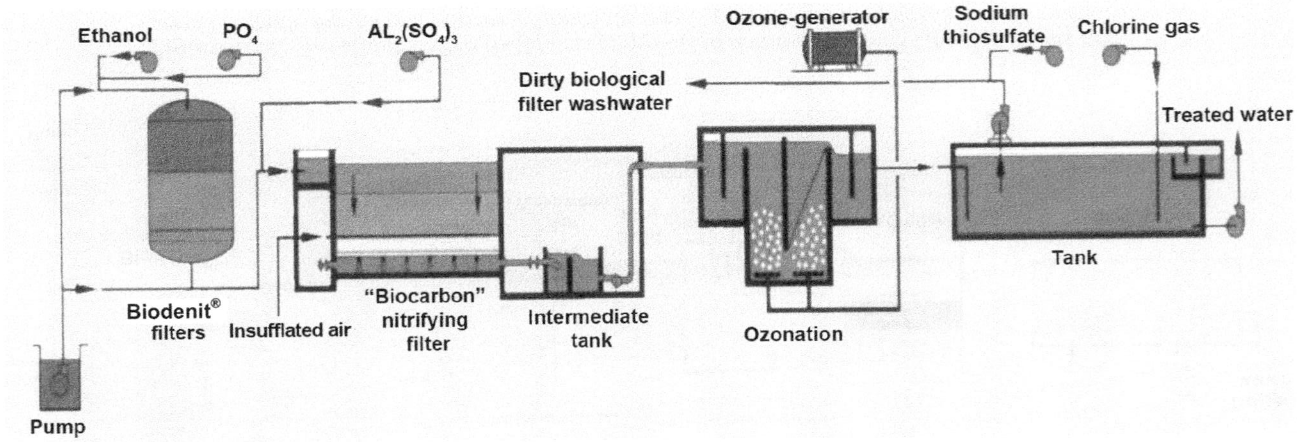

Figure 13.7. *Treatment station removing nitrates and ammonia (including disinfection). For a color version of this figure, see www.iste.co.uk/gaid/watertreatment3.zip*

If the water used for washing the filters is pumped into the tank, the neutralization of the residual ozone (or chlorine) is required to avoid killing the biomass during the washing. When possible, an intermediate tank for non-chlorinated filtered water is installed in order to backwash the filters.

In all cases, Biodenit® filters, sand filters, dual media filters or activated carbon filters can be:

– the closed-type, operating under pressure;

– the concrete open-type, with a high-water level (F or MD types) as described in Chapter 8.

Figure 13.8. *Open-Biodenit® filters with Biodagen® and closed-filter (Veolia sites). For a color version of this figure, see www.iste.co.uk/gaid/watertreatment3.zip*

Remineralization, aggressiveness correction or pH adjustment are implemented, depending on the water quality.

13.1.6. *Factors affecting biological denitrification*

The Biodenit® process can be applied to surface, underground or industrial water, respecting the following limitations:

– water loaded with nitrates, without suspended solids or toxic products ($NO_3^- < 150$ mg·L^{-1});

– water temperature higher than 4°C;

– free-chlorine concentration lower than 0.2 mg·L^{-1}.

13.1.6.1. *pH*

Optimum denitrification appears within a pH zone of 6 to 9, depending on the bacteria family. The pH affects the biodegradation kinetic of nitric nitrogen, and also the nature of the final product: it is only at a pH higher than 7 that nitrogen gas is obtained. In special cases (acid water with low mineral content), an initial pH correction is necessary for the biological phenomenon to set in.

The denitrification reaction naturally tends to alkalize the environment. The denitrification of alkaline water can lead to slightly scaling water.

13.1.6.1.1. Dissolved oxygen

The dissolved oxygen content of raw water must be lower than 5 $mg \cdot L^{-1}$. Heterotrophic denitrification has to take place in an anoxic medium and the conditions for anoxia are quickly obtained within the biological filter. The dissolved oxygen present in raw water is consumed by aerobic bacteria in the upper 20–30 cm of the material layer. Without available oxygen, bacteria consume the oxygen from nitrates in order to fulfill their respiratory and synthetic needs (Figure 13.9).

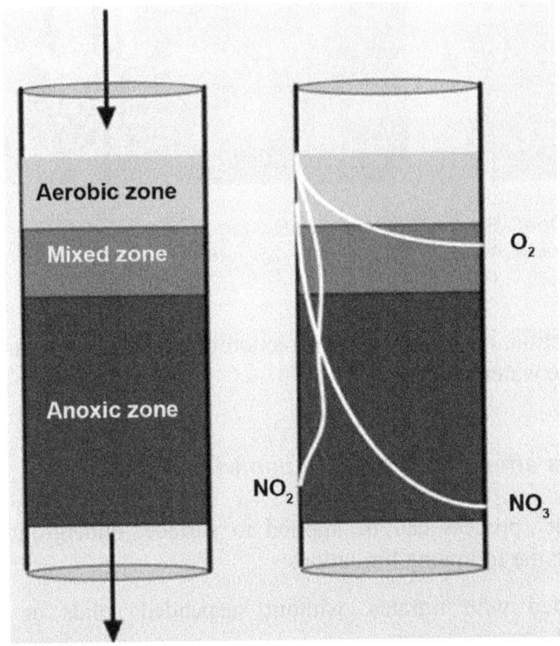

Figure 13.9. *Different identifiable areas in the Biodenit® filter when dissolved oxygen is present in water. For a color version of this figure, see www.iste.co.uk/gaid/watertreatment3.zip*

The seeding period (Figure 13.10) depends on temperature, spanning 3 weeks to 2 months.

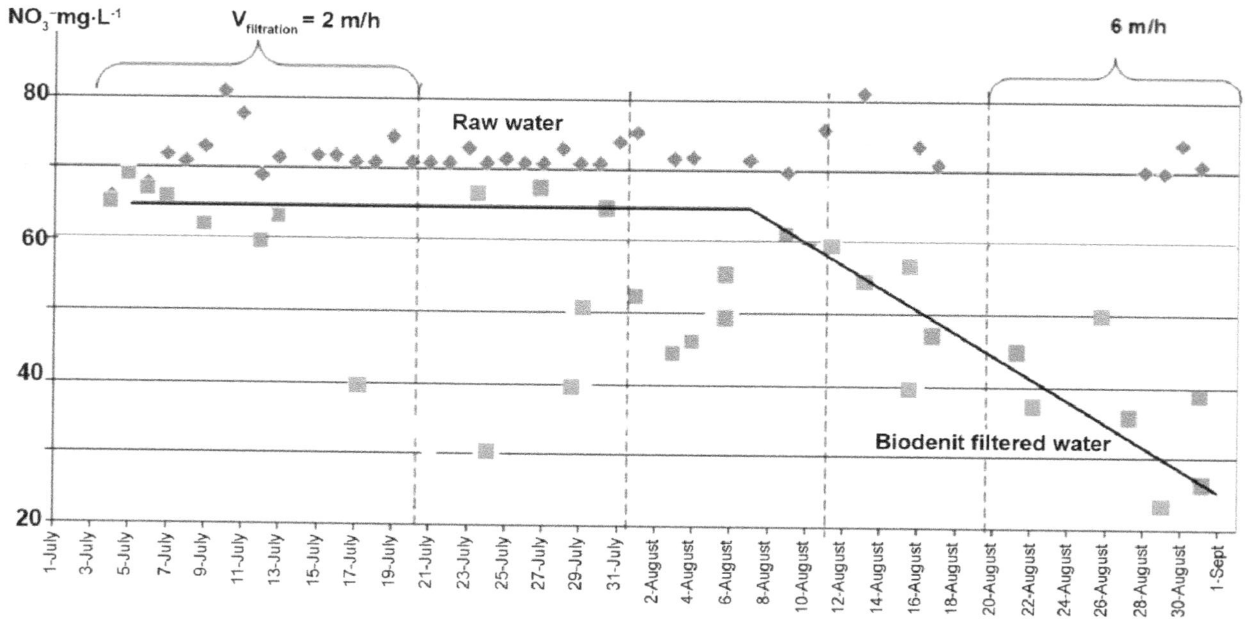

Figure 13.10. *Seeding of a Biodenit® filter (T: 12°C). For a color version of this figure, see www.iste.co.uk/gaid/watertreatment3.zip*

13.1.6.2. *Carbonaceous substrate*

The rate of carbonaceous substrate (ethanol or acetic acid) to be applied is related to the nitrate and dissolved oxygen content in the raw water entering the biological filters.

13.1.6.3. *Phosphorus*

The rate of phosphorus to be applied is related to the quantity of nitrates to be removed.

13.1.6.4. *Temperature*

The impact of temperature on biological systems is well described by the Arrhenius ratio:

$$\text{reaction velocity (T °C): reaction velocity at 20°C } \Theta^{(T-20)}.$$

The denitrification velocity is expressed in mg NO_3^-·g $VSS^{-1}·h^{-1}$. "VSS" stands for volatile suspended solids, which represent the biomass. The coefficient Θ is the Arrhenius constant.

13.1.6.5. *Material*

The support material must have a broad specific surface, large macroporosity, low density and a high resistance to abrasion. Bacteria preferentially get attached to porous materials such as Biodagen®, which promote bacterial adhesion.

The choice of grain size is a compromise, knowing that small-sized materials result in better efficiency, but lead to a faster clogging of the free interstices. Particle size varies between 3 and 6 mm for Biodagen®.

13.1.7. Design parameters: applied volumic load

The applied volumic load (Cv) is expressed in kg of nitrates (NO_3) per m^3 of material per day ($kg·m^{-3}·d^{-1}$). Table 13.3 summarizes the recommended values.

We have Cv (N-NO_3) = Cv (NO_3)/4.428.

Temperature (°C)	3–5	5–8	8–11	>11
Cv (NO_3)	2.8	3.5	4.5	8
Cv (N)	0.63	0.79	1.02	1.58

Table 13.3. *Applied volumic loads as a function of temperature*

13.1.8. *Design parameters: minimum contact time (tc min)*

Table 13.4 summarizes the recommended values for temperatures above 3°C:

$$tc = \frac{60\ vf}{Q}$$

where *vf* is the maximum filtration velocities ($m \cdot h^{-1}$).

Temperature (°C)	3–5	5–8	8–10	10–12	12–14	14–16	>16°C
Contact time (min)	32	22	17	12	10	8	6

Table 13.4. *Contact time (min) as a function of temperature*

Table 13.5 summarizes the values recommended for temperatures above 3°C and those applicable during filter washing (N-1 filters in production):

$$vf = \frac{Q}{S} = \frac{H \times 60}{tc},$$

where:

– Q: flow expressed in $m^3 \cdot h^{-1}$;

– S: area in m^2.

Temperature (°C)	3–5	5–8	8–10	10–12	12–14	>14
Nitrate conc. <100 $mg \cdot L^{-1}$	4.5	5.5	7	9	10	10
Nitrate conc. = 100–150 $mg \cdot L^{-1}$	4.2	5	6.5	8	10	10

Table 13.5. *Filtration velocity ($m \cdot h^{-1}$) as a function of temperature and nitrate concentration*

13.1.9. *Design parameters: height of Biodagen® material (m)*

The height of the media is deduced from velocity and contact time:

$$H = \frac{tcvf}{60}.$$

The minimum height is 1 m and the maximum height is 2.5 m.

13.1.10. *Design parameters: material volume (m³)*

$$Vm = \frac{24\ Q\Delta NO_3}{1{,}000\ Cv}$$

where:

– Q: water flow to be treated in $m^3 \cdot h^{-1}$;

– ΔNO_3: quantity of nitrate to be removed in $g \cdot m^{-3}$;

– Cv: maximum applied volumic load in kg of $NO_3 \cdot m^{-3} \cdot d^{-1}$, depending on the temperature of the water to be treated.

When a 24-h operation is not required, the volume of material then becomes:

$$Vm = \frac{Q\ tf\ \Delta NO_3}{1{,}000\ Cv}$$

where tf is the number of treatment station operating hours (h).

13.1.11. *Partial treatment*

Since most nitrates are reduced to nitrogen gas within the biofilter, the surface area of reactors can be reduced (concomitantly with the investment cost), by treating only a part of raw water at the denitrification stage. At the reactor's outlet, denitrified water is mixed with raw water in accordance with a flow rate ratio resulting in a distribution with a nitrate content $<25\ mg \cdot L^{-1}$, which corresponds to the European standard guidelines.

A fraction of the biological filter's bypass feed rate is then mixed with filtered water before the following stage (filtration with activated carbon, final disinfection, etc.).

If Qin is the nominal flow (m^3-h^{-1}) containing Xin $mg \cdot L^{-1}$ of nitrates in raw water, Q_{DN} is the water flow to be denitrified ($m^3 \cdot h^{-1}$), X_{DN} is the nitrate concentration at the biological filter's outlet (which can decrease to 5 $mg \cdot L^{-1}$), then Xout is the required final nitrate concentration at the treatment station's outlet (after mixing, e.g. 25 $mg \cdot L^{-1}$).

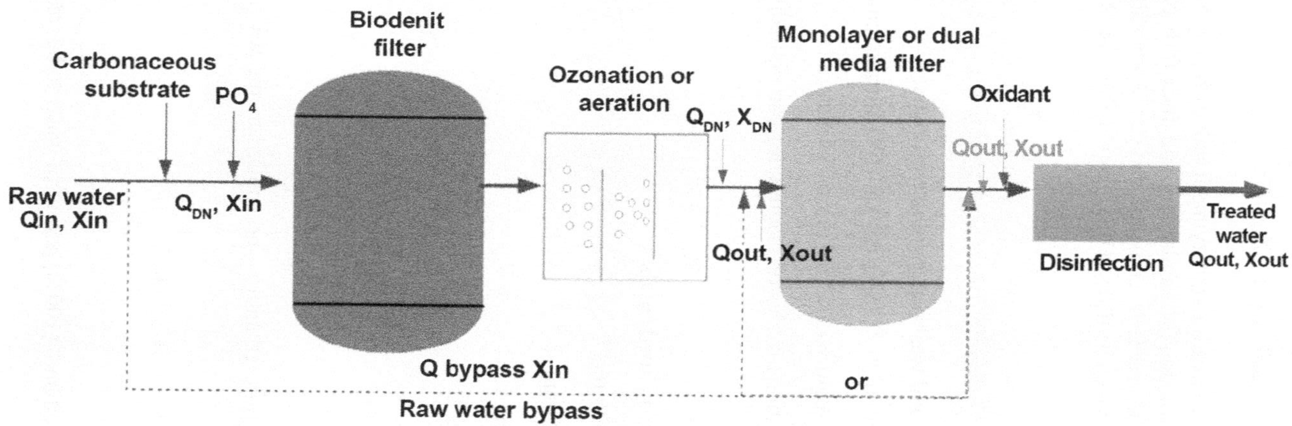

Figure 13.11. *Denitrification with partial treatment of raw water. For a color version of this figure, see www.iste.co.uk/gaid/watertreatment3.zip*

The flow rate Q_{DN} to be denitrified is:

$$Q_{DN} = Q \times (Xin - Xout /(Xin - X_{DN}).$$

It is recommended to avoid denitrifying at values below 5 mg·L^{-1}, at the risk of being unable to control anoxia conditions, and therefore passing into anaerobic conditions. In this case, bacteria consume oxygen from sulfates and form sulfur dioxide (SO_2) or hydrogen sulfide (H_2S), responsible for the odors in the water produced.

13.1.12. *Sludge production*

Sludge production and washwater volume estimate:

– sludge production: 0.09–0.15 kg SS·kg^{-1} NO_3 nitrate removed (with ethanol);

– amount of water required: 2–3 m^3 per m^3 of material;

– sludge concentrations (order of magnitude) corresponding to the first minutes of washing:

- SS: 200–350 mg·L^{-1},
- VM: 95% of suspended solids,
- COD: 700–1,000 mg·L^{-1},
- BDO5: 180–250 mg·L^{-1},
- KTN: 10–35 mg·L^{-1},
- total P: 3–7 mg·L^{-1}.

13.1.13. *The reagents*

13.1.13.1. *Ethanol*

Concentration of pure ethanol to be applied to the water flow passing through Biodenit® filters:

$$Ceth = [Xin \times k1] + [C_{O2} \times k2],$$

where:

– Ceth: g of pure ethanol (100%) per m^3 of water to be treated;

– Xin: NO_3 concentration in raw water (g $NO_3 \cdot m^{-3}$);

– k1: 0.696 (coefficient associated with ethanol/nitrate reactions);

– C_{O2}: dissolved oxygen concentration in the water to be treated ($mg \cdot L^{-1}$);

– k2: 0.55 (coefficient associated with ethanol/oxygen reactions).

Characteristics of alcohol denaturated with sulfuric acid:

– minimum volumetric title at 20°C: approximately 95%;

– density at 20°C: approximately 0.789 $g \cdot cm^{-3}$;

– maximum ether content: 50 $mg \cdot L^{-1}$;

– Qeth hourly flow of denatured alcohol:

$$Qeth = (Ceth \times Q_{DN})/770,$$

where:

– Qeth: pump flow in $l \cdot h^{-1}$;

– Q_{DN}: water flow to be treated on the Biodenit® filter, in $m^3 \cdot h^{-1}$.

In the case of open filters, the introduction of the reagent is done at each filter's inlet, which justifies the need to provide the appropriate flow meters and adjustment valves.

13.1.13.2. *Acetic acid*

Concentration of pure acetic acid (AA) to be applied to the water flow passing through Biodenit® filters:

$$C_{AA} = [Xin \times k1] + [C_{O2} \times k2]$$

where:

– C_{AA}: g of pure AA (100%) per m^3 of water to be treated;

– Xin: NO_3 concentration in raw water (g $NO_3 \cdot m^{-3}$);

– k1: 0.9 (coefficient associated with AA/nitrate reactions);

– C_{O2}: dissolved oxygen concentration in the water to be treated ($mg \cdot L^{-1}$);

– k2: 0.5 (coefficient associated with AA/oxygen reactions).

Characteristics of acetic acid:

– minimum volumetric title at 20°C: >96 %;

– density at 20°C: 1.03 g·cm^{-3};

– Q_{AA} hourly flow:

$$Q_{AA} = (C_{AA} \times Q_{DN})/770$$

where:

– Q_{AA}: pump flow in l·h^{-1};

– Q_{DN}: water flow to be treated on the Biodenit® filter, in m^3·h^{-1}.

In the case of open filters, the introduction of the reagent is done at each filter's inlet, which justifies the need to provide the appropriate flow meters and adjustment valves.

13.1.13.3. *Phosphorus*

The stoichiometric ratio shows that the phosphorus P rate to be introduced and the sought reduction in nitrates are as follows:

$$P = 0.00226 \times Xin + 0.01\ O_2$$

where:

– Xin: NO_3 concentration in raw water (g NO_3·m^{-3});

– P: pure phosphorus rate (100%) in g·m^{-3} for water passing through Biodagen®.

Experience shows that three to four supplementary quantities are required when seeding the biological filter. The final operational dosage is a combination of the stoichiometric dosage and the value injected during start-up.

Phosphorus is obtained from a solution of food-grade dehydrated sodium dihydrogen phosphate, or from monosodium phosphate (NaH_2PO_4, $2H_2O$; MW = 156), where phosphorus P = 19.87%.

Qp hourly flow rate of phosphorus solution:

$$Qp = (5.03 \times P \times Q_{DN}) / C_P$$

where:

– Qp: flow rate of the sodium phosphate pump in l·h^{-1};

– Q_{DN}: raw water flow through the Biodenit® filter, m^3·h^{-1};

– C_P: concentration of the phosphorus solution (from 100 to 600 g·L^{-1}).

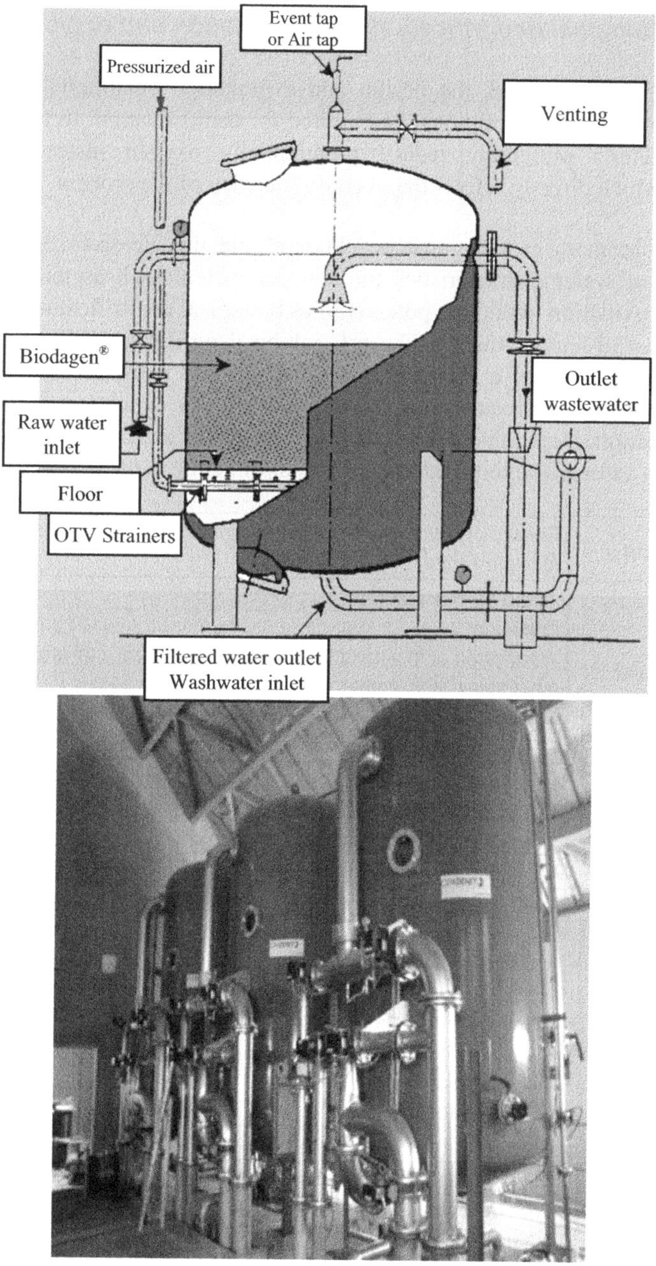

Figure 13.12. *Detail of a closed Biodenit® filter (OTV-Veolia). For a color version of this figure, see www.iste.co.uk/gaid/watertreatment3.zip*

13.1.14. *Biological denitrification implementation and exploitation*

Table 13.6 summarizes the design and exploitation conditions for biological denitrification. Anoxic conditions are required for denitrification. In the presence of oxygen, bacteria preferably reduce preferentially oxygen instead of nitrates, something which directly affects the overall efficiency of the process.

Optimal biomass growth must be obtained and this involves the control and monitoring of water characteristics through parameters such as temperature, pH, salinity and oxidation–reduction potential. For biological denitrification, a pH of 7–8 is preferred, and temperatures below 5°C inhibit denitrification. The pre-treatment includes the addition of a carbonaceous substrate and phosphorus in appropriate doses. Post-processing requirements include aeration or ozonation, filtration and a final disinfection. Depending on the case, the addition of a coagulant, of an activated carbon filter, or even the removal of hardness or of heavy metals may be required.

Pre-treatment	Dosage of organic carbon and phosphorus pH adjustment
Post-treatment	Aeration or ozonation Filtration Adsorption onto activated carbon in the presence of micropollutants or other adsorbable contaminants Disinfection Calco-carbonic equilibrium
Chemical reagents	Contingent pH adjustment Dosage of organic carbon and phosphorus Coagulant/polymer in case of turbidity Oxidizers for disinfection
O&M	Instrumentation for online control: pH, NO_3, DOC, P Permanent management and controls are necessary to ensure chemical and bacteriological performance Necessary control of nitrites and ammonia Management of chemical reagents Management of filter washes, wastewater, sludge and their treatment Odor management
Sewage sludge	Sludge treatment and disposal
Imperatives	Management of chemical reagents Post-treatments required, which can be complex depending on water quality Operator training

Table 13.6. *Summary of design and exploitation conditions for biological denitrification*

Operating conditions should be optimized to ensure thorough denitrification. Nitrite concentrations in denitrified water must be as low as possible in order to avoid their oxidation at the treatment station due to an excessive addition of O_3 or chlorine.

Incomplete denitrification, which may be associated with an excessively high dissolved oxygen content, may lead to the production of greenhouse gases (NO and N_2O). In case of high O_2 concentration, it is recommended to increase the material height to favor the O_2 consumption in the first layers, and to add the required substrate dosages.

Post-treatment is important in order to meet turbidity standards, remove the biomass, residual DOC, to treat odors, micropollutants and any other contaminants present in raw water. It also ensures an optimal disinfection. An excess in the carbonaceous substrate (ethanol, acetic acid or equivalent) can lead to biological development on every structure of the treatment station, including tanks.

Biological denitrification performed with heterotrophic bacteria involves the production of sludge whose volume depends on the quantity of nitrates removed.

Figure 13.13. *Reference examples of the Biodenit® process in France. For a color version of this figure, see www.iste.co.uk/gaid/watertreatment3.zip*

13.2. Treatment with ion exchange resins

Ion exchange resins are ion exchangers of an organic nature, containing a macromolecular network onto which ion-active groups (functional groups) are fixed. The macromolecular network constitutes the resin's backbone, endowing it with its mechanical strength, whereas the functional groups provide the resin with the properties of an ion exchanger. There are two categories of resins depending on the nature of the functional groups attached to the polymer backbone.

Weakly or strongly acidic cation exchange resins fix cations. In this case, the functional group is of the anionic type. Weakly or strongly basic cation exchange resins fix anions. In this case, the functional group is of the cationic type.

The main characteristics of an ion exchanger resin are as follows:

– *type*: the cationic type fixes the cations, whereas the anionic type fixes the anions;

– *ionic form*: it is determined by the nature of the ions bound to the functional groups. For example, the resin is in its chloride form when it fixes chloride ions to its skeleton;

– *granularity*: resins generally come in the form of small balls (400–800 μm diameter);

– *ion exchange capacity*: for a given quantity of resin, there is a precise number of functional groups: this is the resin's exchange capacity. This is the theoretical number of monovalent ion moles which a resin can fix. The exchange capacity is often related to the dry resin's mass or to its volume;

– *density*: resin density is generally between 1.02 and 1.12 for anion exchangers and between 1.13 and 1.38 for cation exchange resins.

13.2.1. *General exchange mechanism (Ecodenit® process)*

The resin exchanges the ions fixed to the functional groups for those present in the water. The ion-fixation equilibrium on a resin differs depending on the ions considered. The resin's affinity for the ions to be fixed must be stronger than the one belonging to the ions initially fixed on the resin. The exchange is stoichiometric for monovalent ions, and there is a possibility of stoichiometry deviation for multivalent ions. The resin's affinity for an ion is significantly higher when the charge density of the ion is important. In that sense, sulfates are better exchanged than nitrates or bicarbonates with an anion exchange resin.

Denitrification via the Ecodenit® process focuses on nitrate removal by percolation of the water to be treated on an ion exchanger resin bed. The resins used are anion exchange resins, regenerated with sodium chloride.

13.2.1.1. *Saturation phase*

During the saturation phase, that is to say, during the production period, not only nitrate ions but also sulfates and bicarbonates are fixed by the resin in exchange for chlorides:

$$\begin{matrix} NO_3^- \\ SO_4^{2-} + R^+\text{-}Cl^- \rightarrow R^+ - \\ HCO_3^- \end{matrix} \left\{ \begin{matrix} NO_3^- \\ SO_4^{2-} + Cl^- \\ HCO_3^- \end{matrix} \right.$$

The water's ionic composition is modified during the treatment.

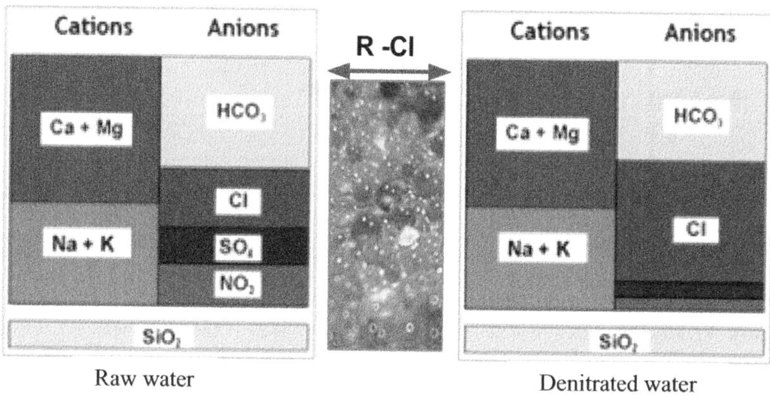

Figure 13.14. *Nitrate and sulfate exchange on an anionic resin (Cl⁻). For a color version of this figure, see www.iste.co.uk/gaid/watertreatment3.zip*

13.2.1.2. *Regeneration phase*

When the resin is exhausted in terms of chloride ions (or saturated in nitrate, sulfate and bicarbonate ions), a leak in nitrate ions is observed in the treated water. Resin regeneration is then triggered by a sodium chloride solution in order to recondition the resin to its initial state.

During this phase, the ions fixed by the resin are eluted by chloride ions and are found in the regeneration eluates.

$$R+ - \left\{ \begin{matrix} NO_3^- \\ SO_4^{2-} + NaCl \rightarrow R^+ - Cl^- + Na \\ HCO_3^- HCO_3^- \end{matrix} \right. \left\{ \begin{matrix} NO_3^- \\ SO_4^{2-} \\ \end{matrix} \right.$$

Such eluates are loaded with nitrate, sulfate and bicarbonate ions, as well as the excess NaCl which was used for regeneration.

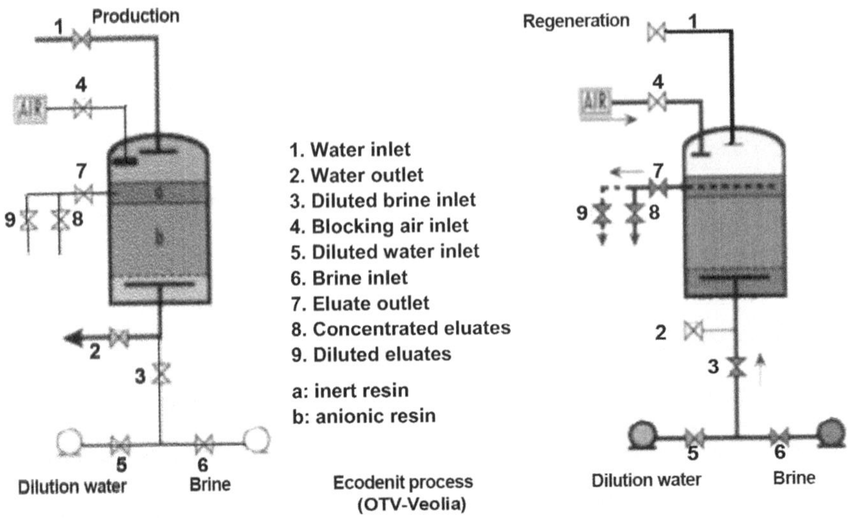

Figure 13.15. *Development of the Ecodenit® process (OTV-Veolia). For a color version of this figure, see www.iste.co.uk/gaid/watertreatment3.zip*

In the Ecodenit® process, the mixing of the resin is performed before putting it back into production, so as to avoid a high nitrate leak in the first water produced. Since the regeneration solution percolates bottom-down, it gradually becomes packed with eluted ions. As a result, nitrate ions are less well eluted in the lower portion of the bed and are later found in the first waters produced. The resin regeneration phase is implemented countercurrently (bottom-up), from a sodium chloride concentrated solution. On the one hand, this flow inversion encourages an excellent regeneration of the lower resin layers (which are generally less saturated by the most concentrated brine), and on the other hand, it improves the anion exchange during the production phase. This is because the descending waters – gradually impoverished in nitrates – come across increasingly better regenerated resins.

To improve this technology, OTV has adapted and improved the air blocking technique, long implemented on demineralization chains destined for the production of high purity water (nuclear or electronics fields). During the regeneration phase, an injection of compressed air into the upper portion of the exchanger enables the formation of a drained resin "cover" which opposes the expansion of the resin bed under the effect of the regenerating flow. In addition, it is possible to proceed with the unclogging and periodic defining of resins by simply raising the water level, without requiring any supplementary equipment. This is the Ecodenit® countercurrent process.

This specifically designed procedure for the Ecodenit® process makes it possible to reach a nitrate leak lower than the 25 mg·L^{-1} guiding value at any given moment of the saturation cycle.

The cocurrent (top-down) regeneration technique is simpler, as it does not require resin blocking. On the other hand, however, the brine and rinsing water volumes required are greater (40–50% more), obtaining a reduction of 85–90%, compared to 95–97% with the countercurrent system (for a given quantity of regenerating agent).

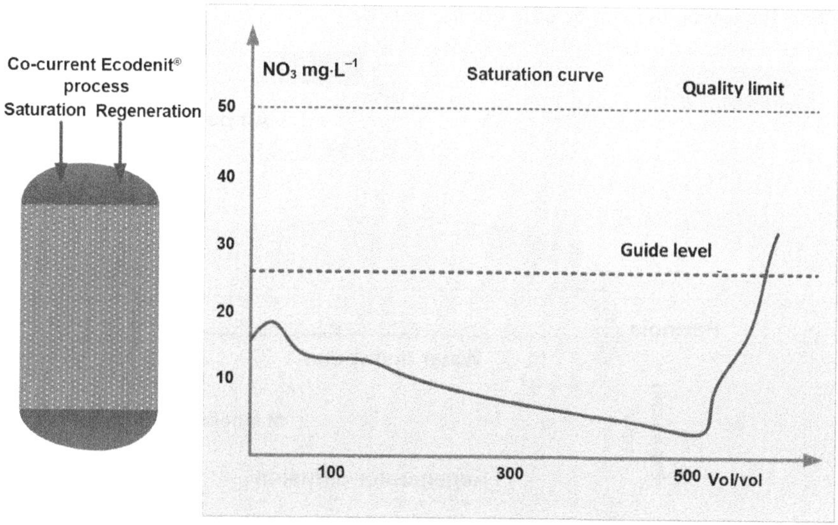

Figure 13.16. *Denitrification saturation curve. For a color version of this figure, see www.iste.co.uk/gaid/watertreatment3.zip*

The operating cycle of an Ecodenit® exchanger takes place in four phases:

– saturation: saturation corresponding to the production phase: the water passes through the bed of the exchanger (downdraft) until it saturates. At the breakthrough point (when the leakage limit value is reached), the production phase comes to a stop;

– water level rising: this phase removes the particles which have settled on the bed's surface using an ascending water current;

– regeneration can be either carried out by introducing a brine by percolation (in cocurrent mode) or by means of an ascending flow (in countercurrent mode). The regeneration process involves displacement;

– rinsing: during this phase, the excess regenerating agent present in the water is removed at a low flow rate until the resin only contains weak traces of regenerating agent: this is the slow rinsing phase. Then comes a rapid rinsing stage with a higher flow rate in order to remove the last traces of regenerating agent.

13.2.2. *Ecodenit*® *process technology*

The exchanger comprises a system device functioning under pressure, made up of a vertical cylinder closed at each end. The upper end has a domed bottom, whereas the lower end can be either domed or flat.

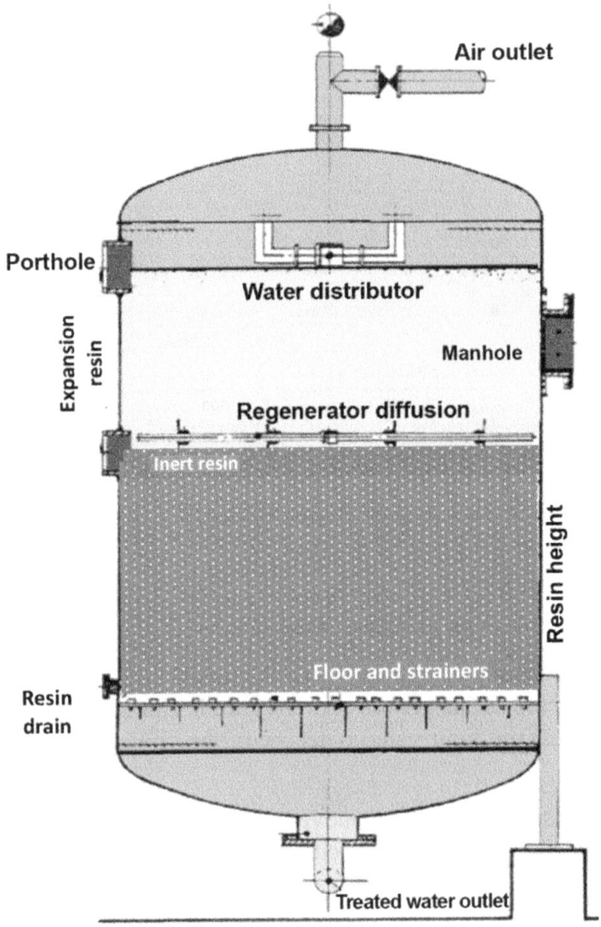

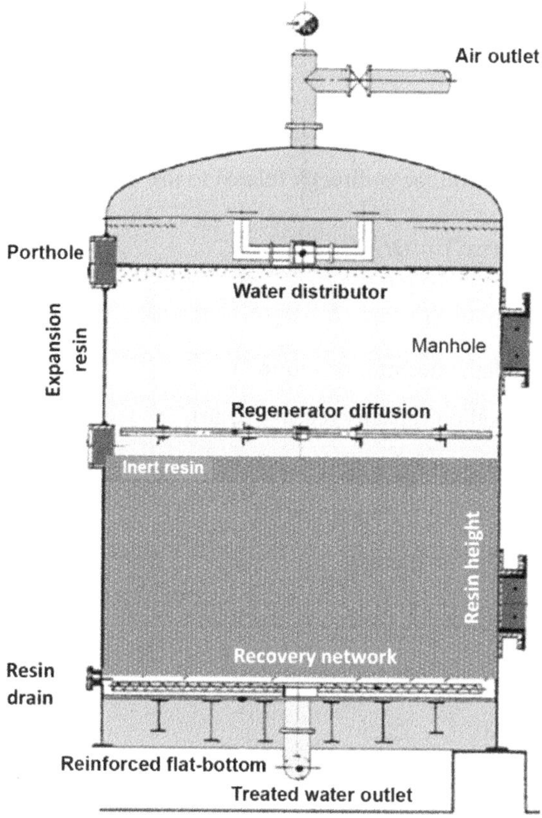

Figure 13.17. *(A) Domed-bottom and (B) flat-bottom exchangers. For a color version of this figure, see www.iste.co.uk/gaid/watertreatment3.zip*

The inner equipment includes:

– a device for distributing water uniformly in the upper portion;

– a branched regenerator diffusion network;

– a treated water recovery device comprising a branched collector (flat bottom) or a floor with strainers (domed bottom).

13.2.2.1. *Design parameters: resin total capacity (RTC)*

The fixing capacity, called "resin total capacity", is about 940 meq·L^{-1} of resin. The regeneration rate is 82.5 g·L^{-1} of resin.

The nitrate fixation capacity (CF_{NO3}) depends on the concentration of anions present (sulfates, bicarbonates and nitrates). It is determined using specific curves for each supplier.

13.2.2.2. *Design parameter: minimum resin volume*

The minimum resin volume is directly related to the nitrates' fixation capacity:

$$\text{Resin volume } (\text{m}^3) = Qt\,\frac{\rho_{NO3}[NO_3]}{CF_{NO3}}\,t$$

where:

- $[NO_3]$: nitrate inlet concentration ($\text{eq}\cdot\text{m}^{-3}$);
- CF_{NO3}: nitrate fixation capacity ($\text{eq}\cdot\text{m}^{-3}$ resin);
- Qt: water flow rate treated with the resin ($\text{m}^3\cdot\text{h}^{-1}$);
- t: expected operating cycle duration (h);
- ρ_{NO3}: denitrification efficiency.

13.2.2.3. *Design parameter: exchanger volumic load*

The exchanger volumic load is given by:

$$\text{Cv}(\text{m}^3\cdot\text{m}^{-2}\cdot\text{h}^{-1}) = \frac{Qt}{\text{Vresin}}$$

It must be between 40 and 60 $\text{m}^3\cdot\text{m}^{-2}\cdot\text{h}^{-1}$. The contact times are usually greater than 2 min.

13.2.2.3.1. Exchanger diameter

The useful diameter of the exchanger is the external diameter. It is related to flow velocity ($\text{m}\cdot\text{h}^{-1}$) and flow rate ($\text{m}^3\cdot\text{h}^{-1}$):

$$D(m) = \sqrt{\frac{4Qt}{\pi v}}$$

Resin height must be between 1 and 2.5 m with a resin H/D ratio which is often between 0.66 and 1.5. The total height of the exchanger must be equal to 2 resin H (in relation to the floor).

13.2.2.4. *Regeneration phase*

- Duration of the rising phase (ts): 10–15 min.

– Water volume used for the rising phase:

$$\text{Water volume (m}^3) = \frac{vs \cdot D^2 \pi ts}{240}$$

where vs is the rising velocity (m·h^{-1}) and ts is expressed in minutes.

The weight of the chlorides required for regeneration is given by:

$$\text{Weight NaCl (kg)} = 58.5 \cdot k \cdot \text{Vresin } CF_{NO3}$$

where:

– resin V: resin volume (m^3);

– CF_{NO3}: nitrate fixation capacity (eq·m^{-3});

– k: regeneration constant (at 150%), that is a k value equal to 1.5.

Brine concentration is: 50–80 g·L^{-1}.

The regeneration time rt is:

$$rt \text{ (min)} = \frac{60\ P}{50\ q\text{NaCl}}$$

where:

– qNaCl: brine flow (m^3·h^{-1});

– P: chloride weight (kg).

The volume used is:

$$\text{Brine volume (m}^3) = \frac{P}{C\ brine}$$

C brine is the concentrated brine (kg·m^{-3}) concentration.

The dilution volume is then:

$$\text{Dilutation volume (m}^3) = \frac{(\text{C brine} - 50)\ P}{50\ \text{C brine}}$$

13.2.2.5. *Eluates volume*

The total amount of eluates produced during regeneration represents the solution's total volume used for the rising, regeneration, displacement, rinsing and leveling phases.

Table 13.7 summarizes the weight (kg) of each anion released and the mean eluates concentration ($g \cdot L^{-1}$).

	Weight rejected (kg)	**Average concentration ($g \cdot L^{-1}$)**
Nitrates	$\rho\, NO_3 \cdot [NO_3] \cdot Qt \cdot t \cdot 10^{-3}$	$\rho\, NO_3 \cdot [NO_3] \cdot Qt \cdot t \cdot 10^{-3}$/eluates V
Sulfates	$\rho\, SO_4 \cdot [SO_4] \cdot Qt \cdot t \cdot 10^{-3}$	$\rho\, SO_4 \cdot [SO_4] \cdot Qt \cdot t \cdot 10^{-3}$/eluates V
Bicarbonates	$\rho\, HCO_3\, [HCO_3] \cdot Q \cdot t \cdot 10^{-3}$	$\rho\, HCO_3 [HCO_3] \cdot Qt \cdot t \cdot 10^{-3}$/eluates V
Chlorides	$\frac{35.5 \cdot P\,(RR - 100)}{58.5\, RR}$	$\frac{35.5 \cdot P\,(RR - 100)}{58.5\, RR.\mathrm{Veluates}}$
Sodium	$\frac{23\, P}{58.5}$	$\frac{23\, P}{58.5 \cdot \mathrm{Veluates}}$

Table 13.7. *Weight and average concentrations of the different anions released in eluates*

The units are the following: Qt is the product flow ($m^3 \cdot h^{-1}$) during the operating time, *t* (time), [NO_3^-], [SO_4^-] and [HCO_3^-] ($mg \cdot L^{-1}$) are anion concentrations, *P* is the chloride weight (in kg), *V* is the eluates volume (m^3), *RR* is the regeneration rate enforced (in %) and ρ is the efficiency for each anion.

13.2.2.6. *Partial treatment*

Since nitrates are fixed on the resin, it is possible to reduce the number of exchangers (and consequently, the investment cost) by treating only part of the raw water at the denitrification stage.

At the reactor's outlet, denitrified water is mixed with raw water in accordance with a flow rate ratio resulting in a distribution with a nitrate content <25 $mg \cdot L^{-1}$, which corresponds to the European standard guidelines. Nevertheless, the exchanger's production must be stopped when the resin reaches a saturation level in which the treated water's nitrate content surpasses the preset concentration (often established at 5 $mg \cdot L^{-1}$).

A fraction of the resins' bypass feed rate is then mixed with denitrified water, before the following stage (filtration with activated carbon, final disinfection, etc.).

If Qin is the nominal flow (m^3-h^{-1}) containing Xin $mg \cdot L^{-1}$of nitrates in raw water, Q_{DN} is the water flow to be denitrified ($m_3 \cdot h^{-1}$), X_{DN} is the nitrate concentration at the exchanger's outlet (which can decrease to 5 $mg \cdot L^{-1}$), Xout is the required final nitrate concentration at the treatment station's outlet (after mixing, e.g., 25 $mg \cdot L^{-1}$), then the Q_{DN} flow to be denitrified is:

$$Q_{DN} = Q \times (Xin - Xout)/(Xin - X_{DN}).$$

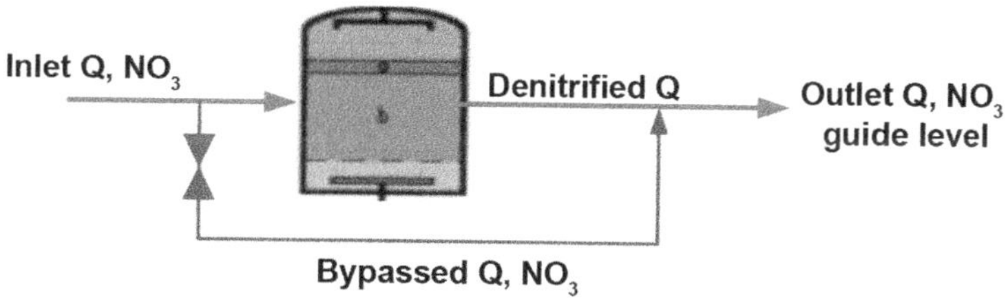

Figure 13.18. *Partial flow treatment. For a color version of this figure, see www.iste.co.uk/gaid/watertreatment3.zip*

13.2.2.7. *Performances*

The performances obtained with the Ecodenit® process are 85% for nitrates, >98% for sulfates and 30% for bicarbonates. Bicarbonates are totally fixed during the first third of the cycle and then gradually released, resulting in a mean efficiency of 30%.

The mean quantity of ions exchanged per m^3 of raw water is given by the ratio:

$$N(eq{\cdot}m^{-3}) = \rho NO_3[NO_3]{\cdot}\rho SO_4[SO_4]{\cdot}\rho HCO_3[HCO_3]$$

ρ is the efficiency for each anion, whereas nitrate, sulfate and bicarbonate concentrations are expressed in $eq{\cdot}m^{-3}$.

We have:

– 1 $eq{\cdot}m^{-3}$ of nitrates: $[NO_3]$ $(mg{\cdot}L^{-1}) \times 0.016129$;

– 1 $eq{\cdot}m^{-3}$ of sulfates: $[SO_4]$ $(mg{\cdot}L^{-1}) \times 0.020833$;

– 1 $eq{\cdot}m^{-3}$ of bicarbonates: $[HCO_3]$ $(mg{\cdot}L^{-1}) \times 0.016393$;

– 1 $eq{\cdot}m^{-3}$ of calcium carbonate: $[CaCO_3]$ $(mg{\cdot}L^{-1}) \times 0.02$.

The nitrate fixation efficiency of conventional resins available on the market depends on the sulfate/nitrate ratio because the competition between these two anions is always to the advantage of sulfates. Water rich in sulfates results in short operating cycles because sulfates will quickly be exchanged for chlorides, to the detriment of nitrates (Figure 13.19). The choice of nitrate removal process may depend on this ratio.

In recent years, DOW resins have appeared. These specifically pertain to nitrates and are less sensitive to the presence of sulfates.

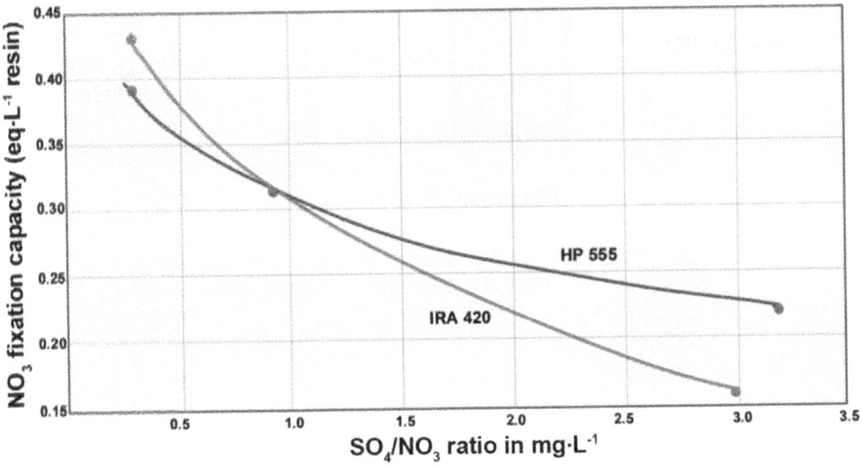

Figure 13.19. *Influence of the SO_4/NO_3 ratio on nitrate fixation capacity. For a color version of this figure, see www.iste.co.uk/gaid/watertreatment3.zip*

13.2.2.8. *Chloride concentration in treated water*

During the resin saturation phase with the fixation of nitrate, sulfate and bicarbonate ions, the treated water becomes enriched with chloride ions. The average chloride concentration is obtained by the ratio:

$$[\mathrm{Cl}]\mathrm{out} = [\mathrm{Cl}]\mathrm{in} + 35.5 \cdot \mathrm{N}.$$

[Cl]in and [Cl]out are the chloride concentrations at the exchanger's inlet and outlet ($mg \cdot L^{-1}$).

When part of the flow is treated with resin (partial treatment), Q_{DN} becomes the resin inlet flow rate and Qin e becomes the overall feed flow rate ($m^3 \cdot h^{-1}$).

$$[\mathrm{Cl}]\mathrm{out} = [\mathrm{Cl}]\mathrm{in} + 35.5 \cdot \mathrm{N} \cdot \frac{\mathrm{QDN}}{\mathrm{Qin}}.$$

Table 13.8 summarizes the concentrations of various anions obtained after passing through the resin. The characteristics of salt and brine are as follows:

– finely refined salt:

- actual density: 2,160 $kg \cdot m^{-3}$;

- apparent density: 1,200 $kg \cdot m^{-3}$;

- volume occupied per 1 ton of salt: 0.463 m^3;

– brine salt:

- actual density: 2,160 kg·m^{-3};
- apparent density: 1,400 kg·m^{-3};
- volume occupied per 1 ton of salt: 0.714 m^3;

– brine per 1 m^3:

- apparent density: 1,200 kg·m^{-3};
- salt weight: 315 kg;
- water weight: 885 kg;
- concentration: around 26%.

	Total treatment	**Partial treatment**
Nitrates	$(1 - \rho\ NO_3)\ [NO_3]$	$(1 - \rho\ NO_3\ Q_{DN}/Qin)\ [NO_3]$
Sulfates	$(1 - \rho\ SO_4)\ [SO_4]$	$(1 - \rho\ SO_4 \cdot Q_{DN}/\ Qin)\ [SO_4]$
Bicarbonates	$(1 - \rho\ HCO_3)\ [HCO_3]$	$(1 - \rho\ HCO_3 \cdot Q_{DN}/\ Qin)\ [HCO_3]$
Chlorides	35.5 N	$35.5\ N \cdot Q_{DN}/\ Qin$

Table 13.8. *Concentrations of various anions following a global treatment and a partial treatment*

13.2.3. *Integration into the water treatment station*

The Ecodenit® process can be integrated into a treatment station, depending on the water quality to be treated.

In the case of low turbid groundwater, it can be placed directly on raw water (after optional pre-filtration). In the event of micropollutant presence, activated carbon filtration can be placed downstream or upstream.

Figure 13.20 shows an example of groundwater with GAC-filtration at the first floor, whose treated water feeds the resins. As nitrates are present only part of the year, the exchangers are bypassed and then restarted when nitrates reappear in the raw water.

In many cases, exchangers are installed directly on raw water, and then denitrified water is disinfected. A pH adjustment may be required prior to disinfection.

In the presence of low bicarbonate concentrations, and insofar as resins remove approximately 30% of bicarbonates, it could be necessary to remineralize denitrified water.

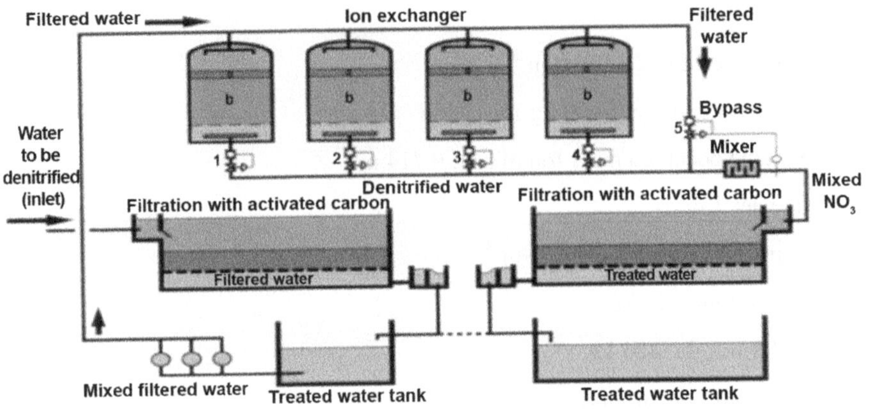

Figure 13.20. *Treatment station with exchangers and granular activated carbon filters. For a color version of this figure, see www.iste.co.uk/gaid/watertreatment3.zip*

13.2.4. *Packaged solutions*

Veolia has packaged solutions for flows of up to 100 $m^3 \cdot h^{-1}$. The installation is connected to the water to be treated, provided that its turbidity is lower than 1 NTU. The basic unit is made up of one or more exchangers mounted on a frame and equipped with the necessary piping and valves, a supply pump and its accessories, an electrical automatic control cabinet and a brine-dissolver silo. As an option, additional equipment can be supplied such as a closed container, insulation with frost-protection heating, a nitrate analyzer, or a remote monitoring system.

13.3. Nitrate removal by high pressure membranes

Currently, nanofiltration (NF) and low pressure reverse osmosis (LPRO) are considered technologies capable of removing a wide range of compounds in a single stage.

The ability to replace many different treatment processes with a single membrane treatment has driven technology development.

The choice of membrane separation procedure is intimately related to the characteristics of the membranes used. Nanofiltration membranes feature a microporous structure with pores having a diameter smaller than 2 nm and negatively charged membrane materials. The high selectivity of the membrane process is generally explained in terms of the size effect, the charge effect (due to the electrostatic interactions between ions and the charged sites in membranes), as well as the differences in diffusivity and solute solubility.

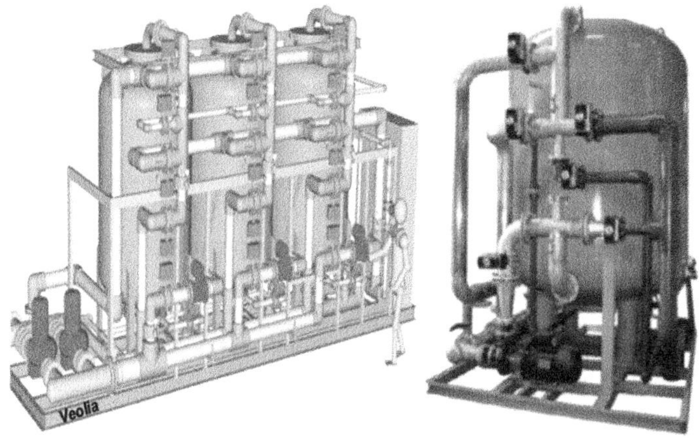

Figure 13.21. *Ecodenit® packaged solution (source: Veolia). For a color version of this figure, see www.iste.co.uk/gaid/watertreatment3.zip*

In low concentrations, the co-ions (same charge as the membrane sites) are repelled outside the membrane due to electrostatic interactions with the membrane charges. In order to ensure the permeate's electroneutrality, counterions (membrane's opposite charge) are also conserved. When the concentration increases, the counterions gradually sieve the membrane charges and co-ions are able to cross the membrane. Therefore, ion retention decreases.

Regarding the retention of nitrates, it sharply decreases with loose NF membranes, and is higher with tight NF membranes or reverse osmosis membranes. Comparatively, sulfate retention is very high for loose NF membranes. On the one hand, the hydrated size of the sulfate ion is greater than the size of all the other ions and on the other hand, the sulfate ion is a divalent anion and is strongly rejected by the electrostatic interactions with the membranes' negative charges, especially since an increase in the anionic charge provokes an increase in electrostatic repulsion with the membranes. The larger the ion's hydrated size and hydration energy, the more ions are retained by the membranes. For this reason, "loose" nanofiltration membranes are generally not used to remove monovalent ions such as nitrates. For tight NF membranes, size exclusion can provide additional ion-rejection.

For loose membranes, the Donnan effect explains it as follows: since divalent sulfate anions are strongly repelled outside the membrane due to a larger size and electrostatic repulsion effects than for monovalent ions, the monovalent anions (and particularly nitrate ions) are forced to cross the membrane so as to ensure the permeate's electroneutrality. In addition, nitrate ions are less retained by the membrane in comparison to sulfates, due to a smaller hydrated size and lower hydration energy (Table 13.9). In the case of very high concentrations, reverse osmosis is recommended for removing nitrates in the production of drinking water. This is because nitrate rejection can be too limited with NF membranes, compared to the outlet desired values.

The reduction of nitrate concentrations and the various existing trials and references (notably Veolia) show that "tight" NF and LPRO are reliable processes for the combined removal of a wide range of compounds, from groundwater and surface water.

Ions	Ion radius (nm)	Hydrated radius (nm)	Hydration energy ($kJ \cdot mole^{-1}$)	Ds diffusion coefficient ($10^{-9} m^2 \cdot s^{-1}$)
Cl^-	0.181	0.332	376	2.032
SO_4^{2-}	0.290	0.379	1,138	1.065
NO_3^-	0.264	0.335	329	1.902
Na^+	0.102	0.358	407	1.333
K^+	0.097	0.331	322	1.957
Ca^{2+}	0.100	0.412	1,584	0.718
Mg^{2+}	0.072	0.428	1,830	0.706

Table 13.9. *Hydrated radius and hydration energy for various ions*

Significant nitrate removal (76%) is achieved with the NF70 membrane (Dow/Filmtec), meeting the expectations for a tight-pore NF membrane. This may suffice to meet nitrate regulation concerning drinking water.

With a tight NF membrane like the NF 90, nitrate retention is important (~88%) and remains almost constant (Figure 13.22), regardless of the concentration. The operating conditions are as follows: 15°C temperature, 6.8 pH, $[NO_3]$ concentration of 150 $mg \cdot L^{-1}$, and 24 $l \cdot m^{-2} \cdot h^{-1}$ (lmh) flow.

Figure 13.23 shows the influence of nitrate retention (NO_3: 100 $mg \cdot L^{-1}$) in the presence of sulfates on the same NF 90 membrane. Sulfate rejection is always >95%. The efficiency of this membrane is illustrated on an operating plant treating surface water loaded with organic matter, pesticides and nitrates. Figure 13.24 shows that the nitrate concentration at the outlet of nanofiltered water is always lower than 10 $mg \cdot L^{-1}$, which corresponds to an 80% reduction on average. The pores of this membrane are smaller than those of NF200 and NF70. As a consequence, the retention is higher, due to the sieving effect and a greater size effect. With this membrane, retention remains constant regardless of the salt concentration, which means that the charge effect is not critical.

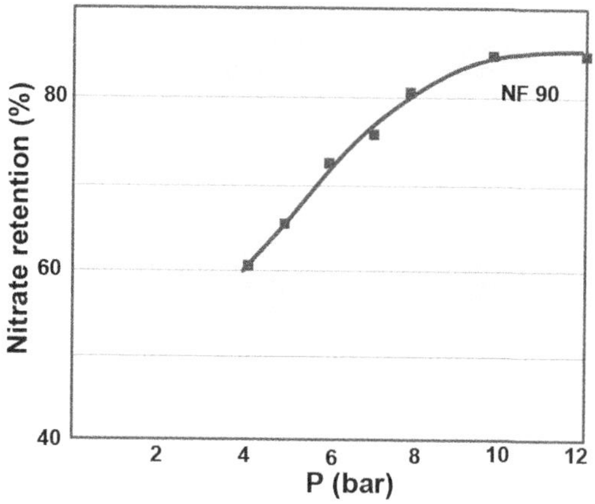

Figure 13.22. *Nitrate retention on an NF 90 membrane*

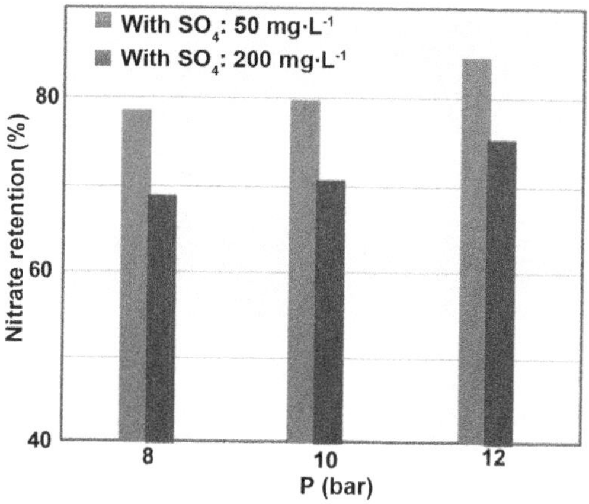

Figure 13.23. *Nitrate retention as a function of sulfate concentration. For a color version of this figure, see www.iste.co.uk/gaid/watertreatment3.zip*

Tight nanofiltration membranes (small pore sizes and characteristics close to reverse osmosis membranes) enable a significant, but not thorough, removal of nitrates.

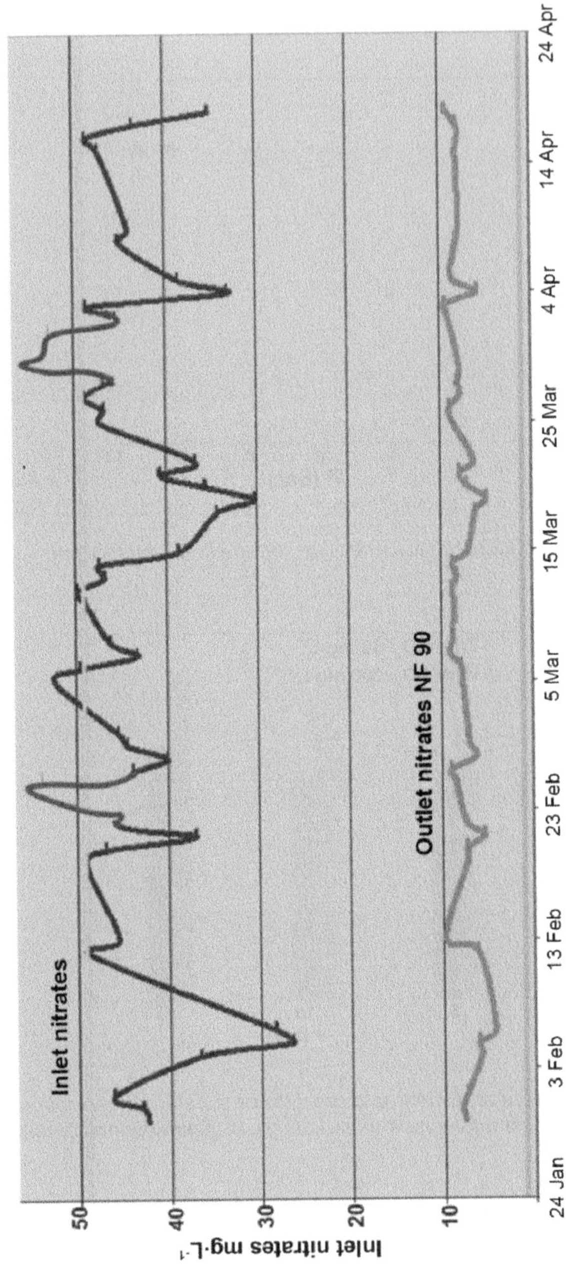

Figure 13.24. *Nitrate reduction monitoring with an NF 90 membrane on an operating plant (source: Veolia). For a color version of this figure, see www.iste.co.uk/gaid/watertreatment3.zip*

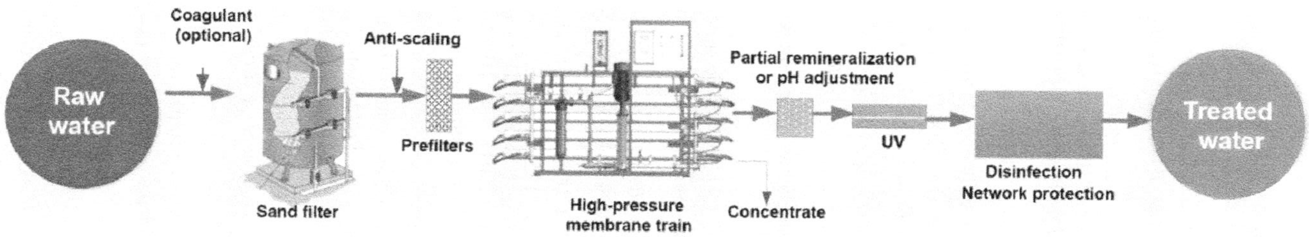

Figure 13.25. *Example of a treatment station for nitrates and other compounds with a groundwater high pressure membrane. For a color version of this figure, see www.iste.co.uk/gaid/watertreatment3.zip*

13.4. References

Alslan, Ş. (2005). Combined removal of pesticides and nitrates in drinking waters using biodenitrification and sand filter system. *Process Biochem.*, 40, 417–424.

Alslan, Ş. and Türkman, A. (2003). Biological denitrification of drinking water by using various natural organic solid substrates. *Water Sci. Technol.*, 48, 11–12.

Amouha, M.A., Bidhendi, G.R.N., Hooshyari, B. (2011). Nanofiltration efficiency in nitrate removal from groundwater: A semi-industrial case study. In *2nd International Conference on Environmental Engineering and Applications*. IPCBEE/IACSIT Press, Singapore, 17.

Archna, Sharma, S.K., Sobti, R.C. (2012). Nitrate removal from ground water: A review. *J. Chem.*, 9(4), 1667–1675.

Bae, B.U., Jung, Y.H., Han, W.W., Shin, H.S. (2002). Improved brine recycling during nitrate removal using ion exchange. *Water Res.*, 25, 3330–3340.

Bandpi, M.A., Elliott, D.J., Memeny-Mazdek, A. (1999). Denitrification of groundwater using acetic acid as a carbon source. *Water Sci. Technol.*, 40(2), 53–59.

Bilidt, H. (1985). The use of reverse osmosis for removal of nitrate in drinking water. *Desalination*, 53, 225–230.

Bohdziewicz, J., Bodzek, M., Wasik, E. (1999). The application of reverse osmosis and nanofiltration to the removal of nitrates from groundwater. *Desalination*, 121, 139–147.

Cameron, S.G. and Schipper, L.A. (2010). Nitrate removal and hydraulic performance of organic carbon for use in denitrification beds. *Ecol. Eng.*, 36, 1588–1595.

Chabani, M., Amran, A., Bensmaili, A. (2006). Kinetic modeling of the adsorption of nitrates by ion exchange resin. *Chem. Eng. J.*, 125, 111–117.

Choi, S., Yun, Z., Hong, S., Ahn, K. (2001). The effect of co-existing ions and surface characteristics of nanomembranes on the removal of nitrate and fluoride. *Desalination*, 133, 53–64.

Clifford, D. and Weber Jr, W.J. (1978). Nitrate removal from water supplies by ion exchange. Report, United States Environmental Protection Agency/Office of Research and Development, Cincinnati.

Clifford, D., Sorg, T., Ghurye, G. (2010). Ion exchange and adsorption of inorganic contaminants. In *Water Quality & Treatment: A Handbook on Drinking Water*, Edzwald, J.K. (ed.). McGraw Hill, New York.

Dahab, M.F. and Sirigina, S. (1994). Nitrate removal from water supplies using biodenitrification and GAC-sand filter system. *Water Sci. Technol.*, 30(9), 133–139.

Darbil, D., Viraraghavan, T., Butler, R., Corkal, D. (2002). Batch studies on nitrate removal from potable water. *Water SA*, 28, 319–322.

De Vet, W.W.J.M. (2011). Biological drinking water treatment of anaerobic groundwater in trinkling filters. PhD Thesis, Delft University of Technology, Delft.

Della Rocca, C., Belgiorno, V., Meric, S. (2006). A heterotrophic/autotrophic denitrification (HAD) approach for nitrate removal from drinking water. *Process Biochem.*, 41, 1022–1028.

Della Rocca, C., Belgiorno, V., Meric, S. (2007). Heterotrophic/autotrophic denitrification (HAD) of drinking water: Prospective use for permeable reactive barrier. *Desalination*, 210, 194–204.

Dördelmann, O., Buchta, P., Panglisch, S., Klegraf, F., Moshiri, A., Emami, A. (2006). Heterotrophic denitrification in drinking water treatment – Results from pilot plant experiments in Mashhad/Iran. In *Recent Progress in Slow Sand and Alternative Biofiltration Processes*, Gimbel, R. (ed.). IWA Publishing, London.

Gabelich, C.J., Gerringer, F.W., Franklin, J.C., Gao, J., Cohen, Y., Suffet, I.H. (2004). Reverse osmosis pretreatment: Challenges with conventional treatment. In *Proceedings of the 2004 AWWA ACE Meeting*, Orlando.

Gaid, K. and Raguenes, I. (1997). L'élimination biologique de l'azote dans les eaux à potabiliser. In *Conference Pollutec*, 1 October, Paris.

Gaid, K., Moreaud, H., Lemarchand, D., Martin, G. (1980a). Elimination biologique des nitrates sur un mélange soufre – carbonates. In *Int. Water Supply Association*, 1–4 September, Paris.

Gaid, K., Moreaud, H., Martin, G. (1980b). Dénitrification autotrophique sur un mélange soufre carbonates. *Techniques & Sciences municipales. L'eau*, 1(75), 39–46.

Gaid, K., Sibony, J., Ravarini, P. (1997). L'élimination biologique et physico-chimique des nitrates dans les eaux à potabiliser. In *Rencontre internationale sur les fluorures, nitrates et pesticides dans les eaux*, 24–25 April, Kenitra.

Garcia, F., Ciceron, D., Saboni, A., Alexandrova, S. (2006). Nitrate ions elimination from drinking water by nanofiltration: Membrane choice. *Sep. Purif. Technol.*, 52, 196–200.

Gomez, M.A., Gonzales-Lopez, J., Hantorie-Garcia, E. (2000). Influence of carbon source on nitrate removal of contaminated groundwater in a denitrifying submerged filter. *J. Hazard. Mater.*, B80, 69–80.

Guter, G.A. (1995). Nitrate removal from contaminated groundwater by anion exchange. In *Ion Exchange Technology: Advances in Pollution Control*, Sengupta, A.K. (ed.). Technomic Publishing, Lancaster.

Hamlin, H.J., Michaels, J.T., Beaulaton, C.M, Graham, W.F., Dutt, W., Steinbach, P., Losordo, T.M., Schrader, K.K., Main, K.L. (2008). Comparing denitrification rates and carbon sources in commercial scale up flow denitrification biological filters in aquaculture. *Aquacult. Eng.*, 38, 79–92.

Holl, W. (1995). CARIX process – A novel approach to desalination by ion exchange. In *Ion Exchange Technology: Advances in Pollution Control*, Sengupta, A.K. (ed.). Technomic Publishing, Lancaster.

Holló, J. and Czakó, L. (1987). Nitrate removal from drinking water in a fluidized-bed biological denitrification bioreactor. *Acta Biotechnol.*, 7, 417–423.

Jensen, B.V., Darby, J.L., Seidel, D.C., Gorman, C.S. (2012). Addressing nitrate in California's drinking water with a focus on Tulare Lake Basin and Salinas Valley groundwater. Report, Center for Watershed Sciences/University of California/Davis Groundwater Nitrate Project, Davis.

Kabay, N., Yuksel, M., Samatya, S., Arar, O., Yuksel, U. (2007). Removal of nitrate from ground water by a hybrid process combining electrodialysis and ion exchange processes. *Sep. Sci. Technol.*, 42, 2615–2627.

Kapoor, A. and Viraraghavan, T. (1997). Nitrate removal from drinking water – Review. *J. Environ. Eng.*, 123, 371–380.

Klapwizk, A. (1987). Nitrate removal from groundwater. *Water Res.*, 21(8), 989–997.

Kurt M., Dunn, I.J., Bourne, J.R. (1987). Biological denitrification of drinking water using autotrophic organisms with H2 in a fluidized-bed biofilm reactor. *Biotechnol. Bioeng.*, 29, 493–501.

Lee, D.U., Lee, S., Choi, D., Bae, J. (2001). Effects of external carbon source an empty bed contact time on simultaneous heterotrophic and sulfur-utilizing autotrophic denitrification. *Process Biochem.*, 36, 1215–1224.

Liu, X. and Clifford, D.A. (1996). Ion exchange with denitrified brine reuse. *J. Am. Water Works Assoc.*, 88, 88–99.

Mimouni, O. and Gaid, K., (1988). Nitrate pollution of Mitidja ground-water (Algiers). In *Int. Symp. Hydrological Processes and Water Management in Urban Areas*, 24–29 April, Duisburg.

Mimouni, O., Chibane, B., Gaid, K. (1988). Predictions on nitrates contaminants of Mitidja plain ground water (Northern Algeria). In *Int. Conf. on Computer Methods and Water Resources*, 21–23 March, Rabat.

Mohseni-Bandpi, A., Elliott, D.J., Momeny-Mazdeh, A. (1999). Denitrification of groundwater using acetic acid as a carbon source. *Water Sci. Technol.*, 40, 53–59.

Mohseni-Bandpi, A., Elliott, D.J., Zazouli, M.A. (2013). Biological nitrate removal processes from drinking water supply – A review. *J. Environ. Health Sci. Eng.*, 11, 35.

Paugam, L., Taha, S., Cabon, J., Dorange, G. (2002). Elimination of nitrate ions in drinking waters by nanofiltration. *Desalination*, 152, 271–274.

Paugam, L., Taha, S., Dorange, G., Jaouen, P., Quemeneur, F. (2004). Mechanism of nitrate ions transfer in nanofiltration depending on pressure, pH, concentration and medium composition. *J. Membr. Sci.*, 231, 37–46.

Pontie, M., Buisson, H., Diawara, C.K., Essis-Tome, H. (2003). Studies of halide ions mass transfer in nanofiltration – Application to selective defluorination of brackish drinking water. *Desalination*, 157, 127–134.

Philipot, J.M. (1982). Elimination des nitrates présents dans les eaux souterraines, T.S.M. *L'eau*. 69, 27–36.

Raguenes, I. (1994). Cours de l'eau : élimination de l'azote. Report, OTV, Paris.

Rittmann, B.E. and McCart, P.L. (2001). *Environmental Biotechnology: Principles and Applications*. McGraw-Hill, New York.

Rogalla, F., Ravarini, P., De Larminat, G., Couttelle, J. (1990). Large-scale biological nitrate and ammonia removal. *Wat. Environ. J.*, 4, 311–401.

Santafé-Moros, A., Gozálvez-Zafrilla, J.M., Lora-García, J. (2005). Performance of commercial membranes in the removal of nitrate ions. *Desalination*, 185, 281–288.

Santafé-Moros, A., Gozálvez-Zafrilla, J.M., Lora-García, J. (2007). Nitrate removal from ternary ionic solutions by a tight nanofiltration membrane. *Desalination*, 204, 63–71.

Santafé-Moros, A., Gozálvez-Zafrilla, J.M., Lora-García, J. (2010). Nanofiltration study of the interaction between bicarbonate and nitrate ions. *Desalination*, 250, 773–777.

Schaep, J., Van der Bruggen, B., Vandecasteele, C., Wilms, D. (1988). Influence of ion size and charge in nanofiltration. *Sep. Purif. Technol.*, 14, 155–162.

Schoeman, J.J. and Steyn, A. (2003). Nitrate removal with reverse osmosis in a rural area in South Africa. *Desalination*, 155, 15–26.

Shon, Z.H., Lee, G., Moon, B., Noh, B., Sung, N. (2001). Parametric studies on the performance of anion exchange for nitrate removal. *Korean J. Chem. Eng.*, 18, 170–177.

Soares, M.I.M. (2000). Biological denitrification of groundwater. *Water, Air, Soil Pollut.*, 123, 183–193.

Soares, M.I.M. (2002). Denitrification of groundwater with elemental sulfur. *Water Res.*, 36, 1392–1395.

Suratt, W.B. and Burke, J.E. (1995). Nitrate removal and water quality improvements with reverse osmosis for Brighton, Colorado. *Desalination*, 103, 101–111.

Tang, Y., Zivel, M., Zhou, C., Shin, J.H., Ahn, C.H., Meyer, K., Rittman, B.E. (2011). Using carrier surface loading to design heterotrophic denitrification reactors. *J. Am. Water Works Assoc.*, 103(3), 68–78.

Torabian, A., ShokouhiHarandi, M., NabiBidhendi, G.R., Ghadimkhani, A. (2009). Nitrate removal from drinking water by using commercial nanofiltration. *Asian J. Chem.*, 21(1), 666–672.

Upadhyaya, G., Jackson, J., Clancy, T.M., Hyun, S.P., Brown, J., Hayes, K.F., Raskin, L. (2010). Simultaneous removal of nitrate and arsenic from drinking water sources utilizing a fixed-bed bioreactor system. *Water Res.*, 44, 4958–4969.

Van der Bruggen, B., Everaert, K., Wilms, D., Vandecasteele, C. (2001). Application of nanofiltration for removal of pesticides, nitrate and hardness from ground water: Rejection properties and economic evaluation. *J. Membr. Sci.*, 193, 239–248.

Van der Hoek, J.P. and Klapwijk, A. (1987). Nitrate removal from ground water. *Water Res.*, 21, 989–997.

Van der Hoek, J.P., Van der Ven, P.J.M., Klapwijk, A. (1988). Combined ion exchange/biological denitrification for nitrate removal from ground water under different process conditions. *Water Res.*, 22, 679–684.

Van Ginkel, S.W., Ahn, C.H., Badruzzaman, M., Roberts, D.J., Lehman, S.G., Adham, S.S., Rittmann, B.E. (2008). Kinetics of nitrate and perchlorate reduction in ion exchange brine using the membrane biofilm reactor (MBfR). *Water Res.*, 42, 4197–4205.

Vasaliadou, I.A., Karanasios, K.A., Pavlou, S., Vayenas, D.V. (2009). Hydrogenotrophic denitrification of drinking water using packed bed reactors. *Desalination*, 248, 859–868.

WQTS (2013). Application of biological denitrification for nitrate removal, from District 37 groundwater. Report, Los Angeles County Department of Public Works Water Works Division, Los Angeles.

14

Removal of Perchlorates

The presence of perchlorate ion traces has been observed in certain resources used for the production of drinking water. Although the presence of this oxy-anion in water is not currently regulated, such analyses raise the question about this parameter's treatability. Perchlorates are found in many industrial applications, particularly in the military and aerospace fields, where they are used for a variety of purposes: as an oxidant for rocket propellants, for manufacturing pyrotechnic devices, flares, the powder of certain firearms, the production of artificial fog, etc.

They can also take part in the composition of certain industrial processes (temporary adhesives for metal–metal bonds, adjusting the ionic strength of electrolytic baths, regulating the impurities of some industrial hypochlorite solutions used for water disinfection, etc.).

The toxicity of perchlorate ions is characterized by their interference with the thyroid gland's iodine absorption, as well as their effects on reproduction and immunity. Certain population groups (pregnant women and infants) might be more vulnerable to the effects of perchlorate ions. Although the mechanisms of action of perchlorates are fairly well known at therapeutic doses, an evaluation of their impact on health at environmental concentrations is less known. In recent years, the health authorities of certain countries or states have begun to address recommendations or regulations regarding limit concentrations of perchlorate ions in drinking water. In the United States, the USEPA recommends a value of 15 $\mu g \cdot L^{-1}$, whereas the state of California has set a value of 6 $\mu g \cdot L^{-1}$ and Massachusetts has determined a 2 $\mu g \cdot L^{-1}$ value.

Until the present moment, there have been no recommendations from the World Health Organization or the European Union regarding the permissible concentration of perchlorate ions in drinking water.

14.1. General aspects

From a strictly chemical point of view, the term "perchlorate" designates the combination of a chlorine atom surrounded by four oxygen atoms: ClO_4^- (valence +7). Perchlorate ion is the most oxidized form of chlorine (Figure 14.1), making it a strong oxidant. Despite its oxidation potential, ClO_4^- is stable due to its tetrahedral shape and requires high activation energy for its reduction to chlorides or chlorates:

$$ClO_4^- + 8\ H^+ + 8\ e \leftrightarrow Cl^- + 4\ H_2O\ E^o = 1{,}287\ V$$

$$ClO_4^- + 2\ H^+ + 2\ e \leftrightarrow ClO_3^- + H_2O\ E^o = 1{,}201\ V$$

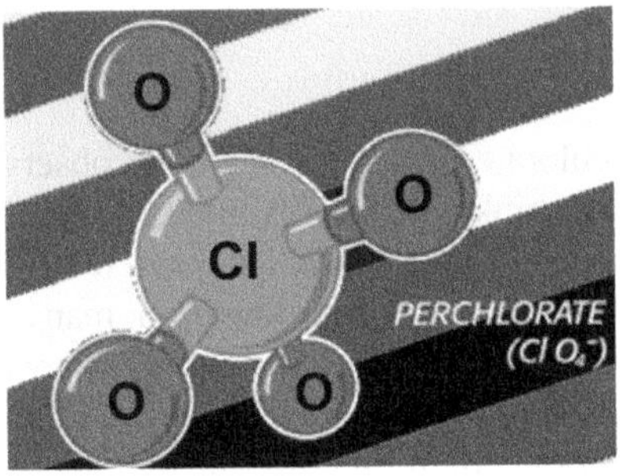

Figure 14.1. *Perchlorate ion. For a color version of this figure, see www.iste.co.uk/gaid/watertreatment3.zip*

The solubility of perchlorate salts is relatively high. For example, it corresponds to 200 g/L (25°C.) for ammonium perchlorate. Its dissociation in water leads to:

$$NH_4ClO_4\ (s) \rightarrow NH_4^+\ (liq) + ClO_4^-\ (liq)$$

14.2. Main processes for removing perchlorate ions

Perchlorate ions can be removed using the following methods:

– non-specific and regenerable ion exchange resins;

– specific non-regenerable resins;

– nanofiltration and reverse osmosis;

– biofilters and fluidized biological reactors.

Perchlorate salts are very soluble and chemically stable. As a result, conventional coagulation–flocculation–settling treatment has no effect on them.

While non-specific and regenerable ion exchange resins are used for treating water contaminated with nitrates and perchlorate ions, they lead to different results in terms of efficiency.

As for ion exchange resins which are specific for perchlorate ions, they are very effective though difficult to regenerate. In point of fact, the ClO_4^- perchlorate is a monovalent ion (often in the NH_4ClO_4 form) whose strong affinity for this selective resin does not make it subject to the exchange reversibility with NaCl brine traditionally used for regeneration. Still, regeneration can be performed using $FeCl_4^-$ (solution of $FeCl_3$, C_2H_5OH and HCl), but cannot be used on site due to the disposal of eluates. These resins can treat 100,000–150,000 v/v of water containing 50 $\mu g \cdot L^{-1}$ of ClO_4^-.

High-pressure membranes such as nanofiltration and reverse osmosis are undoubtedly efficient, but involve the continuous discharge of a concentrate whose destination must be defined for each application.

The use of biofilters makes it possible to reduce perchlorate ions to chlorides, although it entails the risk of by-product formation and the contamination of the water produced.

14.2.1. *Ion exchange resins*

Four types of resins are tested, whose characteristics are summarized in Table 14.1.

	Resin A, specific for perchlorate ions	**Resin B, specific for perchlorate ions**	**Resin C, specific for nitrate ions**	**Resin D, specific for Boron**
Grain size	600 ± 50 μm	0.3–1.2 mm	0.3–1.2 mm	0.3–1.2 mm
Uniformity coefficient	1.3	–	–	–
Density $kg \cdot L^{-1}$	1.04	1.10	1.10	1.09
Ion exchange form	Cl^-	Cl^-	Cl^-	–
Functional group	Bifunctional quaternary ammonium group	Quaternary amine styrene/divinylbenzene copolymer	Quaternary amine styrene/divinylbenzene copolymer	Anion exchange polymer (weak base)
Regeneration	No	No	Yes	Yes

Table 14.1. *Characteristics of the resins tested*

14.2.1.1. *Experimental device*

A specific setup was developed for laboratory tests by VEOLIA-VERI. It features columns containing 15 mL of resin to be tested (Figure 14.2). These columns are supplied with the water to be treated by means of a Master Flex® multichannel peristaltic pump. The samples chosen for each test enable the regular monitoring of perchlorate, nitrate, chloride, carbonate, bicarbonate and sulfate ions. The pH and absorbance in the visible UV spectrum are also analyzed.

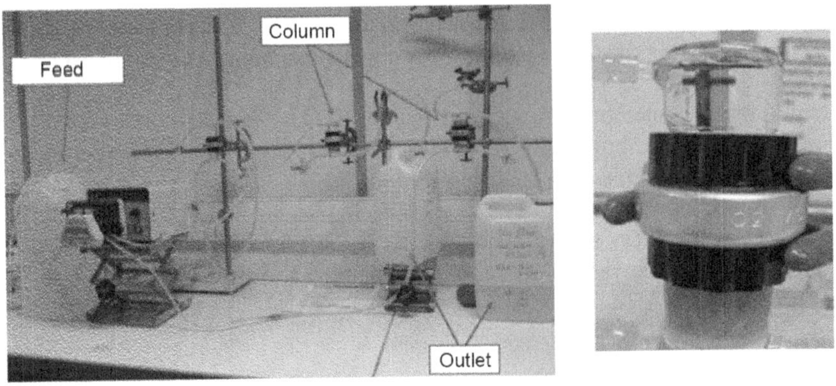

Figure 14.2. *Setup for laboratory tests (VEOLIA-VERI). For a color version of this figure, see www.iste.co.uk/gaid/watertreatment3.zip*

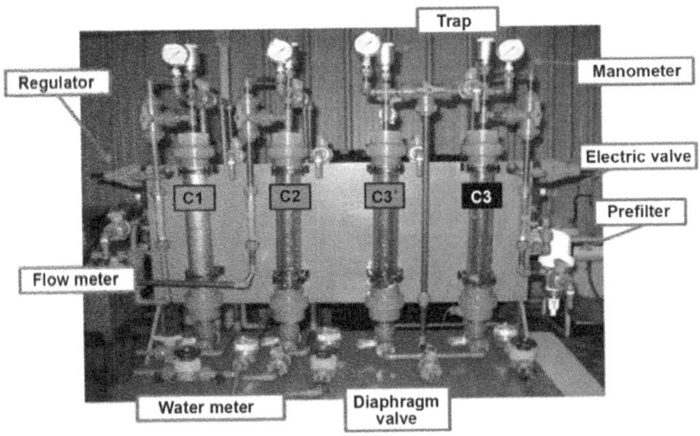

Figure 14.3. *Device for on-site testing of the various selected resins. For a color version of this figure, see www.iste.co.uk/gaid/watertreatment3.zip*

The perchlorate ion treatment pilot using ion exchange resins compares the performances of different resins with water, which could potentially contain other interfering ions, such as nitrates.

The pilot has four columns (Figure 14.3), fed by two types of water:

– Natural water containing perchlorate ions and pre-denitrified water by regularly regenerated anionic resins (NaCl) feeds two columns in parallel. It feeds a first column C1 which contains anionic resins used in denitrification, and then, a second column C2, which contains specific resins for non-regenerated perchlorate ions.

– Raw water containing nitrate and perchlorate ions feeding two non-regenerated columns, laid out in series (C3 & C'3). The procedure comes down to comparing the performances of the two devices and the different resins tested in order to deduce the most efficient configuration for the treatment of water containing nitrate ions and perchlorate ions.

The functional diagram of the treatment pilot is shown in Figure 14.4.

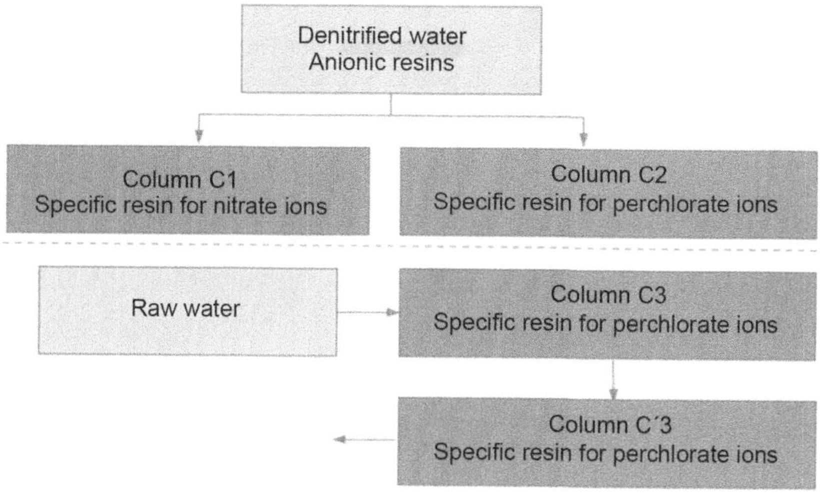

Figure 14.4. *Functional diagram of the pilot. For a color version of this figure, see www.iste.co.uk/gaid/watertreatment3.zip*

The ion exchange mechanism is a physicochemical process whereby ions are transported electrostatically onto a solid's surface and are then exchanged for ions with an equivalent charge (Figure 14.5). The use of this technique for the specific removal of perchlorate ions engages anionic resins of the strong base type. The effectiveness of these resins in relation to the various anions is reported as follows:

$$ClO_4^- > SO_4^{2-} > NO_3^- > Br^- > HAsO_4^{2-}, CO_3^{2-} > Cl^- > H_2AsO_4^-, HCO_3^- > F^-$$

When water containing perchlorate ions passes through an ion exchanger, the Cl^- ions of the chloride resin are exchanged for ClO_4^-, due to the strong affinity for the resin's quaternary amine group. Cl^- ions are cleared with the water produced, whereas the ClO_4^- ions are retained by the resin's functional groups. When these are all bound to ClO_4^- anions, the resin is said to be saturated. It is then regenerated using brine (NaCl), along which chloride ions are fixed onto the resin once again, thus becoming available for exchange with the ClO_4^- anions at restart. It should be noted that in water contaminated with perchlorate ions, nitrates are often present and found in concentrations 1,000–5,000 times higher than perchlorate ions. This influences the removal kinetics.

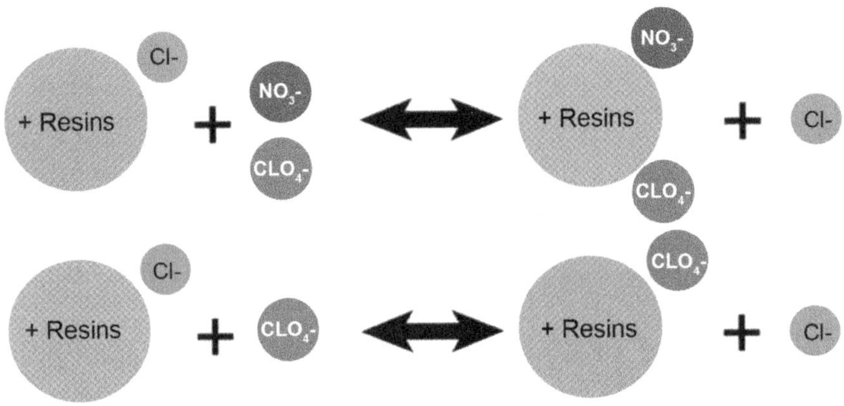

Figure 14.5. *Schematic diagrams of the ion exchange between perchlorate ions, nitrate ions and resin. For a color version of this figure, see www.iste.co.uk/gaid/watertreatment3.zip*

While the use of selective resins targets perchlorates, it does not allow for regeneration with conventional products such as brine, for example. Resins are replaced by new resins after saturation.

The raw water used for treatment trials (Table 14.2) has an average perchlorate ion concentration of 32 $\mu g \cdot L^{-1}$, a nitrate concentration of 50.5 $mg \cdot L^{-1}$ and a pH of 7.94.

Parameters	Units	Values
pH		7.94
Conductivity	$\mu S \cdot cm^{-1}$	854
ClO_4^-	$\mu g \cdot L^{-1}$	32
Cl^-	$mg\ Cl \cdot L^{-1}$	81.2
SO_4^{2-}	$mg\ SO_4 \cdot L^{-1}$	<40
NO_3^-	$mg\ NO_3 \cdot L^{-1}$	50.5
Alkalinity	$mg \cdot L^{-1}\ CaCO_3$	273
Turbidity	NTU	<0.2

Table 14.2. *Physicochemical qualities of the raw water used during the tests*

14.2.1.2. *Results of laboratory tests*

The evolution in the reduction of perchlorate ions depending on the different resins tested is represented on a graph (Figure 14.6).

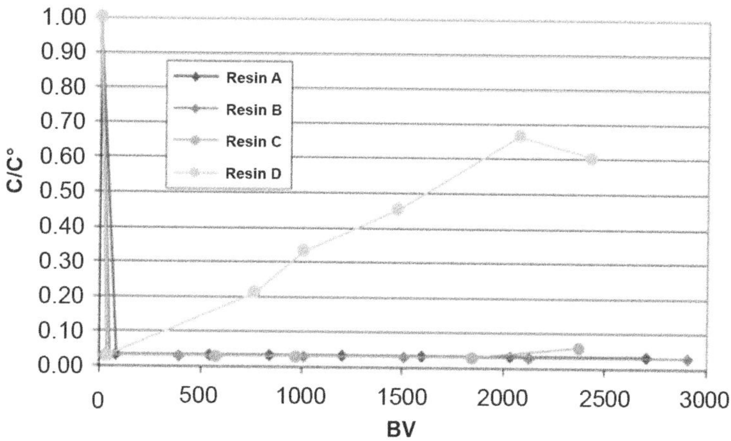

Figure 14.6. *Reduction of perchlorate ions using ion exchange resins. For a color version of this figure, see www.iste.co.uk/gaid/watertreatment3.zip*

As illustrated in Figure 14.6, the specific perchlorate resins (R-A and R-B) are highly efficient, with reductions close to 100%. The specific resin for nitrates (R-C) offers similar results, despite an unexplained loss of efficiency around 1,500 BV (bed volume). The specific resin for Boron (R-D) behaves differently from the other anionic resins tested. Its effectiveness is lower and quickly degrades. This resin has a low affinity for perchlorate and nitrate ions, so it cannot be used effectively to treat those parameters.

The retention of perchlorate ions is accompanied by other modifications in water quality due to the anionic exchanges within the resin bed. At the beginning of the cycle, the chloride-saturated resin will exchange these ions for hydrogen carbonate ions (HCO_3^-), nitrate, sulfate and perchlorate ions. This causes an increase in the chloride content of the treated water and a decrease in all the other parameters, in particular, in perchlorate ions and nitrate ions. The saturation and discharge of nitrates onto resins A, B and C approximately takes places from 400 to 600 BV. Beyond this concentration, resins no longer retain nitrates, but are still efficient for the treatment of perchlorate ions.

Apart from the first BV passed onto resins, the water's pH and conductivity are stable. No modification of the UV absorbance parameter is observed.

14.2.1.3. *Pilot testing*

The treatment pilot is supplied with raw water, whose chemical composition was regularly monitored throughout the trial period. The quality remained stable, presenting the average characteristics reported in Table 14.2. Before use, resins are rinsed at 10 BV for 1 h, with mechanical paddle stirring at 100 rpm. Resins are tested under dynamic conditions at a flow rate set at 30 BV/h, that is, 0.48 $L \cdot h^{-1}$. All trials are carried out at 20°C.

Two configurations were tested:

– treatment of perchlorate ions on previously denitrified water (Table 14.3);

– treatment of perchlorate ions on raw water (Table 14.4).

14.2.1.3.1. First setup

Denitrified water → C1 specific resin for nitrate ions (Column C1) → C2 specific resin for perchlorate ions (Column C2)

The denitrified water which feeds the pilot's C1 and C2 columns has nitrate levels lower than 3 $mg \cdot L^{-1}$, and perchlorate contents between 9 and 20 $\mu g \cdot L^{-1}$. The regenerations regularly carried out on this denitrification column cause variations in the chemical composition of the treated water, in terms of chlorides and Alk, at the beginning of each cycle. The specific resins for nitrate ions placed on the second floor lead to 3 $\mu g \cdot L^{-1}$ perchlorate concentrations. This reduction in perchlorate ions is confirmed over more than 50,000 BV. Specific resins for perchlorate ions show the same reduction performances in the 50,000 BV tested at present.

For the characteristics of the denitrified water used throughout these tests, the perchlorate reduction is highly satisfactory on the specific resins for nitrate (C1) and perchlorate (C2) ions. The perchlorate treatment duration before resin saturation exceeds 50,000 BV.

	Denitrified water Nitrate ion resin				Nitrate-ion resin				Perchlorate ion resin			
BV	ClO_4^- (μg·L^{-1})	Alkalinity mg·L^{-1}	NO_3^- (mg·L^{-1})	Cl^- (mg·L^{-1})	ClO_4^- (μg·L^{-1})	Alkalinity mg·L^{-1}	NO_3^- (mg·L^{-1})	Cl^- (mg·L^{-1})	ClO_4^- (mg·L^{-1})	Alkalinity mg·L^{-1}	NO_3^- (mg·L^{-1})	Cl^- (mg·L^{-1})
0	20	200	<1	168								
30	20	200	<1	168	2	58	<1	266	3	112	<1	222
425	20	200	<1	168	<1	304	<1	78	<1	304	<1	81.6
2,342	11	110	<1	129	<1	160	2.8	195	<1	171	1.3	171
4,231	13	130	<1	80.8	<1	249	2.1	118	<1	258	2	114
10,919	18	180	2.6	192	2	311	1.2	68.2	2	308	<1	70.4
14,358	12	120	2.1	81.6	3	225	1.8	116	3	245	1.6	115
19,871	14	140	9.3	69.4	<1	303	1.3	76	<1	310	<1	74
23,478	9	90	<1	106	<1	218	1.7	143	<1	234	1.8	132
28,153	16	160	<1	302	<1	304	<1	75.4	<1	306	<1	73.6
32,546	17	170	1.5	263	<1	277	1.9	92.2	<1	222	4.7	126
36,841	11	110	1.1	87.8	<1	221	4.1	141	<1	239	3.8	121
43,546	10	100	1.9	77.3	<1	267	2	105	<1	282	1.6	91.8
53,947	10	100	<1	152	<1	131	2.7	199	<1	145	2.6	181

Table 14.3. *Pilot testing results on previously denitrified water*

14.2.1.3.2. Second setup

Raw water → C3 specific resin for perchlorate ions → C'3 specific resin for perchlorate ions

In global terms, the quality of raw water is constant for all parameters over time, with an average perchlorate value around 32 μg·L^{-1}. The specific perchlorate ion resins installed in series are supplied with raw water.

The results obtained (Table 14.4) confirm very good perchlorate ion removal performances by the specific resins. Rates remained below 3 µg $ClO_4^-\cdot L^{-1}$ throughout the overall trial duration. These resins have a lower affinity for hydrogen carbonates and nitrates, which they retain at the cycle's beginning and discharge afterwards. No notable modification of the anion content (other than perchlorates) is observed after 50,000 BV.

	Raw water				**Perchlorate-ion resin**				**Perchlorate-ion resin**			
BV	ClO_4^- (µg·L^{-1})	Alkalinity mg·L^{-1}	NO_3^- (mg·L^{-1})	Cl^- (mg·L^{-1})	ClO_4^- (µg·L^{-1})	Alkalinity mg·L^{-1}	NO_3^- (mg·L^{-1})	Cl^- (mg·L^{-1})	ClO_4^- (µg·L^{-1})	Alkalinity mg·L^{-1}	NO_3^- (mg·L^{-1})	Cl^- (mg·L^{-1})
0	38	264	55	49.5								
30	38	264	55	49.5	3	47	<1	297	5	142	<1	207
425	38	264	55	49.5	<1	297	<1	85.6	<1	267	5.7	59.3
2,342	38	264	55	49.5	<1	264	51.7	49.5	<1	272	51	48.5
4,231	32	262	49.1	45.6	<1	267	57.8	52.5	<1	262	58	54.2
10,919	31	252	50.4	46.7	3	263	50.3	45.2	3	260	53.1	49.5
14,358	31	261	52.7	48.4	2	262	51.2	46.7	2	263	50.4	46.3
19,871	29	261	53.1	48.1	<1	260	52	47.3	<1	261	50.1	47.2
23,478	30	264	53.3	48	<1	264	54.2	49.6	<1	262	52.7	49.2
28,153	29	264	50.6	49.3	<1	261	51.9	48.5	<1	261	54.2	47.6
32,546	26	252	57.2	51.8	2	258	48.3	47.1	<1	255	51.1	49.8
36,841	33	262	48.3	43.7	<1	260	50.8	47.5	<1	265	48.6	46.9
53,947	31	262	48.3	43.7	<1	260	50.8	47.5	<1	265	48.6	46.9

Table 14.4. *Pilot testing results on raw water*

14.2.2. *Nanofiltration membranes*

Nanofiltration is a membrane treatment solution capable of retaining low molecular weight elements present in water. The purpose of these trials is to test the efficiency of NF membrane treatments for the removal of perchlorate ions contained in water resources. Two distinct phases: laboratory tests on a filtration bench with an Osmonics filtration cell (Figure 14.7) and a pilot treatment on a nanofiltration spiral membrane (Figure 14.8) DOW NF 200 and NF 90.

14.2.2.1. *Experimental device*

The water used for these tests is doped with perchlorates from a sodium perchlorate monohydrate commercial solution (Sigma Aldrich).

The sodium perchlorate solution is injected at the pilots' inlet under the experimental conditions shown in Table 14.5.

Experimental parameters	Osmonics cell	NF 200 3-stage pilot	Monotube pilot 6 NF 90
Feed flow ($L \cdot h^{-1}$)	54	6,400	2,000
Inlet-pressure (bar)	4.7	2.7	7.7
Tangential velocity ($m \cdot s^{-1}$)	0.2	0.13	0.16
Membrane area (m^2)	0.014	7.6 × 42 = 319.2	7.6 × 6 = 45.6
Temperature (°C)	25	8.5	8.5

Table 14.5. *Experimental parameters*

Salt rejection tests are carried out to ensure the integrity of the membrane and the system, with the following concentrations:

– NF200: 2000 ppm $MgSO_4$, 70 psi (0.5 MPa), 770 $mg \cdot L^{-1}$ $CaCO_3$ (25°C) alkalinity, and 15% conversion rate;

– NF90: 2000 ppm NaCl, 70 psi (0.5 MPa), 770 $mg \cdot L^{-1}$ $CaCO_3$ (25°C) alkalinity, and 15% conversion rate.

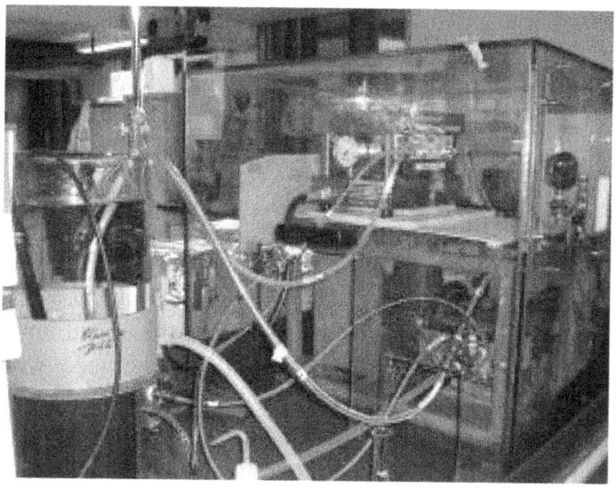

Figure 14.7. *Filtration bench. For a color version of this figure, see www.iste.co.uk/gaid/watertreatment3.zip*

The rejection rates are calculated and monitored by measuring the salt content and conductivity. The system is then rinsed with ultrapure water, before carrying out the tests.

The integrity of NF 200 membranes (pilot scale) was validated, before and after doping, with a $MgSO_4$ retention over 99%. The NF 200 coupon tested at the laboratory scale was also intact, with a $MgSO_4$ retention over 97%. Similarly, the NF 90 membranes remained intact. They, respectively, showed a salt retention over 97% before and after doping for the pilot, a retention in conductivity over 87% and a retention in NaCl over 89%, for the NF 90 coupon tested at the laboratory.

a) b)

Figure 14.8. *Pilots used for perchlorate ion removal trials with NF 90 (a) and NF 200 (b). For a color version of this figure, see www.iste.co.uk/gaid/watertreatment3.zip*

14.2.2.2. *Results*

Laboratory tests have shown that the retention of perchlorate ions on NF 200 membranes does not exceed 27%, with a permeate residual close to 25 $\mu g \cdot L^{-1}$ (± 5 $\mu g \cdot L^{-1}$). At the level of the three-stage pilot, the retention is 21%, with a permeate residual close to 45 $\mu g \cdot L^{-1}$ (± 9 $\mu g \cdot L^{-1}$). The feed water is doped at 30 $\mu g \cdot L^{-1}$.

At the laboratory, the retention of perchlorate ions on NF 90 membranes reaches 88%, with a permeate residual close to 4 $\mu g \cdot L^{-1}$ (± 1.2 $\mu g \cdot L^{-1}$). These retention rates are confirmed on a pilot scale with perchlorate retention rates reaching 90% and a total permeate residual close to 3 $\mu g \cdot L^{-1}$ (± 0.9 $\mu g \cdot L^{-1}$).

Membrane treatment tests to assess the efficiency of nanofiltration membranes in removing perchlorate ions show that these are removed with a retention rate over 87% with the NF 90 membrane. These results lead to satisfactory performances, resulting in a permeate quality containing less than 4 μg·L^{-1} perchlorate ions, compared to the 30 μg·L^{-1} injected. In addition, the trials showed a good correlation between the different scales tested: flat cell and single-tube or three-stage pilot.

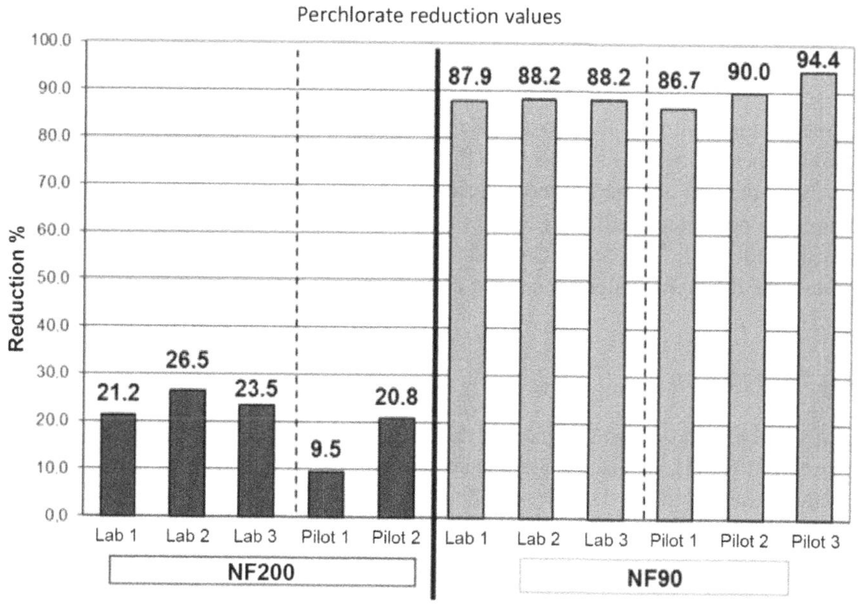

Figure 14.9. *Rejection rate results for lab and pilot scale testing. For a color version of this figure, see www.iste.co.uk/gaid/watertreatment3.zip*

14.3. Conclusions regarding the removal of perchlorates

Trials on specific perchlorate resins have shown good performances for the retention of perchlorate ions. With raw water containing nearly 32 μg·L^{-1}, these resins lead to concentrations lower than 3 μg·L^{-1} on the treated water. These results are confirmed for a treatment of 50,000 volumes of water per resin volume; that is to say, well beyond the resins' saturation concentrations by nitrates (approximately 500 BV under trial conditions).

Due to the very strong affinity of these resins for perchlorate ions under experimental conditions, their retention takes place beyond the resins′ saturation concentration by the main anions present in the water. This makes it possible to pursue with the treatment of perchlorate ions over a long period, without modifying the main anionic characteristics of the water.

Specific resins for perchlorate ions and for nitrate ions treat denitrified water containing perchlorate ions very effectively.

Specific resins for boron showed no affinity for nitrate ions or perchlorate ions.

Membrane treatment tests to assess the efficiency of nanofiltration membranes in removing perchlorate ions show that these are removed with a retention rate over 87% with the NF 90 membrane. These results lead to satisfactory performances, obtaining a permeate quality with less than 4 $\mu g \cdot L^{-1}$ perchlorate ions, in comparison with the 30 $\mu g \cdot L^{-1}$ injected. In addition, the trials showed a good correlation between the different scales tested: flat cell and single-tube or three-stage pilot.

14.4. References

Agence Nationale de sécurité sanitaire et de l'alimentation, de l'environnement et du travail (ANSES) (2011). Evaluation des risques sanitaires liés à la présence d'ions perchlorate dans les eaux destinées à la consommation humaine. Report, Saisine, Maisons-Alfort.

CalDPH (2007). State adoption of a perchlorate standard. Division of Drinking Water and Environmental Management. Report, CalDPH, Sacramento.

Cang, Y., Roberts, D.J., Clifford, D.A. (2004). Development of cultures capable of reducing per-chlorate and nitrate in high salt solutions. *Water Res.*, 38, 3322–3330.

Coppola, E.N. and McDonald, G.R. (2000). Bio-degradation of ammonium perchlorate, nitrate, hydrolysates and other energetic materials. Patent, Applied Research Associates Inc, US6077432A.

Environmental Protection Agency (1998). Perchlorate environmental contamination: Toxicological review and risk characterization based on emerging information. Report, EPA/IRIS, Washington.

Gaid, K., Mazeas, G., Leger, J., Jacob, S., Dassonneville, S., Denis, J., Gromand, V., Gaborit, S. (2011). Les ions perchlorate : quel traitement pour ces nouveaux contaminants des eaux ? *L'Eau, L'Industries, Les Nuisances*, 352, 76–82.

Gingras, T.M. and Batista, J.R. (2002). Biological reduction of perchlorate in ion exchange regenerant solutions containing high salinity and ammonium levels. *J. Environ. Monit.*, 4, 96–101.

Gu, B.H. (2001). Regeneration of perchlorate loaded anion exchangeresins by a novel tetrachloro-ferrate displacement technique. *Environ. Sci. Technol.* 35(16), 3363–3368.

Gu, B.H., Ku, Y.K., Brown, G.M. (2002). Treatment of perchlorate-contaminated groundwater using highly selective, regenerable ion exchange technology: A pilot-scale demonstration. *Remediation J.*, 12(2), 51–68.

Gullick, R.W., Le Chevallier, M.W., Barhorst, T.S. (2001). Occurrence of perchlorate in drinking water sources. *J. Am. Water Works Assoc.*, 93, 66–77.

Massachusetts Department of Environmental Protection (2020). Perchlorate in public drinking water, Bureau of Resource Protection. Report, Massachusetts Department of Environmental Protection, Boston.

Stanbury, J.B. and Wyngaarden, J.B. (1952). Effect of perchlorate on human thyroid gland. *Metabolism*, 1, 533–539.

Wolff, J. (1998). Perchlorate and the thyroid gland. *Pharmacol. Rev.*, 50(1), 89–105.

15

Water Softening

A very high level of hardness in some waters poses practical problems to users, be they domestic or industrial. By way of example, we may mention:

– scaling of boilers, water heaters and hot water circuits;

– long-term scaling of drinking water pipes;

– overconsumption of soaps and detergents;

– difficulty for cooking vegetables;

– oversizing of demineralization facilities.

Water softening achieved by means of carbonate removal undoubtedly constitutes an improvement in user comfort.

15.1. Water hardness

A distinction is made between carbonate hardness (temporary) and non-carbonate hardness (permanent). Carbonate hardness in related to the quantity of carbonates and bicarbonates, responsible for the scaling of pipes and other equipment (kettles, water heaters, boilers, washing machines, etc.). It is mainly associated with Ca^{2+} and Mg^{2+} cations.

Non-carbonate hardness is due to the association of (Ca^{2+} and Mg^{2+}) cations with sulfates, chlorides and nitrates. It is also called "permanent hardness" because it cannot be removed by boiling. When hard water is heated (60°C), an insoluble precipitate forms: tartar or limestone. Water hardness is the marker for mineralization. It indicates the calcium or magnesium ion content. It is expressed in $mg \cdot CaCO_3 \cdot L^{-1}$ or in French degrees; one French degree corresponds to:

$$1\ °f = 4\ mg\ of\ Ca^{2+} \cdot L^{-1}$$

$$= 2.4\ mg\ Mg^{2+} \cdot L^{-1}$$

$$= 10\ mg\ of\ CaCO_3 \cdot L^{-1}$$

Hardness ($mg \cdot L^{-1}$ $CaCO_3$)	**Water quality**
$0 < TH < 80\ mg \cdot L^{-1}\ CaCO_3$	Very soft water
$80 < TH < 150\ mg \cdot L^{-1}\ CaCO_3$	Soft water
$150 < TH < 250\ mg \cdot L^{-1}\ CaCO_3$	Moderately hard water
$250 < TH < 400\ mg \cdot L^{-1}\ CaCO_3$	Hard water
$TH\ 400\ mg \cdot L^{-1}\ CaCO_3$	Very hard water

Table 15.1. *Hardness concentrations and corresponding water quality*

Hardness is defined as the sum of the calcium and magnesium ions. These are both expressed in $mg \cdot L^{-1}$ calcium carbonate ($CaCO_3$). Hardness can consist of both carbonate and non-carbonate hardness. If the hardness is equal to or less than the alkalinity (also expressed in $mg \cdot L^{-1}$ calcium carbonate) then all the hardness is referred to as "carbonate" hardness. If the hardness is greater than the alkalinity, the difference between the hardness value and alkalinity value is referred to as non-carbonate hardness.

Carbonate hardness is due to the presence of carbonate and bicarbonate salts of magnesium and calcium, while non-carbonate hardness arises from various alkaline salts of calcium and magnesium (e.g. $CaCl_2$, $CaSO_4$). Since carbonates, bicarbonates and others compounds that contribute to non-carbonate hardness are alkaline, it is means that a hard water will present a $pH > 7$. Borehole waters are typically very hard due to the dissolution of calcium and magnesium into the water from the soil.

15.2. Alkalinity

Alkalinity is closely related to hardness. Alkalinity is produced mostly by anions such as hydroxides, bicarbonates and carbonates. Alkalinity is expressed in $mg \cdot L^{-1}$ of $CaCO_3$. Alkalinity can be divided into two parts: on the one hand, the alkalimetric title (AT) and, on the other hand, the total alkalinity (Alk).

The AT is characterized by the first neutralization point at pH 8.3. It corresponds to the neutralization of free bases, soda and lime, and the ½ neutralization of carbonates. The total alkalinity (Alk) is characterized by the second neutralization point at pH 4.5. It corresponds to the neutralization of bicarbonates and free bases.

AT: $OH^- + ½CO_3^{2-}$ and Alk: $OH^- + CO_3^{2-}$

If AT and Alk are equal, all the alkalinity is due to hydroxides. If AT is lower than Alk, this means that carbonates and hydroxides are present.

Let CO_3^{2-} be = 2(Alk – AT)

If AT is half the Alk, all the alkalinity is due to carbonates. If AT is less than half the Alk, carbonates and bicarbonates are present. This means that AT is equal to half the carbonate-related alkalinity, and Alk is equal to bicarbonate-related alkalinity plus carbonate-related alkalinity.

Let CO_3^{2-} be = 2AT and HCO_3^- = Alk – 2AT

If AT is null, Alk is only due to bicarbonates.

The alkalinity thus translates the quantity of protons it is necessary to add to those existing in the medium in order to balance the carbonate, bicarbonate and hydroxide ions. Table 15.2 summarizes the different cases.

	OH^-	**CO_3^{2-}**	**HCO_3^-**
AT = Alk	Alk	0	0
2AT > Alk	2AT – Alk	2(Alk – AT)	0
2AT = Alk	0	Alk	0
2AT < Alk	0	2AT	Alk – 2TA
AT = 0	0	0	Alk

Table 15.2. *Carbonate and bicarbonate identification using AT and Alk*

Alkalinity is defined as the sum of carbonate ions. Alkalinity is the sum of all the titratable bases, and is measured by titrating a water sample with an acid to a particular end-point (i.e. pH value). Therefore, reference is made to phenolphthalein alkalinity (phenolphthalein is an indicator that changes color at a pH of around 8.3) and total alkalinity (titration to a pH of 4.5). Generally, the alkalinity of surface water is a function of the carbonates, bicarbonates and hydroxides, although other compounds, such as borates, phosphates and silicates can also contribute to alkalinity. But alkalinity is predominantly due to carbonates, bicarbonates and hydroxides. The most predominant water soluble carbonates and bicarbonates are those of calcium and magnesium. Therefore hard waters are generally also high in alkalinity and the pH of water, which is high in hardness and alkalinity, is therefore generally high[WRC].

In the surface water pH range, AT is most often equal to zero. We can directly calculate permanent calcium hardness because of the difference between THCa-Alk. When Alk is higher than THCa, it means that all the calcium is in the form of calcium bicarbonate. When Alk is higher than the total TH, it means that all the calcium and magnesium are in the form of bicarbonates.

First example:

Water at pH 8.02 containing 682 mg·L^{-1} Ca^{2+} (in $CaCO_3$), total hardness (TH) equal to 1,120 mg·L^{-1} (in $CaCO_3$) and an alkalinity of 130 mg·L^{-1} (in $CaCO_3$) lets calcium be available only partially in the form of bicarbonate (Figure 15.2).

The Mg concentration is 1,120 – 682 = 438 mg·L^{-1} in $CaCO_3$.

The carbonate hardness is therefore 130 mg·L^{-1}.

Non-carbonate hardness or permanent hardness is equal to 0 mg·L^{-1}

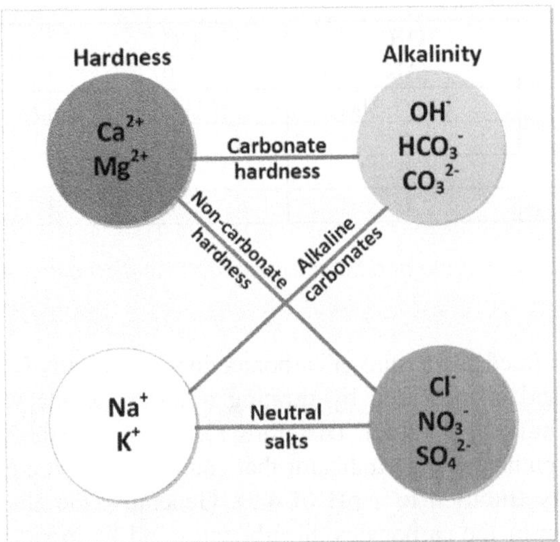

Figure 15.1. *Relationship between hardness, alkalinity and neutral salts. For a color version of this figure, see www.iste.co.uk/gaid/watertreatment3.zip*

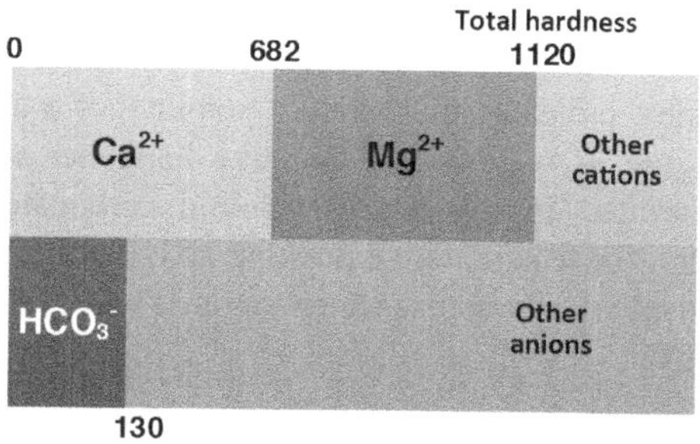

Figure 15.2. *Example of low bicarbonate water in relation to total hardness. For a color version of this figure, see www.iste.co.uk/gaid/watertreatment3.zip*

Second example:

Water at pH 7.72 containing 200 mg·L^{-1} Ca^{2+} (in $CaCO_3$), TH of 332 mg·L^{-1} (in $CaCO_3$) and an alkalinity of 311 mg·L^{-1} (in $CaCO_3$) enables calcium to be totally in the form of bicarbonate and a part of Mg^{2+} to be also available in that form (Figure 15.3).

The Mg concentration is 332 – 200 = 132 mg·L^{-1} in $CaCO_3$.

Carbonate hardness is therefore 311 mg·L^{-1}.

Non-carbonate hardness (or permanent hardness) is 332 – 311 = 21 mg·L^{-1}.

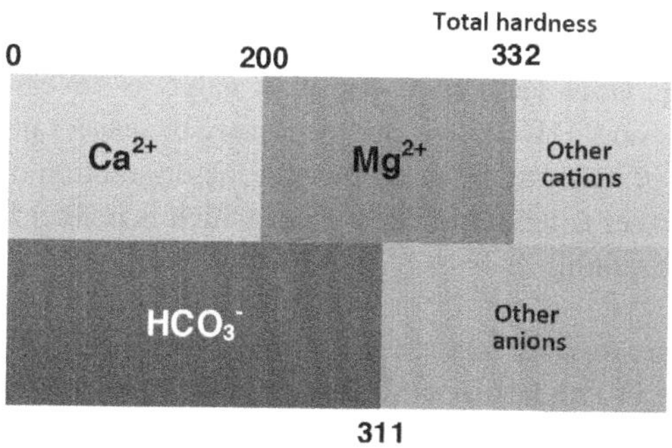

Figure 15.3. *Example of water with calcium totally available in the form of calcium bicarbonate. For a color version of this figure, see www.iste.co.uk/gaid/watertreatment3.zip*

For more details, see Chapter 20, Volume 5 of this series.

15.3. Langelier index (LI or LSI)

The Langelier index indicates the level of stability of water in relation to its $CaCO_3$ saturation degree. If water has a negative Langelier saturated index (LSI) value, it is undersaturated in relation to calcium carbonate and is potentially aggressive. Conversely, for waters with a positive LSI, a protective layer of calcium carbonate can be formed when water is supersaturated with $CaCO_3$. Saturated water has a LSI equal to zero.

The pH at which water is saturated with $CaCO_3$ is known as saturation pH (pHs):

$$LSI = pH - pHs$$

$$pHs = 1.022\ (11.5 - \log\ (Ca^{2+}) - \log\ (alkalinity))^*$$

(*used for a rapid calculation).

Langelier index	Trends
LSI # 0	Water at equilibrium
LSI < 0	Aggressive water
LSI > 0	Water with a tendency to scaling

Table 15.3. *Langelier index correspondences*

15.4. Drinking water hardness goals

The goal for water leaving the water treatment plant is to be at equilibrium, even slightly encrusting. However, since the pHs value is inversely proportional to mineralization (Alk and THCa), and it is necessary to respect pH values compatible with the potability of water at pH < 8.5 or with the dissolution of other salts (lead at pH < 7.5), the engineer is led to produce water which is neither too soft nor too hard in terms of mineralization.

It is a known fact that the "three eights" rule (pH: 8, Alk: 80 $mg \cdot L^{-1}$ $CaCO_3$ and TH: 80 $mg \cdot L^{-1}$ $CaCO_3$) makes it possible to maintain a disinfection compatible with chlorine and to promote the formation of the protective Tillmans layer against the corrosion of iron pipes.

However, this hardness value is minimum in order to avoid falling in the case of aggressive waters, in the event of a decrease in pH. Moreover, water softening performed at lower values is often economically unattractive.

A value of 100 $mg \cdot L^{-1}$ $CaCO_3$ is associated with a saturation pH between 7.8 and 8, while an increase in the final TH ($mg \cdot L^{-1}$ $CaCO_3$) between 120 and 150 $mg \cdot L^{-1}$ $CaCO_3$ leads to an equilibrium pH around 7.5, which is optimal for a proper disinfection.

This is the reason why, for hard waters, the goals in terms of hardness, corrosive potential and water equilibrium are:

$$\text{Alk} < 120\text{–}150 \text{ mg} \cdot L^{-1} \text{ } CaCO_3$$

$$\text{TH} < 120\text{–}150 \text{ mg} \cdot L^{-1} \text{ } CaCO_3$$

$$\text{pH} \geq \text{pHs} + 0.2$$

For each quality of softened water, one first has to check that the calco-carbonic equilibrium (Langelier index) has been reached, and if necessary, to readjust the final pH. The post-treatment goal is to obtain water with a calco-carbonic equilibrium.

15.5. General principles of water softening

The goal of water softening is to condition water by precipitating calcium and carbonates in the form of calcium carbonate. This results in a decrease in alkalinity and hardness. The precipitation of calcium carbonate is possible when the saturation index (Langelier index) is positive.

Saturation index:

$$(\text{LSI}) = \log (Ca^{2+})(CO_3^{2-})/\text{Ks} = \text{pH} - \text{pHs}$$

$$\text{Ks} = -171.9065 - 0.077993 \text{ T} + (2{,}839.319/\text{T}) + 71.595 \log\text{T}$$

This saturation index, related to the temperature, translates the ion concentration, which will trigger the precipitation reaction. For scaling waters, and at usual temperatures (close to 12°C), the Langelier index is positive.

To ensure optimal precipitation, the solubility of $CaCO_3$ must be taken into account as well as various reactions involving the dissociation of carbonic acid:

$$CO_2 + H_2O \rightarrow H_3O^+ + HCO_3^-$$

$$HCO_3^- + H_2O \rightarrow H_3O^+ + CO_3^{2-}$$

$$H_3O^+ + OH^- \rightarrow 2H_2O$$

$$Ca^{2+} + CO_3^{2-} \rightarrow CaCO_3$$

For example, lime softening removes temporary hardness, in particular, calcium-related hardness. Simple lime softening has no impact on permanent hardness (non-carbonate), as demonstrated by the following reactions:

$$Ca(OH)_2 + Ca(HCO_3)_2 \rightarrow 2\ CaCO_3 + 2H_2O$$

$$Ca(OH)_2 + Mg(HCO_3)_2 \rightarrow MgCO_3 + CaCO_3 + 2H_2O$$

Magnesium carbonate only precipitates in cases where pH is above 9.5. With the addition of lime, this reaction continues until reaching a residual alkalinity of around 15–30 mg·L^{-1} $CaCO_3$.

Chemical reagents	**Molecular weight**	**$CaCO_3$ equivalents multiplied by**
Calcium carbonate, $CaCO_3$	100.1	1.00
Lime, $Ca(OH)_2$	74.1	1.35
Quicklime, CaO	56.1	1.78
Carbon dioxide, CO_2	44.0	1.14
Sodium carbonate, Na_2CO_3	106.0	0.94
Soda (sodium hydroxide), NaOH	40.0	1.25
Sodium bicarbonate, $NaHCO_3$	84.0	0.60

Table 15.4. *Correspondences between various reagents and calcium carbonate*

Water softening reactions are regulated by pH adjustment. First, all free acids (CO_2) have to be neutralized. Then the pH is increased to provoke the precipitation of $CaCO_3$.

If necessary, the pH is further increased in order to precipitate the magnesium to $Mg(OH)_2$. Finally, sodium carbonate can be added in order to precipitate non-carbonate hardness.

– *Neutralization of carbonic acid (H_2CO_3)*: in order to raise the pH, the free acids first have to be neutralized. CO_2 is the main acid found in unpolluted natural water. Hardness is not removed during this stage:

$$CO_2 + Ca\ (OH)_2 \rightarrow Ca(HCO_3)_2$$

– *Precipitation of carbonate hardness related to calcium*: in order to precipitate calcium carbonate, all the natural bicarbonate must be converted to carbonate. Carbonate then serves as a common ion for the precipitation reaction:

$$Ca(HCO_3)_2 + Ca\ (OH)_2 \rightarrow 2CaCO_3 + H_2O$$

– *Precipitation of carbonate hardness related to magnesium*: to remove the carbonate hardness resulting from the presence of magnesium, more lime must be added. The reaction can be considered as taking place in two stages.

The first stage occurs when the bicarbonate from stage 2 is converted to carbonate:

$$Mg(HCO_3)_2 + Ca\ (OH)_2 \rightarrow MgCO_3 + CaCO_3 + 2H_2O$$

Water hardness does not change because $MgCO_3$ is soluble (at least 70 mg·L^{-1}). With the supplementary addition of lime, hardness related to magnesium is removed:

$$MgCO_3 + Ca\ (OH)_2 \rightarrow Mg(OH)_2 + CaCO_3$$

– *Removal of non-carbonate hardness related to calcium*: to remove the non-carbonate hardness due to calcium, an additional carbonate in the form of sodium carbonate must be added:

$$Ca^{2+} + Na_2CO_3 \rightarrow CaCO_3 + 2Na^+$$

– *Removal of non-carbonate hardness related to magnesium*: in order to remove the non-carbonate hardness due to magnesium, it is necessary to add both lime and soda. Lime provides the hydroxyl ion for the precipitation of magnesium:

$$Mg^{2+} + Ca(OH)_2 \rightarrow Mg(OH)_2 + 2Ca^{2+}$$

Despite the removal of magnesium, hardness does not change because calcium is still in solution. To remove calcium, sodium carbonate must be added:

$$Ca^{2+} + Na_2CO_3 \rightarrow CaCO_3 + 2Na^+$$

Soda can replace lime.

The effect of pH on the solubility of cations is important. For example, assuming there are no other compounds in solution to react with the calcium hydroxide, the solubility product (Ksp) is:

$$Ksp = [Ca^{2+}]\ [OH^-]^2$$

$$\log\ [Ca^{2+}] = \log Ksp - 2\ \log[OH^-]$$

As $[OH^-]$ is a function of pH:

$$pH = pKw - pOH = pKw + \log[OH^-]$$

$$\log[OH^-] = pH - pKw$$

$$\text{let } \log [Ca^{2+}] = \log Ksp - 2(pH - pKw)$$

Not only does the pH affect the solubility of metal hydroxides, but also other equilibria, which in turn affect the cation's solubility. Carbonated species both have an influence on and are influenced by pH. The sum of carbonate species can be specified as a total concentration of inorganic carbon (TIC):

$$TIC = [H_2CO_3] + HCO_3^-] + [CO_3^{2-}]$$

During the softening of carbonated Ca^{2+}, the pH is controlled by adding lime or soda. To precipitate $CaCO_3$, the water's pH must be >8.3. To precipitate magnesium, the pH must be within a range between 11 and 11.3.

15.5.1. *The main chemical reactions with lime and soda*

15.5.1.1. *With lime*

The softening reaction tends to reduce bicarbonates and calcium hardness, with a significant demand for bicarbonates and calcium hardness. At drinking water stations where the pH of raw water is less than 8.3, and in the presence of lime, water softening leads to the following chemical reactions:

– *calcium carbonate hardness reduction* (no reduction of magnesium TH, which requires a pH > 10):

$$Ca(OH)_2 + Ca(HCO_3)_2 \rightarrow 2CaCO_3 \downarrow + 2H_2O$$

– this reaction is accompanied by the *CO_2 neutralization reaction*:

$$2CO_2 + Ca(OH)_2 \rightarrow Ca(HCO_3)_2$$

The bicarbonate thus formed is added to the temporary hardness.

Thus, for one mole of $Ca(OH)_2$, two moles of $CaCO_3$ are precipitated:

– 1 mole of $Ca(OH)_2$ lime reduces 1 mole of Ca and 2 moles of HCO_3;

– 2 meq of $Ca(OH)_2$ reduce 2 meq of Ca and 2 meq of HCO_3;

– 74 ppm of $Ca(OH)_2$ reduce 40 ppm of Ca and 122 ppm of HCO_3;

– 74 ppm of $Ca(OH)_2$ reduce 100 ppm of Ca (expressed as $CaCO_3$) and 100 ppm of HCO_3 (expressed as $CaCO_3$);

– 74 ppm $Ca(OH)_2$ reduce 100 $mg \cdot L^{-1}$ $CaCO_3$ of Ca, and 100 $mg \cdot L^{-1}$ $CaCO_3$ of HCO_3.

In the presence of magnesium bicarbonate and depending on a suitable pH, the following reaction is:

$$2Ca(OH)_2 + Mg\ (HCO_3)_2 \rightarrow 2CaCO_3 \downarrow + Mg(OH)_2 \downarrow + 2H_2O$$

15.5.1.1.1. Quick estimation of the quantity of lime

First case

Consider a water with the following chemical composition:

– An alkalinity of 160 mg $CaCO_3 \cdot L^{-1}$;

– a calcium hardness of 175 mg $CaCO_3 \cdot L^{-1}$;

– a magnesium hardness of 92 mg $CaCO_3 \cdot L^{-1}$;

– a sulfate concentration of 115 mg $CaCO_3 \cdot L^{-1}$.

The pH is 7.8 and the T °C is 18°C.

The bar graph is illustrated in Figure 15.4.

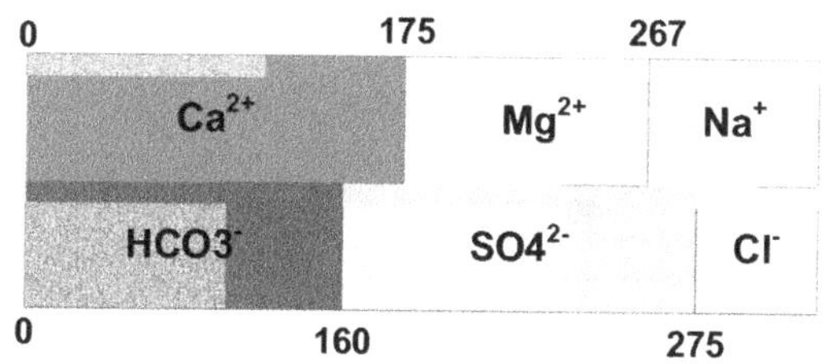

Figure 15.4. *Bar graph for the first case. For a color version of this figure, see www.iste.co.uk/gaid/watertreatment3.zip*

The desired final calcium hardness is 55 mg $CaCO_3 \cdot L^{-1}$. The quantity of lime required is calculated as in Table 15.6.

pH	7.8
Total hardness, mg $CaCO_3 \cdot L^{-1}$	267
Alkalinity, mg $CaCO_3 \cdot L^{-1}$	160
Carbonate hardness, mg $CaCO_3 \cdot L^{-1}$	160
Non-carbonate Ca hardness, mg $CaCO_3 \cdot L^{-1}$	175 – 160 = 15
Mg carbonate hardness, mg $CaCO_3 \cdot L^{-1}$	0
Non-carbonate Mg hardness, mg $CaCO_3 \cdot L^{-1}$	267 – 175 = 92

Table 15.5. *Physicochemical characteristics of the water to be treated (first case)*

Final calcium hardness, mg $CaCO_3 \cdot L^{-1}$	55
Final calcium carbonate hardness, mg $CaCO_3 \cdot L^{-1}$	55 – 15 = 40
Ca hardness removed, mg $CaCO_3 \cdot L^{-1}$	175 – 55 = 120
Residual alkalinity, mg $CaCO_3 \cdot L^{-1}$	40
Removed alkalinity, mg $CaCO_3 \cdot L^{-1}$	160 – 40 = 120
Total residual hardness, mg $CaCO_3 \cdot L^{-1}$	147
Pure lime dose mg $Ca(OH)_2 \cdot L^{-1}$	74 × 120/100 = 88.8
95% purity lime dose, mg·L^{-1}	88.8/95 = 93.47

Table 15.6. *Rapid estimation of lime quantities for water softening (first case)*

The final bar graph is illustrated in Figure 15.5.

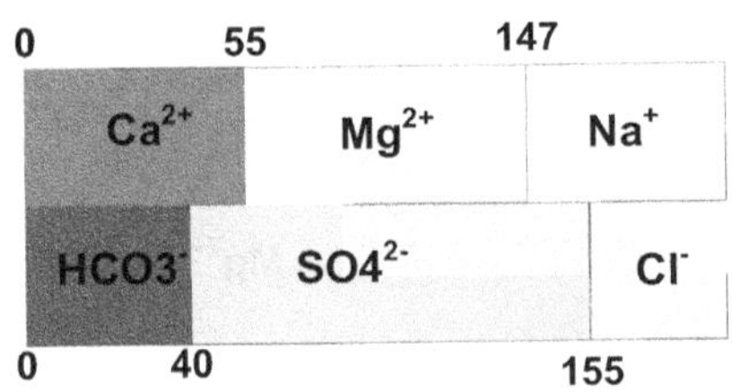

Figure 15.5. *Softened water bar graph (first case). For a color version of this figure, see www.iste.co.uk/gaid/watertreatment3.zip*

This approach does not take into account the CO_2 content in raw water, which creates a demand for lime in order to neutralize the carbonate component. The CO_2 concentration is calculated with specialized software or using the equations detailed in Chapter 20. Then, the real demand for lime is the sum of the dose required to neutralize CO_2 and the one required for the precipitation of carbonate hardness.

Second case

Consider a water with the following chemical composition:

– an alkalinity of 246 mg $CaCO_3 \cdot L^{-1}$;

– a calcium hardness of 187 mg $CaCO_3 \cdot L^{-1}$;

– a magnesium hardness of 164 mg $CaCO_3 \cdot L^{-1}$;

– a sulfate concentration of 29 mg $CaCO_3 \cdot L^{-1}$.

The pH is 7.8 and the T °C is 18°C.

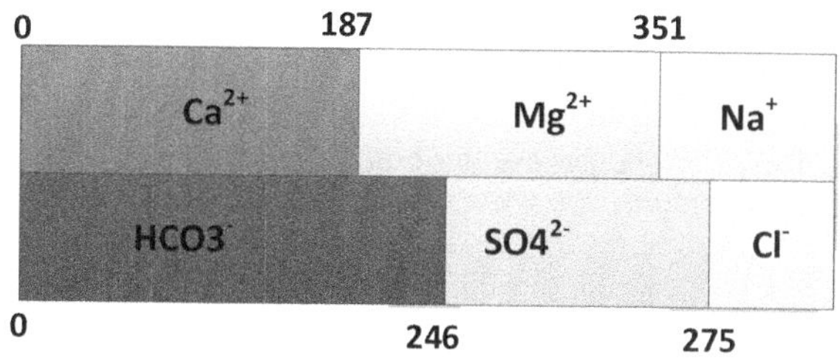

Figure 15.6. *Bar graph for the water to be treated (second case). For a color version of this figure, see www.iste.co.uk/gaid/watertreatment3.zip*

The desired final calcium hardness is 90 mg $CaCO_3 \cdot L^{-1}$. The quantity of lime required is calculated as in Table 15.7.

pH	7.8
Total hardness, mg $CaCO_3 \cdot L^{-1}$	187 + 164 = 351
Ca hardness, mg $CaCO_3 \cdot L^{-1}$	187
Mg hardness, mg $CaCO_3 \cdot L^{-1}$	351 – 187 = 164
Alkalinity, mg $CaCO_3 \cdot L^{-1}$	246
Total carbonate hardness, mg $CaCO_3 \cdot L^{-1}$	246
Total non-carbonate hardness, mg $CaCO_3 \cdot L^{-1}$	351 – 246 = 105
Mg carbonate hardness, mg $CaCO_3 \cdot L^{-1}$	246 – 187 = 59
Non-carbonate Mg hardness, mg $CaCO_3 \cdot L^{-1}$	164 – 59 = 105
Ca carbonate hardness, mg $CaCO_3 \cdot L^{-1}$	187
Total non-carbonate hardness, mg $CaCO_3 \cdot L^{-1}$	0

Table 15.7. *Physicochemical characteristics of the water to be treated (second case)*

The final bar graph is illustrated in Figure 15.7.

The reaction of lime with magnesium bicarbonate is the following:

$$Ca(OH)_2 + Mg(HCO_3)_2 \rightarrow MgCO_3 + CaCO_3 + 2H_2O$$

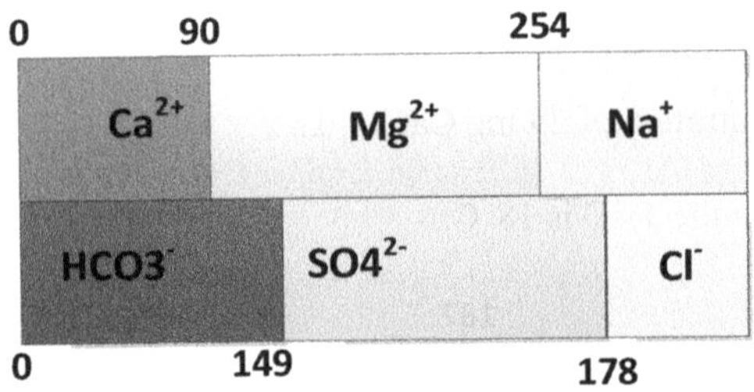

Figure 15.7. *Softened water bar graph (second case). For a color version of this figure, see www.iste.co.uk/gaid/watertreatment3.zip*

Final Ca hardness, mg $CaCO_3 \cdot L^{-1}$	90
Ca hardness removed, mg $CaCO_3 \cdot L^{-1}$	187 – 90 = 97
Removed alkalinity, mg $CaCO_3 \cdot L^{-1}$	97
Residual alkalinity, mg $CaCO_3 \cdot L^{-1}$	246 – 97 = 149
Pure lime dose, mg$\cdot L^{-1}$	71.8
95% purity lime dose, mg$\cdot L^{-1}$	75.6

Table 15.8. *Rapid estimation of lime quantities for water softening (second case)*

As magnesium carbonate is relatively soluble (around 70 mg$\cdot L^{-1}$), a surplus of lime is required to precipitate the Mg:

$$Ca(OH)_2 + MgCO_3 \rightarrow CaCO_3 + Mg(OH)_2$$

On this example, the injected lime precipitates the calcium bicarbonate.

This calculation does not take into account the CO_2 concentration which – depending on its concentration – may require a supplementary addition of lime.

15.5.1.2. *With caustic soda*

The overall softening reaction tends to reduce bicarbonates and calcium hardness in different molar proportions (half the quantity of bicarbonates). Calcium hardness reduction reactions with caustic soda are as follows:

$$Ca(HCO_3)_2 + 2NaOH \rightarrow CaCO_3 \downarrow + Na_2CO_3 + 2\ H_2O$$

Followed by:

$$Ca(HCO_3)_2 + Na_2CO_3 \rightarrow CaCO_3 \downarrow + 2NaHCO_3$$

where:

– 1 mole of NaOH soda reduces 1 mole of Ca and 1 mole of HCO_3;

– 1 meq of caustic soda reduces 2 meq of Ca and 1 meq of HCO_3;

– 40 ppm of caustic soda reduce 40 ppm of Ca and 61 ppm of HCO_3;

– 40 ppm of caustic soda reduce 100 ppm of Ca (expressed as $CaCO_3$) and 50 ppm of HCO_3 (expressed as $CaCO_3$);

– 40 ppm of NaOH increases the Na concentration by 23 ppm.

Consider a water with the following chemical composition:

– an alkalinity of 160 mg $CaCO_3 \cdot L^{-1}$;

– a calcium hardness of 175 mg $CaCO_3 \cdot L^{-1}$;

– a magnesium hardness of 92 mg $CaCO_3 \cdot L^{-1}$;

– a sulfate concentration of 115 mg $CaCO_3 \cdot L^{-1}$.

The pH is 7.8 and the T °C is 18°C.

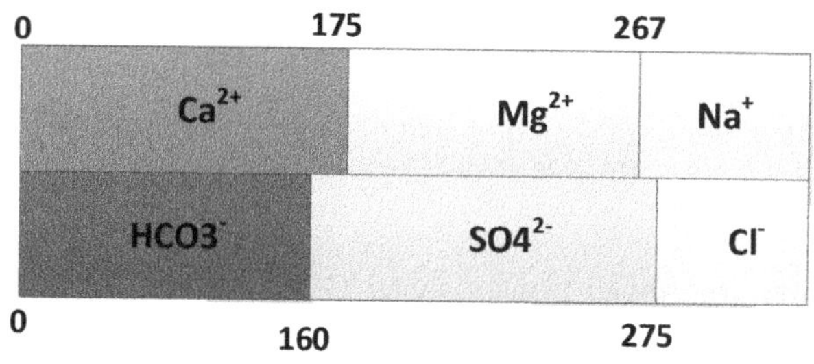

Figure 15.8. *Bar graph for the water to be treated. For a color version of this figure, see www.iste.co.uk/gaid/watertreatment3.zip*

The desired final calcium hardness is 55 mg $CaCO_3 \cdot L^{-1}$. The quantity of caustic soda required is calculated as follows:

pH	7.8
Total hardness, mg $CaCO_3 \cdot L^{-1}$	267
Alkalinity, mg $CaCO_3 \cdot L^{-1}$	160
Carbonate hardness, mg $CaCO_3 \cdot L^{-1}$	160
Non-carbonate Ca hardness, mg $CaCO_3 \cdot L^{-1}$	175 – 160 = 15
Mg carbonate hardness, mg $CaCO_3 \cdot L^{-1}$	0
Non-carbonate Mg hardness, mg $CaCO_3 \cdot L^{-1}$	267 – 175 = 92

Table 15.9. *Physicochemical characteristics of the water to be treated*

Final Ca hardness, mg $CaCO_3 \cdot L^{-1}$	55
Ca hardness removed, mg $CaCO_3 \cdot L^{-1}$	120
Removed alkalinity, mg $CaCO_3 \cdot L^{-1}$	60
Residual alkalinity, mg $CaCO_3 \cdot L^{-1}$	100
Total residual hardness, mg $CaCO_3 \cdot L^{-1}$	147
Pure soda dose, mg$\cdot L^{-1}$	54
33 % purity soda dose, mg$\cdot L^{-1}$	163.6

Table 15.10. *Rapid estimation of caustic soda quantities for water softening*

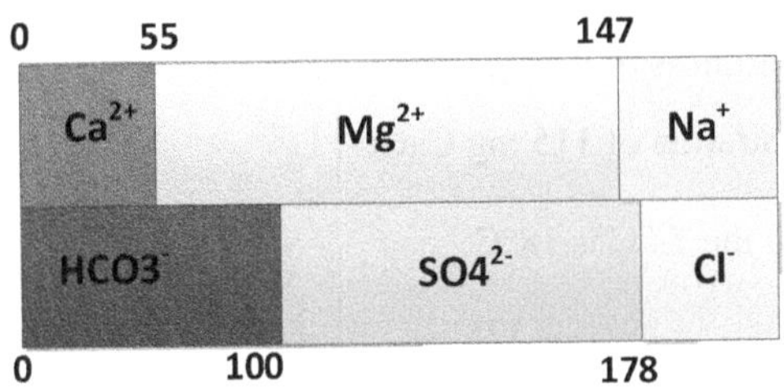

Figure 15.9. *Bar graph of water softened with caustic soda. For a color version of this figure, see www.iste.co.uk/gaid/watertreatment3.zip*

15.5.1.2.1. Same remark as previously for CO_2

The addition of an equivalent of caustic soda leads to a decrease in water hardness (TH) by two equivalents, and a decrease in alkalinity by one equivalent. This results in a TH decrease of 20 mg$\cdot L^{-1}$ $CaCO_3$ and an Alk decrease of 10 mg$\cdot L^{-1}$ $CaCO_3$.

There is no reduction in magnesium TH, because the removal of magnesium TH is triggered at higher pH values (≥11).

This precipitation reaction should be accompanied by the neutralization of free CO_2. The reaction is as follows:

$$CO_2 + NaOH \rightarrow NaHCO_3$$

This reaction leads to the formation of hydrogen carbonates, therefore producing residual. Softened water is always slightly scaling, and therefore, acidity stabilization should be considered.

Compared to water softening with lime, catalytic water softening with soda decreases alkalinity to a lesser extent and adds sodium.

In the presence of magnesium bicarbonate and depending on a suitable pH, the following reaction is:

$$4NaOH + Mg\,(HCO_3)_2 \rightarrow 2Na_2CO_3 \downarrow + Mg(OH)_2 \downarrow + 2H_2O$$

15.5.1.3. *With sodium carbonate*

Sodium carbonate (Na_2CO_3) is used for removing non-carbonate calcium (e.g. $CaCl_2$, $CaSO_4$). The softening reaction does not lead to a reduction in bicarbonates:

$$Na_2CO_3 + CaCl_2 \rightarrow CaCO_3 \downarrow + 2NaCl$$

$$Na_2CO_3 + CO_2 + H_2O \rightarrow 2NaHCO_3$$

$$Na_2CO_3 + CaSO_4 \rightarrow CaCO_3 \downarrow + Na_2SO_4$$

$$Na_2CO_3 + Ca(OH)_2 + MgCl_2 \rightarrow CaCO_3 \downarrow + Mg(OH)_2 \downarrow + 2NaCl$$

$$Na_2CO_3 + Ca(OH)_2 + MgSO_4 \rightarrow CaCO_3 \downarrow + Mg(OH)_2 \downarrow + Na_2SO_4$$

where:

– 1 mole of Na_2CO_3 reduces 1 mole of Ca and forms 1 mole of $CaCO_3$;

– 2 meq of Na_2CO_3 reduce 2 meq of Ca and form 2 meq of $CaCO_3$;

– 106 ppm of Na_2CO_3 reduce 40 ppm of Ca and form 100 ppm of $CaCO_3$;

– 106 ppm of Na_2CO_3 reduce 100 ppm Ca (expressed as $CaCO_3$).

Parameters to take into account:

– maintenance and upkeep costs and constraints;

– (liquid) caustic soda handling is easier than that of lime;

– the use of caustic soda and sodium carbonate as softening agents provokes an increase in the sodium content in the treated effluent.

15.5.1.4. *CO_2 neutralization*

The CO_2 in raw water behaves as H_2CO_3, and its removal is the first stage in water softening. It can be neutralized by adding lime or soda. When the concentration exceeds 10 mg·L^{-1} in CO_2 (22.7 mg·L^{-1} of $CaCO_3$ or 0.45 meq·L^{-1}), aeration or stripping removal processes must be considered. No hardness is removed at this stage of the process.

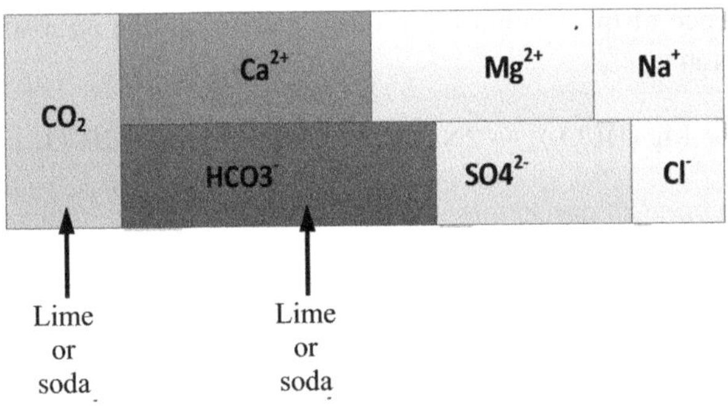

Figure 15.10. *CO_2 neutralization, followed by carbonated calcium water softening. For a color version of this figure, see www.iste.co.uk/gaid/watertreatment3.zip*

The concentration in CO_2 can be estimated using the equilibrium expressions related to the dissociation of carbonic acid. Water pH and alkalinity must be known. The equilibrium expressions for carbonic acid are:

$$H_2CO_3 \rightarrow H^+ + HCO_3^- \; pKa_1 = 6.35 \text{ at } 25°C$$

$$Ka_1 = \frac{[H^+][HCO_3^-]}{[H_2CO_3]} \; Ka_1 = 10^{14.8453 - 3404.71/T - 0.032786T}$$

$$HCO_3^- \rightarrow CO_3^{2-} + H^+ \; pKa_2\text{: } 10.33 \text{ at } 25°C$$

$$Ka_2 = \frac{[H^+][CO_3^{2-}]}{[HCO_3^-]} \; Ka_2 = 10^{6.498 - 2909.39/T - 0.02379T}$$

Temperature T is expressed in Kelvin degrees.

The following example calculates the CO_2 concentration for water at pH 7.65 and with an alkalinity of 310 $mg \cdot L^{-1}$ as $CaCO_3$. The temperature is 25°C.

The [H+] concentration is:

$$10^{-7.65} = 2.24 \times 10^{-8} \text{ mol} \cdot \text{L}^{-1}$$

The HCO_3^- concentration is:

$$HCO_3 = 310 \text{ mg} \cdot \text{L}^{-1} \frac{61 \text{ mg} \cdot \text{meq}^{-1}}{50 \text{ mg} \cdot \text{meq}^{-1}} \frac{1}{(61 \text{ g} \cdot \text{mol} \cdot \text{L}^{-1})(1{,}000 \text{ mg} \cdot \text{g}^{-1})} = 6.20 \times 10^{-3} \text{ mol} \cdot \text{L}^{-1}$$

$pKa_1 = 6.35$ at 25°C, that is, $Ka_1 = 10^{-6.35} = 4.47 \times 10^{-7}$ $mol \cdot L^{-1}$ at 25°C

$$[H_2CO_3] = \frac{[2.24 \times 10^{-8}][6.20 \times 10^{-3}]}{4.47 \times 10^{-7}} = 3.11 \times 10^{-4} \text{ mol} \cdot \text{L}^{-1}$$

In other terms:

$$[CO_2] = 3.11 \times 10^{-4} \text{ mol} \cdot \text{L}^{-1} \times 44{,}000 \text{ mg} \cdot \text{mol}^{-1} = 13.7 \text{ mol} \cdot \text{L}^{-1}$$

15.6. Water softening chemical processes

Water softening chemical processes can be divided into two groups:

– processes leading to the precipitation of $CaCO_3$ in the form of liquid sludge;

– processes leading to precipitation of $CaCO_3$ by means of crystallization on a solid support.

The choice between precipitation and crystallization is mainly determined by factors such as investment and operating costs, implementation and exploitation difficulties, the destination of the precipitates obtained and the ground surface. On the other hand, the type of chemical is mainly indicated by the constraints of raw water quality.

Lime, caustic soda and sodium carbonate are the basic reagents. The choice of reagent depends on the ratio between the bicarbonates present in raw water (namely the calcium and bicarbonate concentration) and the desired reduction in calcium hardness.

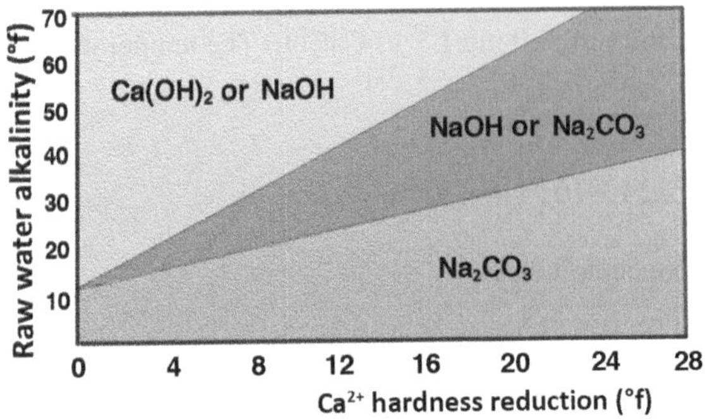

Figure 15.11. *Choice of reagents depending upon raw water quality. For a color version of this figure, see www.iste.co.uk/gaid/watertreatment3.zip*

Water softening is possible in the following cases:

– with lime, when the bicarbonate concentration is high;

– with soda, when the bicarbonate concentration is moderate;

– with sodium carbonate, when the bicarbonate concentration is low (extremely rare case). It is combined with lime or soda to remove permanent hardness.

In water softening chemical processes, the reagents are identical for catalytic and for conventional water softening. Ion exchange resins use NaCl as a regenerating agent, whereas nanofiltration uses mineral acids and anti-scaling in order to prevent the precipitation of salts onto the membrane.

Processes	Reagents
Catalytic water softening	Lime, soda, sodium carbonate
Classic water softening	Lime, soda, sodium carbonate
Ion-exchange resins	Sodium chloride, biocide cleaner
Nanofiltration	Hydrochloric acid, anti-scaling, sulfuric acid

Table 15.11. *Chemical reagents used in various water softening processes*

15.6.1. *Limitations of the process and empirical considerations*

The minimum calcium hardness that can be reached is around 30 $mg{\cdot}L^{-1}$ as $CaCO_3$. Commonly a supplementary quantity of lime is added beyond the stoichiometric value with a 20 $mg{\cdot}L^{-1}$ dosage of $Ca(OH)_2$ expressed as $CaCO_3$.

Reagent addition	*Reactions*
Carbonate hardness	
Lime for CO_2	Conversion into H_2CO_3
Lime for HCO_3^-	pH increase, conversion of HCO_3^- into CO_3^{2-}
Lime for Mg^{2+}	pH increase Precipitation into $Mg(OH)_2$
Non-carbonate hardness	
Sodium carbonate	CO_3^{2-} ion contribution

Table 15.12. *Summary of reagents required for water softening with lime*

15.7. Veolia water softening technologies

15.7.1. *Clariflocculator*

The clariflocculation process used for removing turbidity can also be used for water softening.

The rapid mixing is done in a tank with vertical-type mixers, whereas the flocculation reaction takes place in a concentric settler tank. Part of the precipitate can be recirculated ahead of the process. It accelerates precipitation reactions and the kinetics approach solubility limits as the previously formed $CaCO_3$ crystals become mixed in the reaction tank. The velocity gradient (G) is between 300 and 700 s^{-1} and the minimum mixing time should be 2–10 min, depending on the temperature.

The recirculation rate is typically 10–20% of the raw water inlet flow. The mixer is equipped with a variable speed power drive because of which the operator can regulate the recirculation rate.

– The total water level varies between 4.0 and 5.5 m.

– Overflow velocity (that is to say, the flow divided by the surface, Q/S) is typically measured 0.6 m below the water surface based on the surface between the deflector's wall and the basin's wall. It is between 2.0 and 3.0 $m \cdot h^{-1}$.

– The retention time is between 2 and 3 h.

– The $CaCO_3$ sludge concentration removed is between 60 and 100 $g \cdot L^{-1}$.

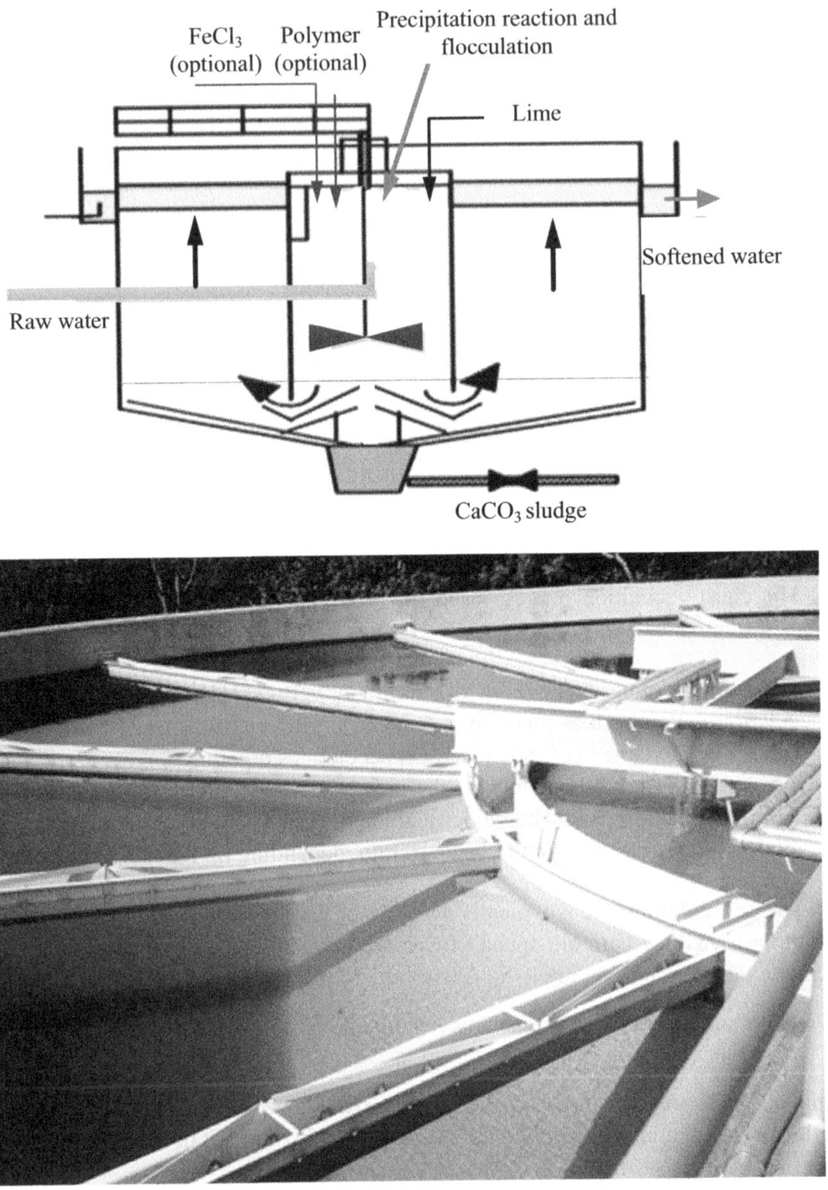

Figure 15.12. *Clariflocculator water softening. For a color version of this figure, see www.iste.co.uk/gaid/watertreatment3.zip*

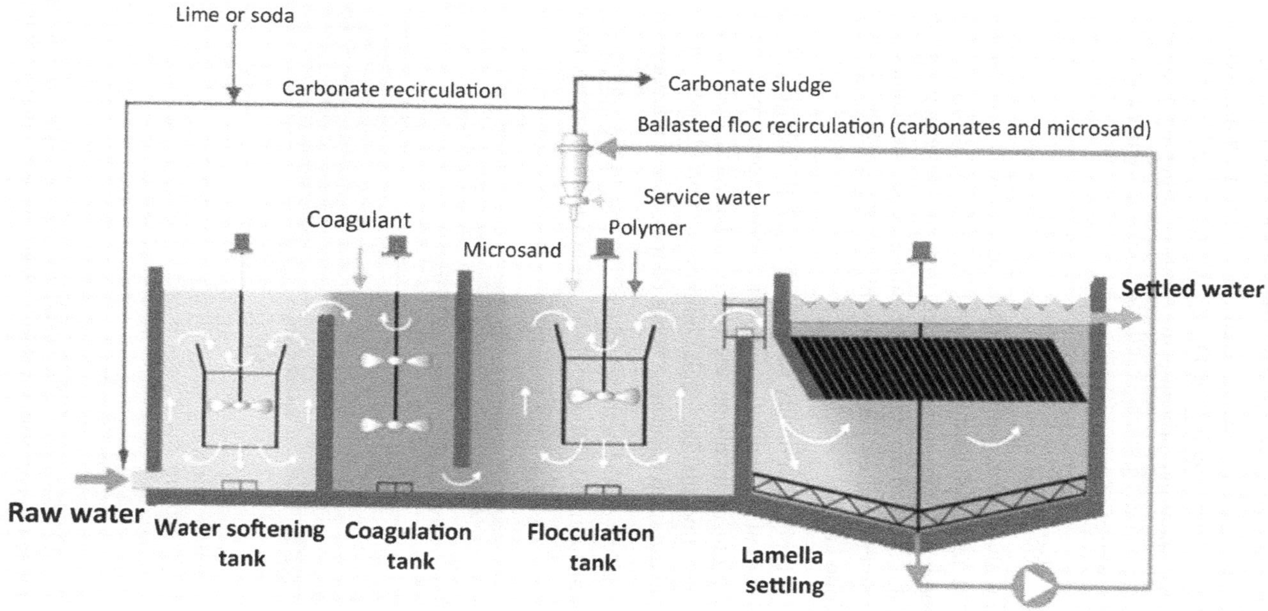

Figure 15.13. *Actiflo® softening process schematic diagram. For a color version of this figure, see www.iste.co.uk/gaid/watertreatment3.zip*

15.7.2. *Actiflo® softening*

The Actiflo® softening process treats raw water turbidity and water hardness at the same time. It is a compact process derived from Actiflo®, which relies on a base as a reagent for the precipitation of $CaCO_3$, a coagulant and a polymer associated with microsand. The operating principle of Actiflo® softening (Figure 15.13) is based on a separation of the chemical reactions resulting from water softening and those related to the removal of colloidal matter.

This process includes:

– A first stage involving water softening (with soda or with lime) for treating raw water exclusively in terms of hardness, regardless of the presence of suspended solids (SS). Depending on the goals assigned to water softening, the dosages of alkaline reagent are adapted to the treatment. Sludge recirculation, including essentially precipitate carbonates, permanently brings precipitation germinates, thus increasing the reaction kinetics.

– A second stage bringing turbidity removal through the addition of a coagulant (Fe or Al) and a natural or synthetic polymer. This stage is part of the classic turbidity treatment protocol involving the use of a ballasted floc. A Turbomix™ incorporated in the flocculation tank optimizes flocculation efficiency by homogenizing the mixture and ensuring a better distribution of retention times.

– A third stage involves the settling of $CaCO_3$ and hydroxide sludge/colloids within a lamellar structure, and associated with a $CaCO_3$ sludge recirculation system via a hydrocyclone.

a) b) c)

Figure 15.14. *Actiflo® softening stages: (a) water softening, (b) flocculation and (c) settling. For a color version of this figure, see www.iste.co.uk/gaid/watertreatment3.zip*

Ferric chloride is the most suitable coagulant for this type of application. Its dosage is directly related to raw water turbidity. Various tests conducted on different raw waters have shown that a dose of $FeCl_3$ (commercial product at 41%) between

10 and 50 ppm is generally sufficient for the treatment of turbidity. The synthetic polymer dosages are between 0.1 and 0.4 mg·L^{-1}.

15.7.2.1. *Performances: removal of water hardness*

The results obtained on a surface water with a TH between 500 and 650 mg·L^{-1} $CaCO_3$, an alkalinity between 150 and 200 mg·L^{-1} $CaCO_3$, large variations in temperature (4–22°C), in turbidity (5–150 NTU) and in organic matter (4–8 mg·L^{-1} of total organic carbon [TOC]) are summarized in Figure 15.15, for a variable mirror velocity between 60 and 80 m·h^{-1}. A constant reduction in alkalinity (between 150 and 180 mg·L^{-1} $CaCO_3$) is obtained with a dosage of 140 g·m^{-3} of lime (96% pure lime).

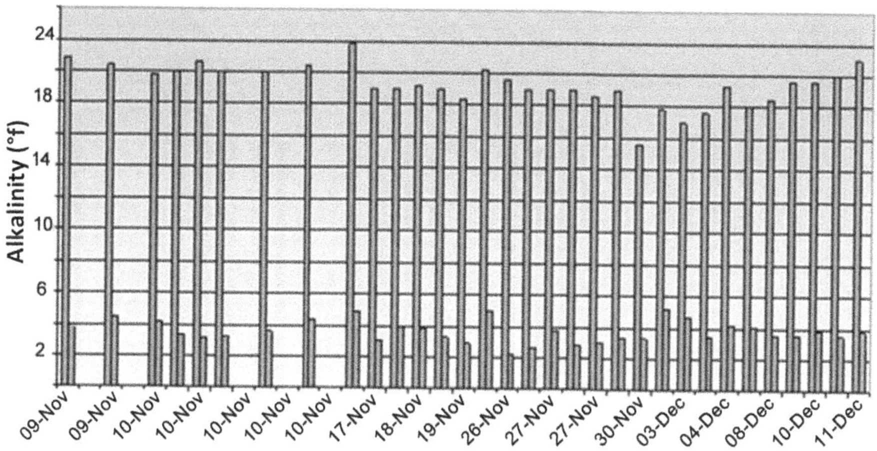

Figure 15.15. *Water softening with lime on Actiflo® softening (1°f = 10 mg·L^{-1} $CaCO_3$). For a color version of this figure, see www.iste.co.uk/gaid/watertreatment3.zip*

A residual turbidity between 1 and 3 NTU is obtained for a $FeCl_3$ dosage (41% com.) of around 10–30 g·m^{-3}. The turbidity of settled water remains practically stable, despite a sudden variation in the quality of raw water (95 NTU turbidity).

The results obtained are for groundwater characterized by a TH between 200 and 270 mg·L^{-1} $CaCO_3$, an alkalinity between 150 and 200 mg·L^{-1} $CaCO_3$ and an organic matter concentration (3.6–5 mg·L^{-1} of TOC). We observe a constant reduction (Figure 15.16) of TH between 100 and 120 mg·L^{-1} $CaCO_3$, obtained with a lime dosage of around 120 g·m^{-3} (96% pure lime).

The operating conditions are:

– water softening contact time: 5 min;

– carbonate load in the reactor: 3–5 $g \cdot L^{-1}$;

– softening pH: 9.5;

– global contact time in the Actiflo® softening unit: 15 min;

– mirror velocity: 80 $m \cdot h^{-1}$;

– hydrated lime (96% purity): 120 $g \cdot m^{-3}$;

– use of an iron salt: 30$g \cdot m^{-3}$ of $FeCl_3$ commercial product (41%);

– use of a synthetic polymer: 0.3 $g \cdot m^{-3}$.

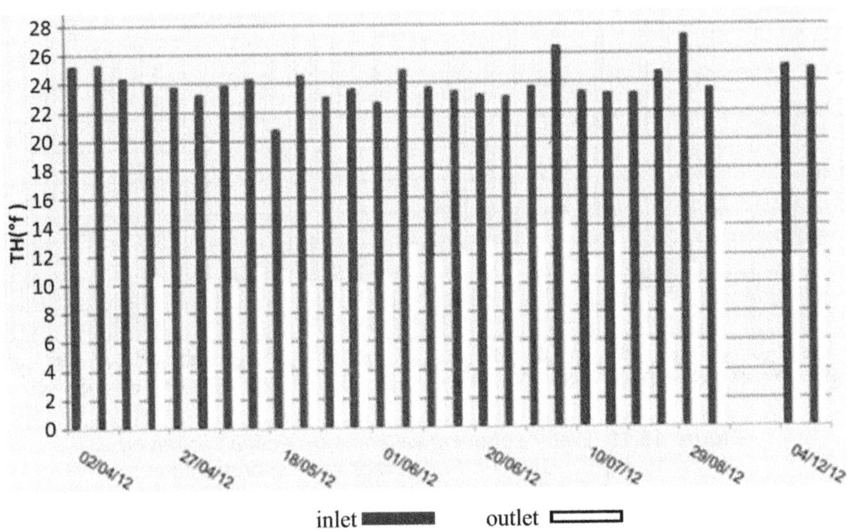

Figure 15.16. *Hardness removal with Actiflo® softening on groundwater (1°f = 10 mg·L^{-1} $CaCO_3$). For a color version of this figure, see www.iste.co.uk/gaid/watertreatment3.zip*

15.7.2.2. *Removal of dissolved organic carbon*

The removal of natural organic matter (NOM) during the water softening process is interesting in order to reduce the potential for the formation of trihalomethanes (THMs) and halogenated acetic acids (HAA5) when chlorine is used as a disinfectant. This removal, nonetheless, depends on the raw water's organic matter composition. However, we rely on the following elements.

The removal of NOM (TOC and dissolved organic carbon [DOC]) between 30 and 40% is observed. The removal of DOC is obtained by co-precipitation into $CaCO_3$ or into $Mg(OH)_2$. Organic substances such as humic acids can be adsorbed onto $CaCO_3$ crystals, associated with precipitation in the form of calcium humates. This is explained by the affinity of certain functional groups (particularly carboxylic acids) for Ca^{2+} ions. But it is mostly $Mg(OH)_2$ which is the most effective because of its presentation in the form of a hydrated solid with a high exchange surface, an amorphous structure and a positive charge. $Mg(OH)_2$ acts as a coagulant, making it a very good adsorbent for high molecular weight organic substances.

	Raw water		**Actiflo® softening outlet**	
	TOC ($mg \cdot L^{-1}$)	**Abs UV254 nm (m^{-1})**	**TOC ($mg \cdot L^{-1}$)**	**Abs UV254 nm (m^{-1})**
Mean	4.5	12.8	3.1	9.5
Max	5	14.7	3.5	10.5
Min	3.6	10.3	2.3	5.4
95%	4.8	14.3	3.3	10.3

Table 15.13. *Removal of TOC with Actiflo® softening*

Calcium carbonate precipitation typically removes 10–30% of color.

Magnesium hydroxide precipitation typically removes 30–60% of TOC and disinfection by-product (DPB) precursors, and 50–80% of color.

Adding iron in the form of ferric chloride or sulfate typically removes an additional 5–15% of color, TOC and disinfection by-products from calcium or magnesium precipitation.

Alum hydroxide is an amphoteric hydroxide. That is, it is soluble either at low or at high pH values. At the pH values normally found during water softening with lime and soda, it dissolves, thereby becoming inefficient to improve the removal of NOM.

15.7.2.2.1. $CaCO_3$ sludge

Microscopic observations (Figure 15.17) of the mixture represented by the $CaCO_3$ precipitate and hydroxide sludge: colloids show a clear distinction between microsand and carbonate grains. The size of the microsand does not increase, which means that there is no progressive coating of these grains with $CaCO_3$ as obtained during catalytic water softening. This also conveys that, when the water softening reaction has been completed during the first treatment stage, there is no longer any risk of subsequent precipitation. The particle sizes produced at the hydrocyclone's outlet show that there is a separation between the microsand (underflow) recycled in the flocculation tank and $CaCO_3$ (overflow), part of which is recycled in the water softening tank (first treatment stage).

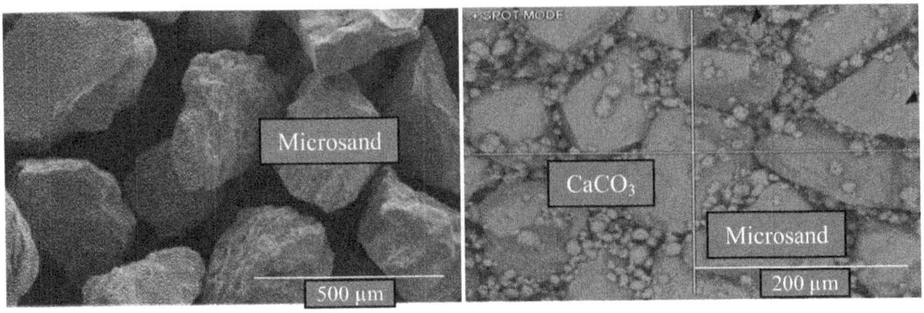

a) b)

Figure 15.17. *(a) Microsand and (b) $CaCO_3$ precipitate. For a color version of this figure, see www.iste.co.uk/gaid/watertreatment3.zip*

The $CaCO_3$ precipitated grains have a mean size of around 18 µm (between 4 and 40 µm), whereas the microsand grains have a size which is around 100 µm. The sludge extracted from $CaCO_3$ (without microsand) has high settling velocities (around 20 $m{\cdot}h^{-1}$) and its concentration after thickening is between 200 and 400 $g{\cdot}L^{-1}$.

Actiflo® softening combines water softening and turbidity removal within the same structure. With more than 1,000 references, this process benefits from Veolia's know-how in terms of ballasted settling, associated with incomparable expertise in softening chemistry. This process is the ideal tool for clarifying "hard" surface water or groundwater requiring softening before sand or membrane filtration. It improves the following physicochemical parameters:

– reduction of network scaling problems;

– decrease in chlorine dosages and risk attenuation of bacterial revival within networks;

– maintenance cost reduction for flow regulation devices and water quality measurement instruments;

– removal of manganese and many other elements (arsenic, silica, selenium, etc.);

– reduction of thermal losses in heat exchangers;

– risk attenuation of legionella development in hot water networks;

– removal of certain micropollutants due to the precipitation effect of carbonates;

– trapping of certain radionuclides, depending on pH and retention time;

– excellent clarification, leading to optimal filtration times;

– increase in UV transmittance enabling an easier disinfection downstream, with this type of technology;

– strong reduction in the need for individual softening systems for end consumers (e.g. dishwasher or washing machine protection).

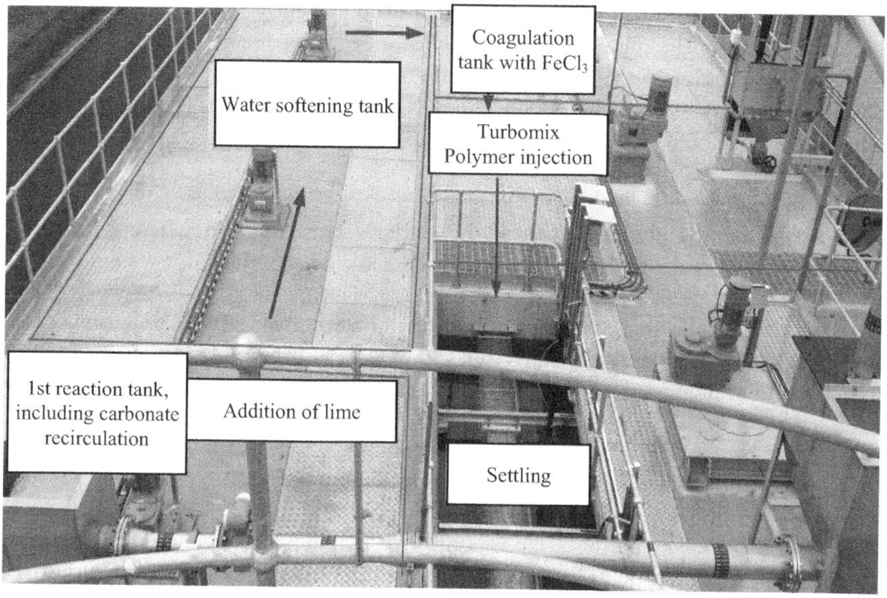

Figure 15.18. *Water circuit in an Actiflo® softening installation (Dublin, Ireland). For a color version of this figure, see www.iste.co.uk/gaid/watertreatment3.zip*

15.7.3. *Multiflo® softening*

Multiflo® softening is a water softening lamella settler derived from Multiflo®. It is made up of several compartments which change depending on raw water turbidity, temperature and hardness treatment goals. For high turbidity waters and whose temperature can be very low during a period of the year, Multiflo® softening includes (Figure 15.19):

– a first rapid mixing tank, mixing the raw water with the alkaline reagent;

– a second tank at high stirring speed (coagulation tank) whose mission is to neutralize the colloidal particles present in raw water. Ferric chloride coagulant is injected;

– a third tank, with a more moderate stirring speed (flocculation-maturation tank) where the calcium carbonate particles formed agglomerate into flocs. This tank is equipped with a flow guide tube called Turbomix™ which prevents hydraulic short-circuits and ensures good flocculation. A natural or synthetic polymer is injected;

– a settling tank: the softened and flocculated water is introduced into the Multiflo® clarification zone, under lamella plates inclined at 60°. The flocculated water is distributed beneath the lamella, along the entire settler's width. It circulates bottom-up, filling the inter-lamellar spaces co-currently and countercurrently. Particles settle on the plates and slide downwards, under the effect of gravity. The uniform distribution of the clarified water flow along the settler's entire surface is ensured by recovery troughs which collect the clarified water. These troughs are equipped with notched weirs granting an appropriate flow distribution within the structure.

Water softening sludge settles at the settler's bottom, from where it is partly pumped and recirculated to the water softening tank, or sent to sludge treatment.

The creation of a sludge bed by sludge recirculation makes it possible to increase the adhesion supports and to densify flocs, therefore improving sludge decantability. The formation of calcium carbonate is the result of the following consecutive steps:

> Supersaturation condition [ionic product > solubility product] followed by nucleation [formation of a calcium carbonate nucleus]

Calcium carbonate particle size is controlled by sludge recirculation. The size of calcium carbonate particles is controlled by sludge recirculation, which improves precipitation kinetics and crystal growth. This is because the recirculation of solids improves precipitation kinetics and crystals growth. The size of the reaction tank and the carbonate recirculation ratios are designed and controlled to provide sufficient hydraulic retention time and crystal surface area to induce an optimized precipitation reaction and to prevent carbonate deposition within the system. In addition, the advantage of recirculation is that it minimizes fluctuations in the quality of influent water as well as post-precipitation effects. It permanently enriches the reaction medium with crystals and its entire mass is engaged in the water softening process.

Water softening systems by Veolia have obtained solids recirculation ratios of up to 20:1 (solids recirculated/solids formed), depending on influent water characteristics. A newly formed calcium carbonate precipitate can be recirculated within the system for up to 18 h. Although particle size is not controlled directly, the long particle retention time in the system ensures optimum crystal size, resulting in faster sedimentation rates and easier dehydration.

For example, the precipitation of calcium sulfate in Multiflo® showed that, without sludge recirculation, the mean $CaSO_4$ particle size is 10 µm. With recirculation, it rises to 80 µm.

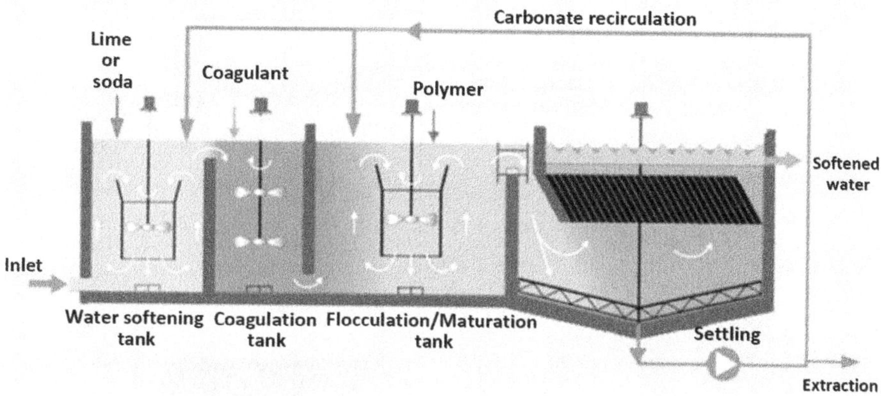

Figure 15.19. *Schematic diagram of the Multiflo® softening process. For a color version of this figure, see www.iste.co.uk/gaid/watertreatment3.zip*

For low turbidities (<5 NTU) or groundwater, the Multiflo® softening only comprises two tanks (Figure 15.20), with and without Turbomix™ in the water softening tank, depending on the carbonate concentration to be removed and the treatment goals.

Multiflo® softening thus enables:

– a highly efficient carbonate settling thanks to sludge recirculation and controlled flocculation;

– constant efficiency, even in the event of raw water quality variations;

– sewage sludge thickening with scraper;

– ease of operation.

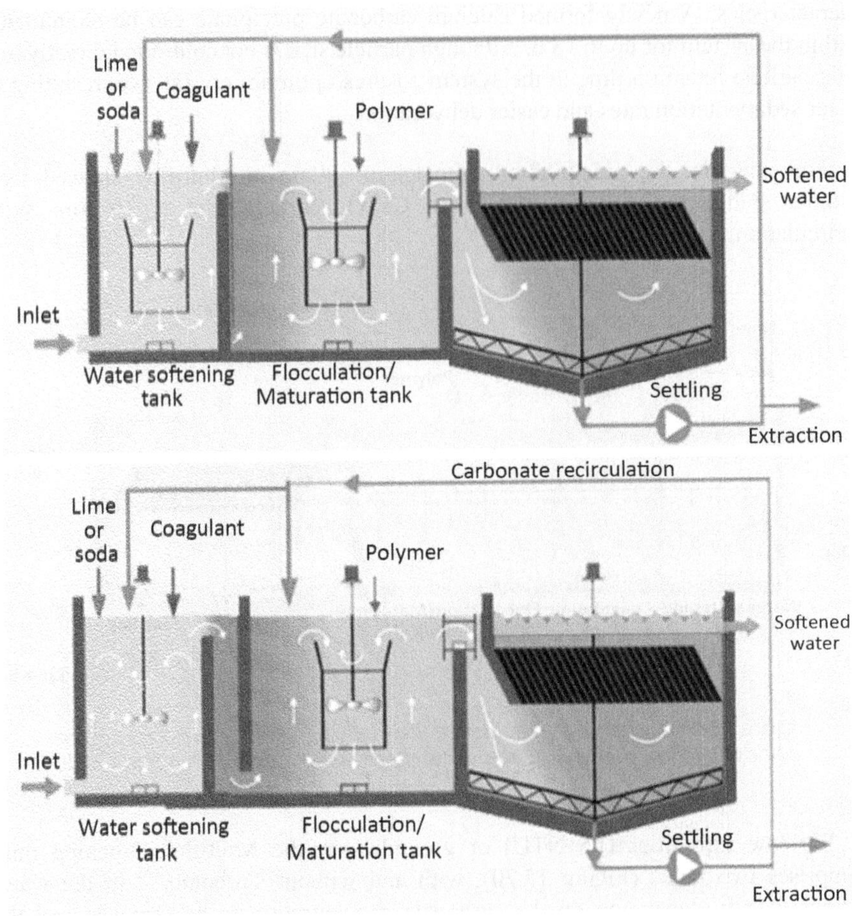

Figure 15.20. *Multiflo® softening in two tanks with and without Turbomix™. For a color version of this figure, see www.iste.co.uk/gaid/watertreatment3.zip*

The contact times in the reaction tank vary between 3 and 5 min, and those in the flocculation-maturation tank between 15 and 20 min. The velocity at the mirror in the clarification zone is 20–25 $m \cdot h^{-1}$. The reagents injected include:

– lime or caustic soda: water softening reagents;

– the rates are calculated directly by an automaton, depending on the hardness to be removed;

– ferric chloride: coagulant. The coagulant rate is supervised by the operator. The automaton gives the instruction to the dosing pumps;

– polymer: flocculation adjuvant. Automatic preparation and injection. The polymer rate is supervised by the operator. With the known polymer concentration, the automaton gives the instruction to the dosing pumps.

Figure 15.21 illustrates an example of performances observed at a plant in France: surface water softening with whitewash on a Multiflo® softening device. The temperature fluctuated between 5 and 10°C. Raw water turbidity was between 3 and 135 NTU before the operation, whereas the Multiflo®'s outlet water varied between 1 and 5 NTU.

In addition, a DOC reduction between 15 and 27% was observed. The recirculated sludge concentration was between 250 and 300 $g \cdot L^{-1}$. The lime/alkalinity ratio ($CaCO_3$) was 1–1.2 mg per $mg \cdot L^{-1}$ $CaCO_3$, whereas the stoichiometric ratio was 0.74 mg per $mg \cdot L^{-1}$ $CaCO_3$.

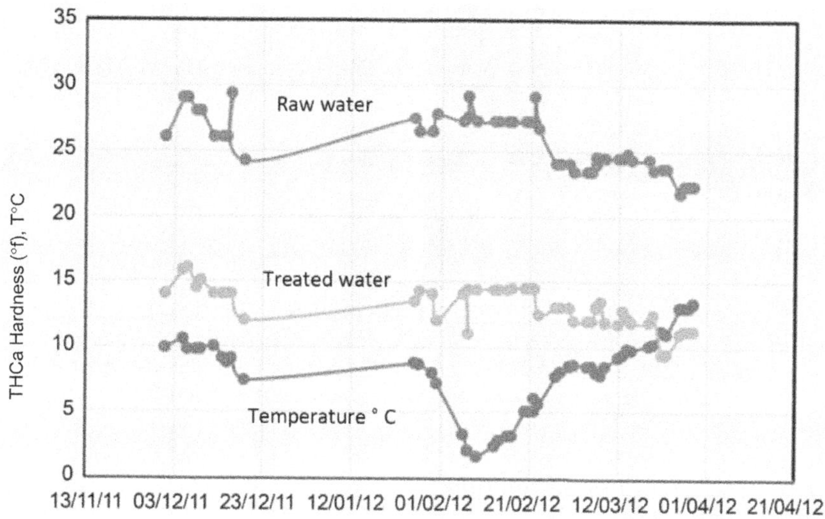

Figure 15.21. *Removal of surface water hardness with whitewash on a Multiflo® softening device (1°f = 10 $mg \cdot L^{-1}$ $CaCO_3$). For a color version of this figure, see www.iste.co.uk/gaid/watertreatment3.zip*

In Veolia processes, the control of operating parameters such as pH, chemical dosages, sludge recirculation rate, process performances makes it possible to obtain calcium hardness and alkalinity concentrations at desired levels.

Figure 15.22. *Some Multiflo® softening references. For a color version of this figure, see www.iste.co.uk/gaid/watertreatment3.zip*

15.7.4. *Catalytic water softening*

Another chemical water softening method provokes the reaction on a support acting as a catalyst. This is known as catalytic water softening.

15.7.4.1. *Calcium carbonate nucleation on a support: underlying principle*

Crystallization can be classified into two processes: nucleation and growth. Nucleation refers to crystal generation or crystal birth. Growth is the process whereby ions are transported to the crystal surface. In the presence of low crystal concentrations, nucleation and growth take place simultaneously. This is called spontaneous homogeneous crystallization or precipitation. In the presence of high crystal concentrations, growth is the dominant process. This is called heterogeneous nucleation.

The crystallization of $CaCO_3$ is mainly determined by the oversaturation degree. According to thermodynamic considerations, it appears that the Gibbs free energy proper to the precipitation of $CaCO_3$ is directly proportional to the saturation index (Langelier index).

In theory, a precipitation reaction is possible if the LSI is positive. In practice, an activation energy is required to form a nucleus, and the precipitation reaction only occurs above a minimum LSI value.

Spontaneous nucleation of $CaCO_3$ occurs with an LSI of 1.7 at 20°C, whereas in the presence of already precipitated $CaCO_3$ crystals (heterogeneous nucleation) onto a support, the LSI is lower than 0.2.

This explains the swift kinetics of water softening on a catalytic support. In general, $CaCO_3$ precipitation processes take a variable time between 15 min and 2 h. The heterogeneous nucleation mechanism only requires less than 5 min, in particular due to a high local pH, which promotes precipitation and nucleation.

The precipitation reaction takes place in a reactor in contact with a sand bed fluidized by an ascending water current. The sand grains are permanently expanding and offer all the specific surface for the adhesion of the calcium carbonate precipitate. It is therefore important for fluidization to reach 100% in order to prevent sand sedimentation at the bottom of the reactor, which would generate irreversible scaling at the structure's level.

The sand grain acts as a catalyst or crystallization seed, which is coated with a calcium carbonate precipitate obtained by the localized and rapid increase in the water's pH.

For the pellet softening process the maximum fluidization velocity is of interest. When the velocity is too high, the finest sand will be suspended instead of fluidized and will leave the column. In practice the limit for the finest sand depends on the expansion of the fine sand in the top of the column and the amount of the finest grain.

With a velocity of 100 $m{\cdot}h^{-1}$, normal quartz sand with a grain size of 0.3 mm will be expanded 4 times and sand with a grain size less than approx. 0.25 mm will be suspended and washed out.

The chemicals introduced into the column will cause building up a coating on the granulate and other inert material inside the column. If the chemicals are introduced into the raw water or the nozzles, there will be a scaling in the pipes and nozzles, blocking the nozzles for a short while.

$CaCO_3$ precipitation kinetics and water softening are characterized by the quantity of supersaturated $CaCO_3$ (the quantity of $CaCO_3$ which can be precipitated until an equilibrium is obtained, in other words, until LSI = 0). The crystallization reaction follows a first-order kinetic law:

$$\frac{dC}{dt} = -k\,A\,(C\,\mathrm{CaCO_3} - \frac{Ks}{f^8})$$

where:

- C_{CaCO3}: quantity of supersaturated $CaCO_3$ over time t;
- k: reaction constant ($m^4 \cdot mol^{-1} \cdot s^{-1}$); $k = 0.0255 \cdot 1.053^{(T-20)}$;
- A: specific material's area (6 (1 – Ɛ)/dp) in m^{-1};
- Ɛ: porosity;
- dp: pellet diameter (m);
- Ks: solubility product of $CaCO_3$ ($mol^2 \cdot m^{-6}$);
- f: activity factor related to ionic strength I.

$$\log f = -0.0002I$$

For calcium carbonate, the ionic strength is equal to 0.184 $mole \cdot m^{-3}$.

The crystallization of calcium carbonate on the support (sand) generates a new chemical equilibrium and the carbonates, bicarbonates and carbon dioxide concentrations are modified. Therefore, pH, ionic strength and activity factor change. The calcium and carbonate concentrations, as well as the activity coefficient, have an influence on crystallization kinetics. Supersaturation is highest in the lower portion of the reactor and most of the calcium is removed in this area.

The basic equations are:

$$CO_2 + 2\,H_2O \rightarrow H_3O^+ + HCO_3^-$$

$$K1 = \frac{(H_3O+)(HCO_3-)}{(CO_2)};\; K'1 = \frac{(H_3O+)(HCO_3-)}{(CO_2)fm^2} = \frac{K1}{fm^2}$$

The molar concentrations can be used if the ionic force f is included for each product. In this case, the constants K_1, K_2, K_w are replaced by K'_1, K'_2 and K'_w

$$HCO_3^- + H_2O \rightarrow H_3O^+ + CO_3^{2-}$$

$$K2 = \frac{(H_3O+)(CO_3{}^{2-})}{(HCO_3-)};\ K'2 = \frac{(H_3O+)(CO_3{}^{2-})}{(HCO_3-)fd} = \frac{K2}{fd}$$

$$H_3O^+ + OH^- \rightarrow 2\ H_2O$$

$$Kw = (OH-)(H_3O^+);\ K'w = \frac{(OH-)(H_3O^+)}{fm^2} = \frac{Kw}{fm^2}$$

K_1, K_2 and K_w are the constants for a given temperature and a given concentration, and K'_1, K'_2 and K'_w are the constants for a given temperature and a given ionic strength (ionic strength is a function of the concentrations and valences of the ions present in the solution).

If the solution is saturated with calcium carbonate, the application of the law of mass action to the heterogeneous equilibrium:

$$CO_3^{2-} + Ca^{2+} \rightarrow CaCO_3\ Ks = (CO_3^{2-})(Ca^{2+})$$

$$K's = \frac{(CO_3^{2-})(Ca^{2+})}{fd^2}$$

$$K's = \frac{Ks}{fd^2}$$

where:

– Ks: $CaCO_3$ solubility product;

– fm and fd: activity coefficients of monovalent (m) and divalent (d) ions.

$$\text{Alkalinity} = 0.5\,[HCO_3^-] + [CO_3^{2-}] + 0.5[OH^-] - 0.5[H_3O^+]$$

$$pH = -\log\{fm[H_3O+]\}$$

The constants K’1, K’2, K’w and K’s are commonly expressed by their decimal cologarithms, respectively, written as pK’1, pK’2, pK’w and pK’s.

Their values can be calculated for temperatures between 0 and 80°C:

$$pK'1 = 6.583 - 646\cdot10^{-9}\,T3 + 163.5\cdot10^{-6}\,T^2 - 12.3\cdot10^{-3}\,T$$

$$pK'2 = 10.627 - 328 \cdot 10^{-9}\, T^3 + 135.3 \cdot 10^{-6}\, T^2 - 15.04 \cdot 10^{-3}\, T$$

$$pK'w = 14.940 - 1.49 \cdot 10^{-9}\, T^3 + 270 \cdot 10^{-6}\, T^2 - 43.44 \cdot 10^{-3}\, T$$

$$pK's = 8.022 + 444 \cdot 10^{-9}\, T^3 - 61 \cdot 10^{-6}\, T^2 + 14 \cdot 10^{-3}\, T$$

The potential for the formation of calcium carbonate precipitation or its dissolution is described by the LSI, defined by the following equation:

$$LSI = pH - pHs$$

The pH is known from laboratory or field measurements. It can also be calculated from the equation:

$$pH = -pK2 - \log (CO_3^{2-}) + \log (HCO_3^-)$$

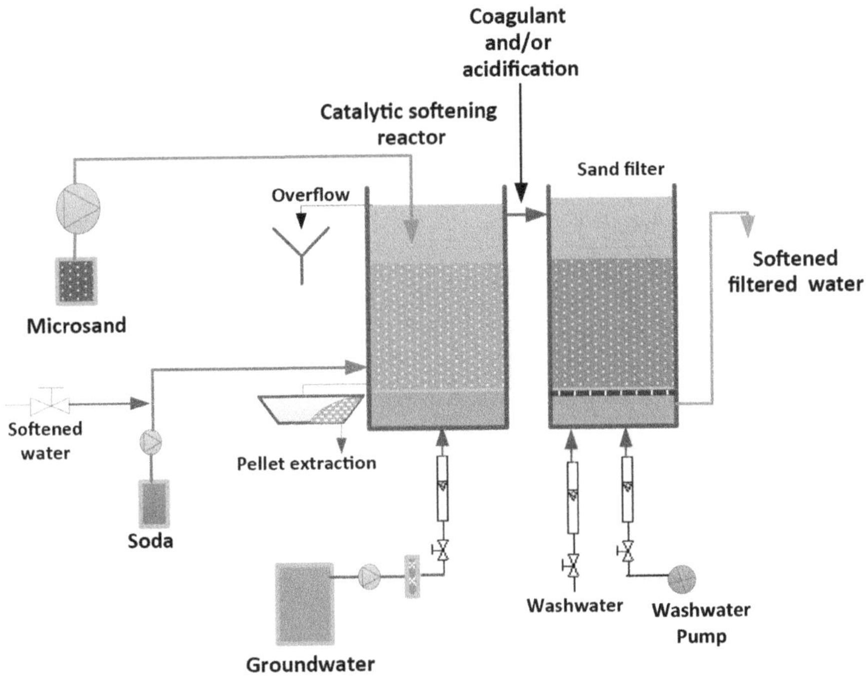

Figure 15.23. *Schematic diagram of catalytic softening process. For a color version of this figure, see www.iste.co.uk/gaid/watertreatment3.zip*

The grain thickens and forms 1–2 mm diameter carbonate pellets. The pellets thus formed are regularly removed from the reactor's base, as soon as they reach optimum size (Figure 15.23). To keep the quantity of sand constant, the reactor is replenished with sand after each removal. The height of a column is depending on the required retention time in the column, the vertical hydraulic velocity of the water and the unity of the distribution of the vertical flow. The retention time is depending on the chemical reaction time, the concentration of the reactants and the specific surface area of the fluidized granulate. The specific surface area is again depending on granulate diameter, specific weight of the granulate and again the vertical velocity (the expansion of the granulate).

Water is injected at the reactor's base by means of a diffusion floor, which ensures a homogeneous flow distribution. Bed expansion is achieved by maintaining a high fluidization velocity (80–100 $m \cdot h^{-1}$).

Due to the species solubility, the limit of water softening is 15 $mg \cdot L^{-1}$ $CaCO_3$.

Calcium salts	**Formulae**	**Solubilities at 15°C**
Calcium bicarbonate	$Ca(HCO_3)_2$	87,000 $mg \cdot L^{-1}$
Calcium carbonate	$CaCO_3$	15 $mg \cdot L^{-1}$
Calcium chloride	$CaCl_2$	745,000 $mg \cdot L^{-1}$
Lime	$Ca(OH)_2$	1,740 $mg \cdot L^{-1}$
Calcium sulfate	$CaSO_4$	2,000 $mg \cdot L^{-1}$
Magnesium carbonate	$Mg\ CO_3$	110 $mg \cdot L^{-1}$
Sodium carbonate	Na_2CO_3	215 $mg \cdot L^{-1}$
Magnesium hydroxide	$Mg(OH)_2$	8.4 $mg \cdot L^{-1}$

Table 15.14. *Solubilities of various salts at 15°C*

15.7.4.1.1. Process limitations

The sodium concentration of the treated water must be lower than 200 $mg \cdot L^{-1}$, so particular attention should be given to the sodium concentration in raw water during water softening with soda. A low sodium content is required for water softening with soda or sodium carbonate. If water softening with soda is rendered impossible, considering the initial sodium or bicarbonate content, we will turn to water softening with lime. The pH of softened water must be less than 8.25 in order to avoid the risk of scaling and to guarantee chlorine disinfection efficient. Beyond that, acidification must be systematically scheduled.

Feed water characteristics recommended:

– turbidity: <5 NTU;

– apparent color (related to iron): < 5 $mg \cdot L^{-1}$ Pt/Co;

– dissolved Fe: <2 $mg \cdot L^{-1}$ (standard = 0.2 $mg \cdot L^{-1}$);

– Mn: <0.5 $mg \cdot L^{-1}$ (standard = 0.05 $mg \cdot L^{-1}$);

– PO_4: <0.5 $mg \cdot L^{-1}$;

– organic matter: <20 $mg \cdot L^{-1}$ $KMnO_4$;

– pesticides: no removal by catalytic softening;

– T °C: lime: >8°C and < 25°C and ; soda: >1°C and <25°C.

In the case of surface water treatment, catalytic water softening is preceded by coagulation–flocculation–settling or dual media filtration with anthracite sand in order to reduce turbidity from the very beginning (inlet turbidity < 5 NTU).

When raw water turbidity reaches exceptional peaks of 15–20 NTU for a few days per year, catalytic softening can be fed directly from raw water. The addition of a coagulant (WAC, $FeCl_3$) can be injected upstream of the filters placed after the catalytic softening stage in order to trap this turbidity.

It is the dissolved iron concentration that is important. There is no limit for (already oxidized) particulate iron, which is easily retained on sand grains. When the iron concentration in raw water exceeds 2.0 $mg \cdot L^{-1}$ in Fe, its removal is recommended before water softening.

The strongly basic pH within the reactor favors the precipitation of a large fraction (>90%) of iron and manganese in the carbonate pellets. The iron thus precipitated colors the pellets with a reddish-brown color. The use of these pellets for soil enrichment should therefore be avoided, because it leads to a redissolution of iron. This iron can contaminate groundwater.

An iron reduction over 95% for iron concentrations of 5 $mg \cdot L^{-1}$ was observed on some installations. However, this reduction percentage is not constant. For this reason, one may only accept dissolved Fe concentrations of 5 $mg \cdot L^{-1}$ when they occur episodically (just a few times a month, for a few hours).

Catalytic softening is followed by filtration (possibly by adhesion), so as to remove any trace of turbidity in raw water. A slight acidification of water before filtration is required to solubilize calcium carbonate (either in supersaturation or entrained with water). It is also necessary to stabilize water's calco-carbonic equilibrium, so as to prevent filter scaling.

There are no constraints as far as the ammonia is concerned, which can be removed after catalytic softening by a biological process in the filters or by breakpoint chlorination.

15.7.4.2. *Velocity within the reactor*

For the design of a specific reactor for pellet softening, it is necessary to know:

1) the chemical process and feed water quality;

2) the amount of base dosing in the softening reactor and the recommended pH level in the process;

3) effluent limits;

4) the retention time in the reactor;

5) vertical velocity of the water in the reactor;

6) max. and min. expansion acceptable in the reactor;

7) particle size and specific weight.

The Actina® reactor is designed for a fluidization velocity between 80 (with lime) and 100 $m{\cdot}h^{-1}$ (with soda). In the event of a significant flow decrease, the recycling of softened water is provided for in order to maintain a minimum fluidization velocity of 50 $m{\cdot}h^{-1}$.

15.7.4.3. *Static material's height (in the case of sand)*

The greater the number of active sites of sand available in the catalytic softeners, the closer the residual concentration of $CaCO_3$ gets to $CaCO_3$ solubility, provided that the soda or lime dosages are sufficient. A very high removal of TH is thus possible. This can be sought for industrial process water or for drinking water, when the TH of raw water is above 800 $mg{\cdot}L^{-1}$ $CaCO_3$.

Static sand height	**Sand in expansion height**	**Raw water TH $mg{\cdot}L^{-1}$ $CaCO_3$**	**Final water TH $mg{\cdot}L^{-1}$ $CaCO_3$**
1.5 m	3 m	700	80–150
2 m	4 m	900	80–150
3 m	6 m	1,000	30

Table 15.15. *Sand height calculation depending upon the hardness to be removed*

However, for drinking water, the final TH must be between 80 and 150 $mg \cdot L^{-1}$ $CaCO_3$. It is therefore advisable to choose a static bed height matching the desired final TH.

Attaining the treated water's final hardness depends on the reagent used (lime or soda) and the alkalinity of raw water. The reactor's total height acknowledges the height of the expanding material, to which is added the water's tranquillizing zone before evacuation and the height below the floor. Therefore, the reactor's total is between 6 and 8 m.

15.7.4.4. *Reactor hydraulics*

The hydraulics of reactors is identical to that of filters during the backwash phase. While in a sand filter, the material's expansion is in the range of 20%, it rises to at least 100% in reactors. Depending on the support material and its density, expansion can reach 200%. Resistance is indicated by the weight of pellets in the water:

$$\Delta H\ max = (1 - \mathcal{E})\ H\ \left(\frac{\rho b - \rho w}{\rho w}\right)$$

where:

– ΔH max: maximum resistance (m);

– ρw: water density ($kg \cdot m^{-3}$);

– ρb: pellet density ($kg \cdot m^{-3}$);

- $\mathcal{E}$: porosity;

– h: material's height.

The bed expansion is given by:

$$\frac{hexp}{ho} = \frac{1 - \mathcal{E}o}{1 - \mathcal{E}exp}$$

ho and hexp are the initial and expanding material height (m), respectively. $\mathcal{E}o$ and $\mathcal{E}exp$ are the initial and expanded bed porosity.

The porosity of the expanded bed is given by the ratio:

$$\frac{\mathcal{E}exp^3}{(1 - \mathcal{E}exp)^{0.8}} = 130\ \frac{\eta^{0.8}}{g}\ \frac{\rho w}{\rho b - \rho w}\ \frac{v^{1.2}}{db^{1,8}}$$

where:

– v: flow rate ($m \cdot s^{-1}$);

– db: pellet diameter (m);

– g: acceleration due to gravity ($m^2 \cdot s^{-1}$);

– η: water kinematic viscosity ($kg \cdot m^{-2} \cdot s^{-1}$).

Minimum velocity to fluidize the material :

$$v_{min}^{1.2} = \epsilon_{exp}^{3}/(1-\epsilon_{exp}) / (\rho_w/(\rho_b-\rho_w)) / (\eta^{0.8}/d^{1.8}) \times (g/130)$$

Maximum fluidization speed, just before suspension :

$$v_{max} = 1/10/\eta^{0.6} \times ((g) \times (\rho_w/(\rho_b-\rho_w)))^{0.8} \times d^{1.4}$$

If the velocity in the reactor is too low (velocities 1 – 2 × v_{min}), upright vertical upright whirls will be created, and the fluidization will be unstable. For pellet softening and crystallization processes a maximum vertical velocity of 100 m/h is recommended.

The expansion height Hexp is then inferred therefrom.

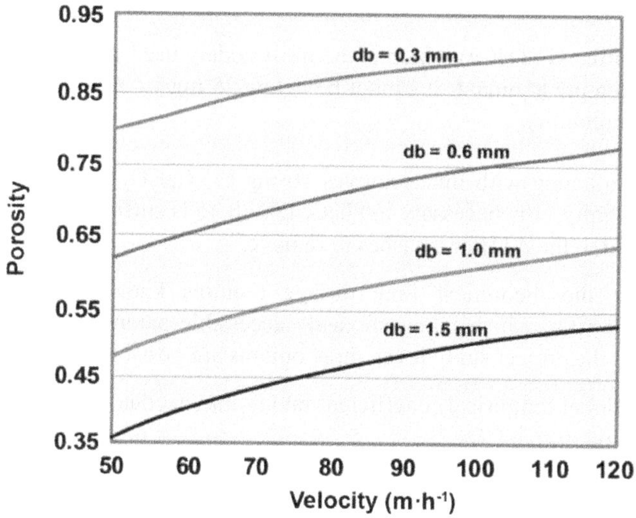

Figure 15.24. *Evolution of the fluidized bed porosity as a function of velocity and pellet diameter. For a color version of this figure, see www.iste.co.uk/gaid/watertreatment3.zip*

15.7.4.5. *Design parameters for catalytic water softening with lime*

Three design parameters have to be considered: the CO_2 concentration, the raw water calcium hardness and the desired final water calcium hardness.

15.7.4.5.1. Determination of the quantity of lime

In terms of chemical reactions, the theoretical lime treatment rate to be implemented depends on the following:

– On the one hand, the difference between the raw water's TH and the treated water's intended TH:

$$[Ca(OH)_2] = 74 \times [\Delta TH\ (mg{\cdot}L^{-1}\ CaCO_3)]\ (g{\cdot}m^{-3})$$

where TH: total hardness.

– On the other hand, the neutralization of free CO_2 from raw water and the removal of hardness formed during this reaction:

$$[Ca(OH)_2] = 2^{*}\ 74/88 \times [\Delta CO_2]\ (g{\cdot}m^{-3})$$

The pure lime treatment rate is then the following:

$$[Ca\ (OH)_2] = 74\ [\Delta TH]\ mg{\cdot}L^{-1}\ CaCO_3 + 2^{*}\ 74/88\ [\Delta CO_2]\ (g{\cdot}m^{-3})$$

Calcium hardness (THCa) which is the most widely used parameter insofar as it is the most often predominant. It cannot be below 25 $mg{\cdot}L^{-1}$ $CaCO_3$ after treatment due to $CaCO_3$ solubility.

As water softening with lime removes 10 $mg{\cdot}L^{-1}$ $CaCO_3$ in TH per 10 $mg{\cdot}L^{-1}$ $CaCO_3$ in alkalinity, it is necessary to check that there is sufficient alkalinity in raw water compared to the ΔTH one wishes to remove.

Calculating the theoretical lime dosage requires knowing the quantity of neutralized free CO_2, which is not an easily accessible parameter. To estimate the lime content at the project study level, three options are possible:

– using a global (empirical) coefficient taking into account the TH decrease and the neutralization of free CO_2:

$$[Ca(OH)_2] = 0.9\text{–}1.2 \text{ mg of lime for 1 mg } CaCO_3 \text{ to be removed;}$$

– in the calculation formula, using the free CO_2 from the raw water, assuming it will be completely neutralized. In this case, the lime rate will be overestimated;

– using a computer calculation tool integrating the calco-carbonic equilibrium's chemical formulas (which will provide information as to the quantity of neutralized CO_2).

In case of lacking information about the free CO_2 in raw water, the following calculation can be made. Consider water whose physicochemical composition is as follows:

pH 7.5,

T °C: 20°C,

Calcium hardness (THCa): 210.5 mg·L^{-1} $CaCO_3$,

alkalinity: 180 mg·L^{-1} $CaCO_3$.

The treatment's goal is to achieve a THCa of 100 mg·L^{-1} $CaCO_3$.

Using a calculation tool, we obtain a free CO_2 of 10.1 mg·L^{-1}.

The quantity of lime is therefore:

0.74 (210.5 – 100) + 2* 10.1*74/88 = 98.75 mg·L^{-1} pure $Ca(OH)_2$.

Without access to the free CO_2 value by a tool, a quick calculation solution would be the following:

– input alkalinity: 180 mg·L^{-1} $CaCO_3$ = 3.60 meq·L^{-1} = 3.60 mM;

– non-carbonate hardness: THCa – Alkalinity = 210.5 – 180 = 30.5 mg·L^{-1} $CaCO_3$;

– carbonate hardness: 180 mg·L^{-1} = 3.6 meq·L^{-1} = 1.80 mM;

– carbonate hardness to be removed: 210.5 – 100 mg·L^{-1} = 110.5 mg·L^{-1} $CaCO_3$;

– the alkalinity to be removed is then equivalent to = 110.5 mg·L^{-1} $CaCO_3$ = 2.21 meq·L^{-1} = 2.21 mM.

The H_2CO_3 concentration is:

$$K1 = \frac{[HCO_3^-][H+]}{[H_2CO_3]}$$

$$[H_2CO_3] = \frac{[HCO_3^-][H^+]}{K1}$$

pH = 7.5, that is, $[H+] = 10^{-7.5}$

K'1 at 20°C: $10^{-6.38}$

We then have:

$$[H_2CO_3] = \frac{[3.60 \cdot 0.001][10^{-7.5}]}{10^{-6.38}} = 0.000273 \text{ M} = 0.273 \text{ mM}$$

Pure $Ca(OH)_2$ lime rate:

$(0.273 \text{ mM} + 2.21 \text{ mM}/2)\ 74 = 101.9 \text{ mg} \cdot L^{-1}$

This slight difference is derived from the value of constant K'1, which is more accurate with the calculation tool in better control of the temperature effect.

The quantity of commercial lime is then equal to:

Commercial lime $(g \cdot m^{-3}) = [Ca(OH)_2] \times 100/\ LP$

where LP is the lime purity percentage.

Quantity of lime $[kg \cdot J^{-1}]$ = quantity of commercial lime $(g \cdot m^{-3}) \times Q\ (m^3 \cdot d^{-1}) \times 10^{-3}$

and:

Required lime volume $(m^3 \cdot d^{-1})$ = quantity of lime $[kg \cdot d^{-1}]$/apparent density

Depending on quality and grain size, the apparent density of lime is between 300 and 450 $kg \cdot m^{-3}$.

For example:

– flow to be treated = 500 $m^3 \cdot h^{-1}$;

– operation = 20 h/24 h;

– raw water calcium hardness (TH) = 420 $mg \cdot L^{-1}$ $CaCO_3$;

– final calcium hardness (TH) = 200 $mg \cdot L^{-1}$ $CaCO_3$;

– neutralized CO_2 concentration = 22 $mg \cdot L^{-1}$;

– purity of commercial lime = 95%;

– density = 350 $kg \cdot m^{-3}$;

– quantity of lime required $[Ca(OH)_2] = 0.74\ (420 - 200) + 74/88\ (22) = 181.3\ g \cdot m^{-3}$;

– Q to be injected = 181.3 × 100/95 = 190.08 $g \cdot m^{-3}$ and 190.08 × 500 × 20 × 10^{-3} = 1,900.8 $kg \cdot d^{-1}$;

– daily lime volume = 1,900.8/350 = 5.43 $m^3 \cdot d^{-1}$;

– lime volume to be stored for 1 week (7 days) = 5.43 × 7 = 38.01 m^3;

– volume to be stored for 1 month (30 d) = 5.43 × 30 = 162.9 m^3.

15.7.4.5.2. Lime milk

The use of micronized (5 μm) lime milk is strongly recommended, or eventually, a well-prepared lime milk, with a particle size between 5 and 10 μm. In point of fact, too high a particle size (>20 μm) does not have enough time to react within the reactor, due to the short contact times. As a result, it is entrained with the outlet water. Turbidities greater than 50 NTU are then obtained.

When using lime, the rise in pH is weaker than with soda, and the water then requires a longer contact time. This is why the contact time corresponding to a velocity of 80 $m \cdot h^{-1}$ is 2.25 min for an expanded bed height of 3 m.

Catalytic softening works properly only if the magnesium concentration is not very high in the raw water. Magnesium hardness (THMg) must be < calcium hardness (THCa)/3. Otherwise, the quickly formed magnesia clogs the filters (it is not retained by the sand grains or by the pellets).

Lime milk can be prepared either from quicklime or from hydrated lime. Quicklime is a white-colored fine dry powder, essentially composed of CaO (calcium oxide).

Hydrated lime is obtained from dry slaking quicklime.

The small-diameter whitewash particles are more reactive and provide improved water softening performances under the high-speed operating conditions determined by the reactor.

The extinction temperature in the reactor is important for the reactivity of produced lime milk. Although the ideal temperature is 80°C, in practice, it is between 50 and 70°C.

15.7.4.5.3. Pellet production

The quantity of carbonates formed during water softening with lime is:

– quantity of carbonates formed = 2 mg per mg $CaCO_3$, to which is added the quantity of $CaCO_3$ formed during the formation of $Ca(HCO_3)_2$ from CO_2

neutralization by lime, that is to say: 1.1 mg $CaCO_3$ formed·mg^{-1} CO_2 neutralized. When mature, the pellets contain 95–98% carbonates. As a safety measure (presence of iron, etc.), a value of 90% is often considered;

– pellet production (kg·d^{-1}) = quantity of carbonates formed (kg·d^{-1})/0.90. The apparent density of pellets is approximately 2,800 kg·m^{-3}. Maturing pellets have an optimal size of 0.8–1 mm;

– daily volume of pellets (m^3·d^{-1}) = pellet production (kg·d^{-1})/2,800 (kg·m^{-3}).

A larger pellet size results in:

– a decrease in the fluidized bed expansion rate;

– a decrease in the water softening cont surface;

– an increase in turbidity at the reactor's outlet;

– a decrease in the efficiency of water softening.

It is advisable not to exceed a pellet size of 1–1.2 mm. A classification of pellets takes place within the fluidized bed: the larger pellets are in the lower section of the water softener, whereas the smaller pellets are in its upper section.

Pellet color may vary depending on the nature of the raw water treated. They can be white without traces of impurities in the calcium carbonate, or acquire a brown-red tarnish when raw water has high levels of iron and manganese, which precipitate in the reactor.

For the same example as before, pellet production will be equal to:

– pellet production = 2.0 × (420 – 200) × 500 × 20 × 10^{-3}/0.9009 = 4,884 kg·d^{-1};

– daily volume = 4,884/1,600 = 3.05 m^3·d^{-1};

– weekly storage volume = 3.05 × 7 = 21.36 m^3;

– weekly storage volume = 3.05 × 30 = 1,602.3 m^3;

15.7.4.6. *Design parameters for catalytic water softening with caustic soda*

15.7.4.6.1. Quantity of caustic soda

The chemical reactions show that the theoretical caustic soda treatment rate to be implemented depends on the difference between raw water's TH and the intended TH for treated water:

$$[NaOH]\ g\cdot m^{-3} = 4 \times [\Delta TH\ (mg\cdot L^{-1}\ CaCO_3)] = 0.4\ [\Delta TH],\ (mg\cdot L^{-1}\ CaCO_3)$$

where ΔTH is the calcium hardness to be removed and the neutralization of free CO_2 from raw water:

$$[NaOH] = 40/44 \times [\Delta CO_2]\ (g\cdot m^{-3})$$

The pure caustic soda treatment rate is then the following:

$$[NaOH]\ g\cdot m^{-3} = 0.4 \times [\Delta TH]\ (mg\cdot L^{-1} CaCO_3) + 40/44 \times [\Delta CO_2\ g\cdot m^{-3}]$$

Calculating the theoretical lime dose requires knowing the quantity of neutralized free CO_2, which can be easily inferred knowing the pH, T °C and alkalinity of raw water.

A first approximation to CO_2 can be obtained using the ratio:

$$\text{free } CO_2\ mg\cdot L^{-1} = 10^{(\text{Log (Alkalinity/1.78)} + 0.2 - \text{pH} + 6.523 - 12.3*(0.001*T\ °C) + 163.5*((0.001*T\ °C)^2) - 646*((0.001*T\ °C)\text{^}3))}$$

Alkalinity is expressed in $mg\cdot L^{-1}$ $CaCO_3$. THCa cannot be below 25 $mg\cdot L^{-1}$ $CaCO_3$ due to $CaCO_3$ solubility.

As water softening with soda removes 2.0 mg $CaCO_3$ in TH for every 1.0 mg $CaCO_3$ in alkalinity, it is necessary to check that there is sufficient alkalinity in raw water compared to the ΔTH one wishes to remove.

The caustic soda supply increases the quantity of sodium in treated water. The quantity of sodium provided is:

$$[Na]\ (g\cdot m^{-3}) = [NaOH]\ (g\cdot m^{-3}) \times 23/40 \text{ (i.e. 57.5\% of the rate)}$$

For the same example used for lime consumption, we find with caustic soda as the reagent used for softening:

– flow to be treated = 500 $m^3\cdot h^{-1}$;

– operation = 20 h/24 h;

– raw water calcium hardness (TH) = 420 $mg\cdot L^{-1}$ $CaCO_3$;

– final calcium hardness (TH) = 200 $mg\cdot L^{-1}$ $CaCO_3$;

– neutralized CO_2 concentration = 22 $mg\cdot L^{-1}$;

– purity of commercial NaOH = 30%;

– quantity of pure NaOH = 0.4 (420 – 200) + 40/44 (22) = 108 $g\cdot m^{-3}$;

– commercial NaOH: 108 × 100/30 = 360 $g\cdot m^{-3}$.

Total soda consumption will be 108 $g \cdot m^{-3}$ in NaOH, that is, that means 62.1 $g \cdot m^{-3}$ in Na^+ will be added in the final water quality.

Soda solutions crystallize easily and their viscosity increases considerably when the concentration increases, or when the temperature decreases.

We preferably use sodium hydroxide at 30%, taking into account the risks of crystallization due to dilution (when its temperature is too low). Note that 50% sodium hydroxide (soda lye) crystallizes at 12°C. The density of 30% sodium hydroxide solution is 1.30–1.33, the solution containing approximately 400 $g \cdot L^{-1}$ of NaOH. The crystallization temperature of 30% NaOH is between –1 and +4°C, which, if necessary, involves a compulsory tracing of pipes up to the point of injection or dilution.

The use of caustic soda produces two times less calcium carbonate than lime. Since this reagent is liquid, it can be injected directly, without prior preparation. This is opposite to lime, which requires an additional whitewash preparation phase.

In the catalytic softener using caustic soda, the contact time is 1.8 min for a 100 $m \cdot h^{-1}$ flow rate and an expanded bed height of 3 m.

15.7.4.6.2. Pellet production

The quantity of carbonates formed during water softening with caustic soda is:

Quantity of carbonates formed = Q (0.9–1.1 kg/per kg $CaCO_3$ removed$\cdot m^{-3}$)
Q is the treated flow ($m^3 \cdot d^{-1}$)

When mature, the pellets contain 95–98% carbonates. As a safety measure (presence of iron, etc.), we recommend using a value of 90% for the calculations:

Pellet production ($kg \cdot d^{-1}$) = quantity of carbonates formed ($kg \cdot d^{-1}$)/0.90

The apparent density of pellets is approximately 2,800 $kg \cdot m^{-3}$. The optimal size of pellets at maturation is 0.8–1 mm:

Daily volume of pellets ($m^3 \cdot d^{-1}$) = pellet production ($kg \cdot d^{-1}$)/2,800 ($kg \cdot m^{-3}$).

The same remarks relating to the size of the pellets to be removed and to pellet classification when using lime are valid for soda.

15.7.4.6.3. Sand consumption

Sand consumption depends on the production of $CaCO_3$ and the size of the pellets removed. It is obtained from the ratios shown in Table 15.16, depending on the chemical reagent used.

Note that 1,500 (kg·m^{-3}) in Table 15.16 represent the apparent density of sand. This is useful for inferring the apparent volume of sand.

	Lime	Caustic soda
Sand % compared to $CaCO_3$ formed	6–10% of $CaCO_3$ production	4–6% of $CaCO_3$ production
Sand consumption (g/mg·L^{-1} $CaCO_3$·m^3)	0.20 g sand·m^{-3}·mg·L^{-1} $CaCO_3$	0.12 g·mg·L^{-1} $CaCO_3$ removed
Sand consumption (kg·d^{-1})	0.2 × [Δ TH] (mg·L^{-1} $CaCO_3$) × Q × 10^{-3}	0.12 × [Δ TH] (mg·L^{-1} $CaCO_3$) × Q × 10^{-3}
Apparent sand volume (m^{-3}·d^{-1})	Sand consumption (kg·d^{-1})/1,500 (kg·m^{-3})	Sand consumption (kg·d^{-1})/1,500 (kg·m^{-3})

Table 15.16. *Sand consumption for the chemical reagents lime and soda*

When loading sand for the first time (static height = 1.5 m), a sand with a higher grain size is used, so as to avoid sand loss during fluidization. The sand used has an effective size (ES) 0.4 mm (absence of mesh smaller than 0.35 mm due to 0.2–0.3 mm strainer slots) and a uniformity coefficient of 1.15. The sand actually used (after each pellet extraction) has an effective size of 0.34 mm (absence of mess smaller than 0.25 mm) and a uniformity coefficient of 1.3. Since the fine sand added is immediately fluidized in the reactor's upper section, there is no risk of it entering into the strainer slots. Its smaller particle size ensures a higher cont surface and faster kinetics.

Garnet can replace sand, which has the advantage of a finer grain size (d_{10}: 0.2 mm, d_{90}: 0.4 mm). This leads to a greater number of active sites than with sand, as well as a lower material consumption, for removing the same hardness.

Material quantities (and volumes) are lower with garnet than with sand. The initial height of garnet is 1.5–2 m.

The physicochemical characteristics of the two materials are shown in Table 15.17.

Strainers with specific slots must be provided in relation to the garnet's grain size.

Parameters	Sand	Garnet
Grain size (mm) d_{10}-d_{90}	0.4–0.6	0.2–0.4
Density kg·m^{-3}	1,600	2,380
Apparent density kg·m^{-3}	2,630	4,100
Mohs hardness	7.5	7.5
Particle volume (Ø sphere) mm^{-3}	0.0654 (0.5 mm)	0.0141 (0.3 mm)
Relative ratio of the 2 volumes	× 4.63	1
Particle surface (Ø sphere) mm²	0.785 (0.5 mm)	0.283 (0.3 mm)
Relative ratio of the two surfaces	× 2.77	1

Table 15.17. *Comparison of the characteristics of the two materials: sand and garnet*

15.7.4.6.4. Sand and reagent consumption

Table 15.18 makes it possible to project the approximate consumption of sand and chemical reagents.

Parameters	Caustic Soda	Lime
Fluidization velocity	80–100 m·h^{-1}	80 m·h^{-1}
Chemical reagent consumption	0.45–0.60 mg NaOH· mg^{-1}·$CaCO_3$ removed	0.90–1.0 mg Ca(OH)2· mg^{-1}·$CaCO_3$ removed
Sand consumption	0.12 mg·mg^{-1}·$CaCO_3$ removed	0.20 mg·mg^{-1}·$CaCO_3$ removed
Sand consumption (kg·d^{-1})	0.12 × [ΔTH] (mg·L^{-1} $CaCO_3$) × Q × 10^{-3}	0.20 × [ΔTH] (mg·L^{-1} $CaCO_3$) × Q × 10^{-3}
Pellet production	0.9–1.1 mg·mg^{-1} $CaCO_3$ removed	2 mg·mg^{-1} $CaCO_3$ removed
Apparent sand volume (m^3·d^{-1})	Sand consumption (kg·d^{-1}) / 1,500 (kg·m^{-3})	Sand consumption (kg·d^{-1})/ 1,500 (kg·m^{-3})

Table 15.18. *Operating parameters with caustic soda and lime*

15.7.4.7. *Parameters common to the two alkaline reagents*

Five design parameters are common to the two alkaline reagents. They are presented as follows.

15.7.4.7.1. Estimation of the volume of pellets located at the reactor's bottom

The previous calculation of pellet production with lime and soda informs us about the global volume of pellets produced from hardness consumption. This volume of pellets is distributed throughout the expanded bed. As the largest pellets are at the reactor's bottom because the flow rate is not sufficient to expand them, it is possible to calculate the height of the pellets to be removed, depending on their diameter.

The total pellet mass is given by:

$$M\ (\text{kg}) = \rho b(\frac{\pi db^3}{6} - \frac{\pi ds^3}{6})N$$

where:

- ρb: pellet density ($kg \cdot m^{-3}$);
- db: pellet diameter in the desired zone (at the reactor's bottom);
- ds: sand diameter (m);
- N: number of pellets in the chosen area.

That is to say, at the bottom of the reactor where the pellets are removed:

$$N = 6\left(\frac{\Delta Ca\ \text{removed}}{\pi\ \rho b(db^3 - ds^3)}\right)MQ$$

where:

- Ca: calcium concentration ($mol \cdot m^{-3}$) removed;
- MM: $CaCO_3$ molar mass ($100\ g \cdot mol^{-1}$);
- Q: water flow ($m^3 \cdot d^{-1}$);
- N: number of pellets in the chosen area;
- ds: initial sand diameter (m).

The volume of pellets to be removed at this height section is given by:

$$Vb = M(kg)/\rho b$$

15.7.4.7.2. Water consumption

Pellet removal requires dilution with treated or mains water. The dilution ratio may vary from 10 to 30, depending on the design of the pellet extraction circuit. Pipe bending or narrowing should be avoided in order to limit the risk of clogging.

15.7.4.7.3. Balancing-acidification

Softened water is always slightly scaling. To avoid the precipitation of carbonates into filters, intermediate acidification is recommended between the water softened from the catalytic softener and the filters, especially if the pH is above 8.25. This acidification is obtained by injecting CO_2 or another acid. This will ensure favorable pH conditions for chlorine disinfection.

15.7.4.7.4. Filtration after water softening

Softened water may present residual turbidity due to calcium carbonate (or magnesium hydroxides) precipitates which did not get fixed onto the grains, lime impurities or suspended matter contained in the water to be treated and, possibly, due to fine sand leaks.

In order to obtain clear softened water free of suspended solids, sand filtration is applied downstream of the softener. The role of this filtration procedure is to ensure the removal of raw water turbidity, since the process accepts raw water with up to 5 NTU, but also to remove possible turbidity due to a leak in carbonates. The filters can remove also ammonia, iron or manganese.

If there is any turbidity in raw water, a ferric chloride unit (2–5 $g \cdot m^{-3}$ in a commercial product) has to be installed (an aluminum-based coagulant should be avoided due to pH and aluminum solubility considerations). The design of sand filters is the one generally adopted for the treatment of drinking water with conventional filtration rates of around 6–7 $m \cdot h^{-1}$ (1 m sand height), or high-rate filters operating at 15 $m \cdot h^{-1}$ in N (2 m sand height).

Another possible design is to install dual media filters (anthracite/sand) to provide longer operating times between two washes (due to the higher retention capacity of anthracite), or when raw water turbidity exceeds 5 NTU.

Finally, GAC/sand dual media filters will be used in the presence of organic micropollutants (pesticides).

15.7.4.7.5. Pellet removal

Pellet removal is done at the reactor's bottom because it is there where pellets are the largest. From reactor operating velocities (80–100 $m \cdot h^{-1}$), the fluidization velocities calculated with Leva, Kozeny or Ergun equations (see section 9.5.4, Chapter 9), indicate the pellets distribution along the reactor (Figure 15.25 and Table 15.19). This makes it possible to estimate the volume to be removed, even if the head loss is already a means for estimating it.

% Pellet expansion	Lime flow rate (80 m·h^{-1})	Caustic soda flow rate (100 m–h^{-1})
d_{10} 0.5 mm	100%	100%
d_{10} 1 mm	85%	85–90%
d_{10} 1.5 mm	35%	50%
d_{10} 2.0 mm	18%	25%

Table 15.19. *Percentage of pellet expansion along the reactor depending on pellet diameter*

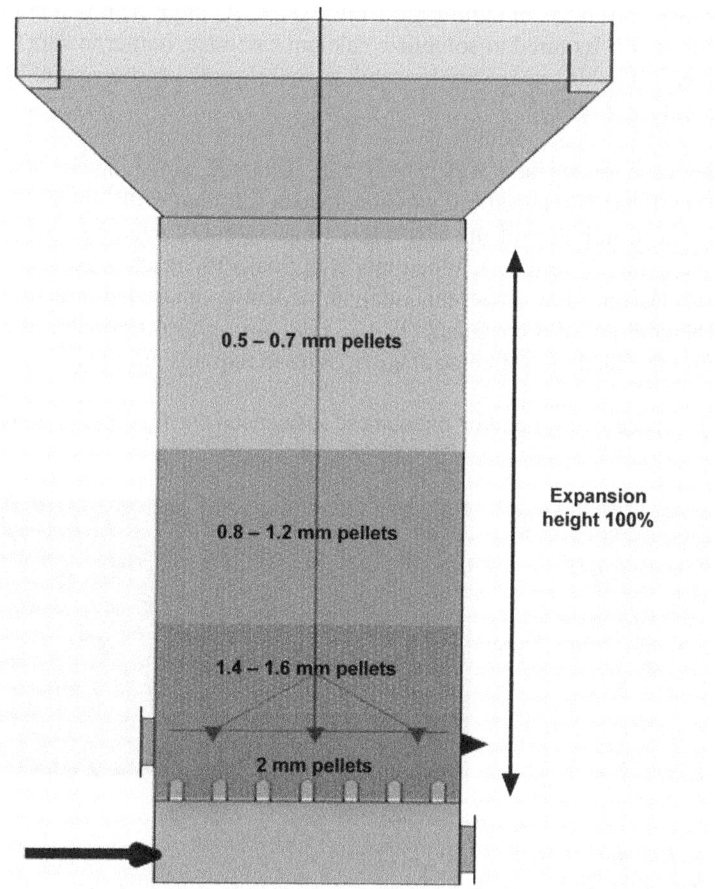

Figure 15.25. *Pellet distribution depending on diameter (Actina® process). For a color version of this figure, see www.iste.co.uk/gaid/watertreatment3.zip*

15.7.4.8. *Factors influencing catalytic water softening*

The experience gained shows a marked preference for sodium hydroxide: it is easy to use (simple dilution), efficient for a wide range of raw water qualities (in case of moderate bicarbonate content), its low pellets production and inexpensive operating cost for installations treating 500–1000 $m^3 \cdot h^{-1}$.

A low sodium content is required for water softening with caustic soda or sodium carbonate.

Catalytic softening is followed by filtration (possible addition of coagulant), so as to remove any trace of turbidity in raw water. A slight acidification of water before filtration is required to solubilize calcium carbonate (either in supersaturation or entrained). It is also necessary to stabilize water's calco-carbonic equilibrium, so as to prevent filter scaling.

In the case of surface water treatment, catalytic softening is preceded by coagulation–flocculation–settling or dual media filtration with anthracite-sand in order to reduce turbidity of the raw water (inlet turbidity must be <5 NTU at the inlet of the catalytic softener). When raw water turbidity reaches exceptional peaks of 15–20 NTU for a few days/year, catalytic softening can be fed directly from raw water. The addition of a coagulant (WAC, $FeCl_3$) is applied upstream of the filters placed after the catalytic-softening stage in order to trap this turbidity.

The presence of iron and/or manganese influences the final pellet color (Figure 15.26).

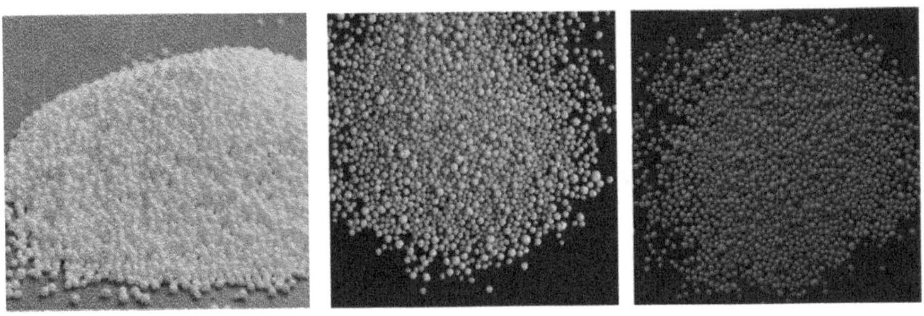

a) b) c)

Figure 15.26. *(a) $CaCO_3$ pellets without iron, (b) with low iron concentration, (c) with high iron concentration. For a color version of this figure, see www.iste.co.uk/gaid/watertreatment3.zip*

Parameters to take into account:

– maintenance and upkeep costs and constraints;

– (liquid) soda handling is easier than that of lime;

– the production of $CaCO_3$ pellets is twice higher with lime than with caustic soda, which results in a higher sand consumption in the first case;

– maintaining an optimal fluidization velocity to compensate for temperature-related viscosity differences that directly affect the bed expansion level;

– the use of caustic soda and sodium carbonate as softening agents provokes an increase in the sodium content in the treated effluent.

15.7.4.9. *The place of chemical water softening at a water treatment plant*

The place of chemical water softening at a water treatment plant depends on the quality of raw water.

15.7.4.9.1. The case of groundwater containing hardness

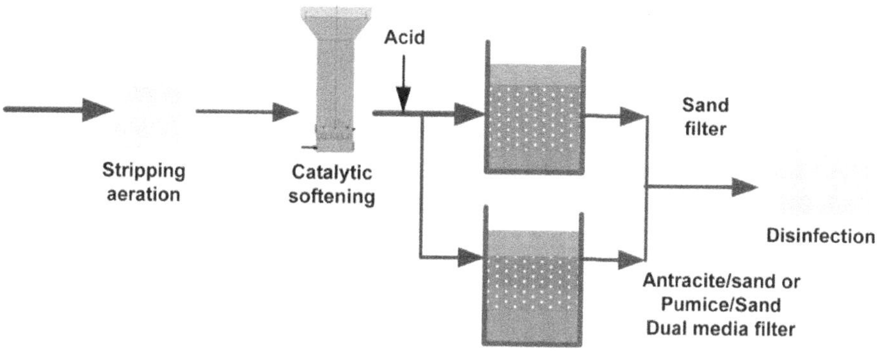

Figure 15.27. *Water treatment plant for water only containing hardness. For a color version of this figure, see www.iste.co.uk/gaid/watertreatment3.zip*

15.7.4.9.2. The case of groundwater containing hardness and micropollutants

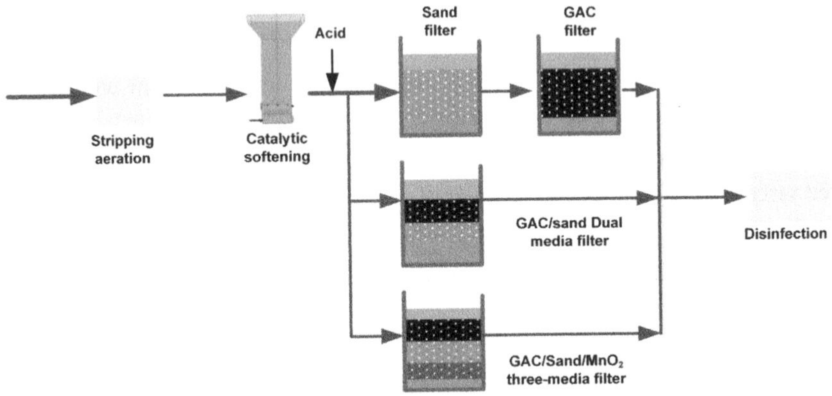

Figure 15.28. *Water treatment plant for water containing hardness and micropollutants. For a color version of this figure, see www.iste.co.uk/gaid/watertreatment3.zip*

15.7.4.9.3. The case of groundwater containing nitrates, hardness and micropollutants

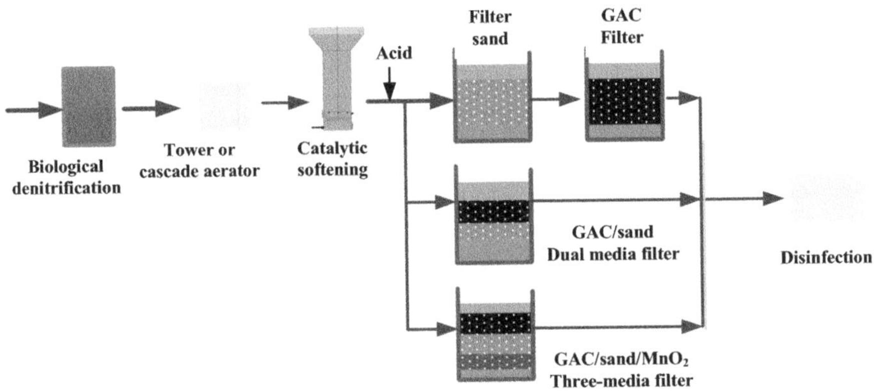

Figure 15.29. *Water treatment plant for water containing nitrates, hardness and micropollutants. For a color version of this figure, see www.iste.co.uk/gaid/watertreatment3.zip*

15.7.4.9.4. The case of surface water containing turbidity, hardness, organic matter and micropollutants

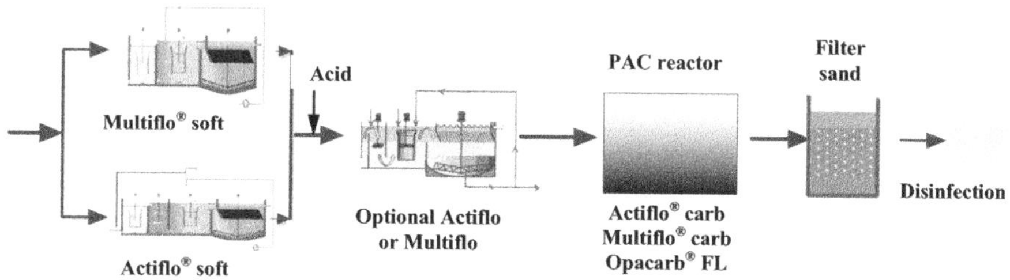

Figure 15.30. *Water treatment plant for water containing turbidity, hardness, organic matter and micropollutants. For a color version of this figure, see www.iste.co.uk/gaid/watertreatment3.zip*

15.7.4.9.5. The case of surface water containing turbidity, hardness and seasonal micropollutants

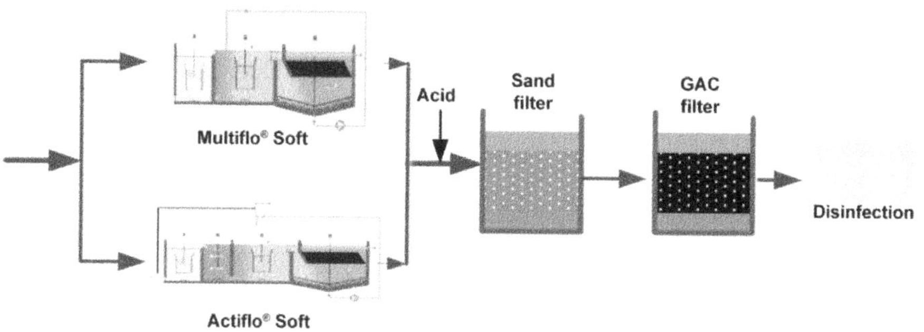

Figure 15.31. *Water treatment plant for water containing turbidity, hardness and seasonal micropollutants. For a color version of this figure, see www.iste.co.uk/gaid/watertreatment3.zip*

15.7.4.10. Performances and guarantee levels

The main guarantee concerns treated water hardness (TH). The target value must correspond either to a value desired by the customer (insofar as the TH/Alk ratio allows it), or to the value dictated by the calco-carbonic equilibrium.

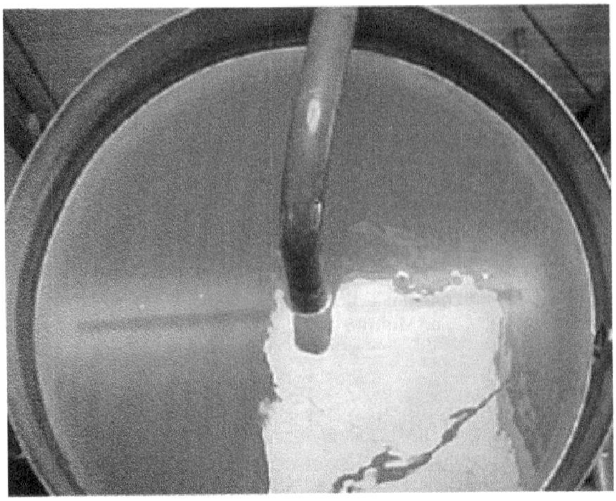

Figure 15.32. *Softened water quality. For a color version of this figure, see www.iste.co.uk/gaid/watertreatment3.zip*

The turbidity of softened water is on average between 2 and 5 NTU on OTV-Veolia installations (Figure 15.32). Figure 15.33 shows a change in turbidity as a function of residual hardness. The lower the residual hardness (50 mg·L^{-1} in $CaCO_3$), the higher the final turbidity. The average value is 3 NTU for a final hardness of 120–150 mg·L^{-1} $CaCO_3$ corresponding to installations usually built by OTV-Veolia.

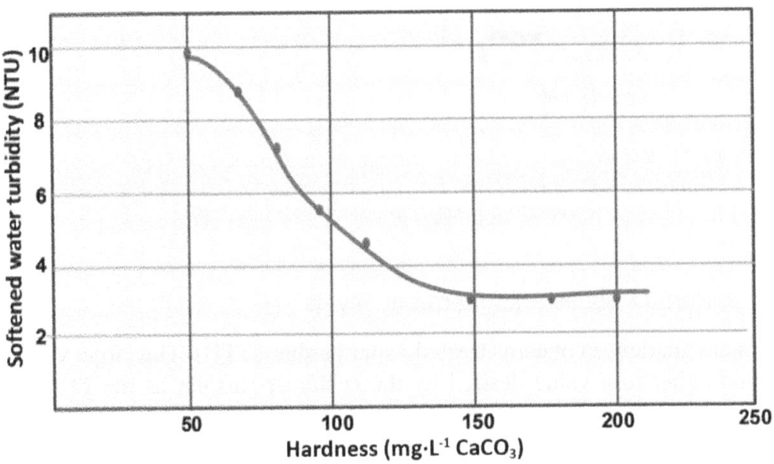

Figure 15.33. *Softened water turbidity as a function of output hardness (catalytic material: sand)*

On the other hand, with garnet as catalytic material (replacing conventional sand), the final turbidities are between 0.5 and 1.0 NTU, regardless of the final hardness.

15.7.4.11. *The case of mitigation*

The alternative to treating the entire flow is to soften part of the raw water, followed by a mitigation with untreated water. An interesting gain in investment cost can thus be obtained. The operating costs are identical insofar as the consumption of alkaline reagents is related to the level of calcium removed. On the other hand, the possible presence of metal ions in raw water requires a pretreatment (iron and Mn) to avoid finding them in the mitigated water. In the event of heavy metals (Ni, Pb, etc.) being present, this solution is only possible after having estimated the final concentrations with proper calculations.

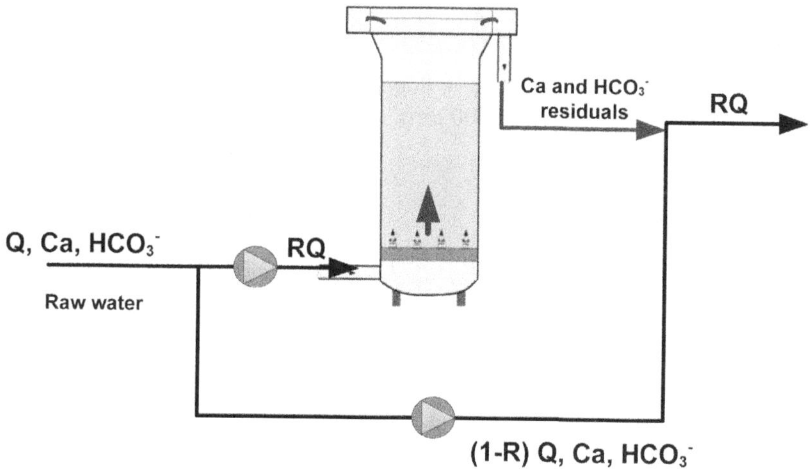

Figure 15.34. *Bypass of a part of raw water, followed by mitigation with treated water. For a color version of this figure, see www.iste.co.uk/gaid/watertreatment3.zip*

15.7.4.12. Magnesium removal

Catalytic softening is poorly adapted to remove magnesium due to the high pH (pH 11) required by the magnesium removal reaction. Table 15.20 lists the various processes for removing calcium and, in some cases, magnesium.

Processes	Calcium removal %	Magnesium removal %
Catalytic softening	95	5
Classic softening	95	90
Ion exchange resins	>99	>99
Nanofiltration	90	90

Table 15.20. *Comparative removal of Ca and Mg by various processes*

15.7.4.13. Actina® process: technology and equipment

The technology and equipment elements involved in the Actina® process are the following:

15.7.4.13.1. Reactor floor

Actina® softening reactors have a floor made up of strainers whose slots are smaller than the sand's grain size (400 μm).

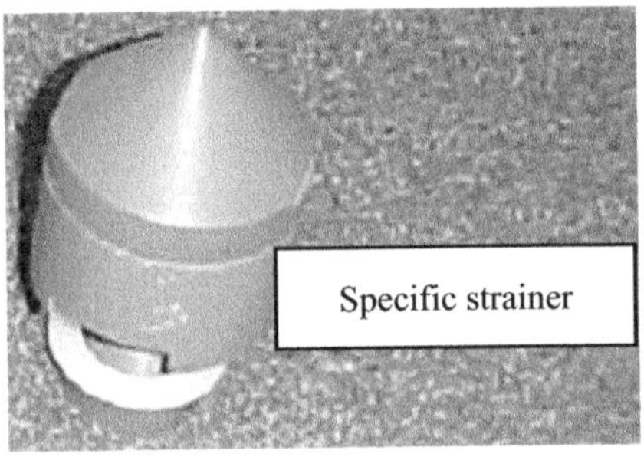

Figure 15.35. *Specific strainer for the floor of Actina® catalytic softener reactors. For a color version of this figure, see www.iste.co.uk/gaid/watertreatment3.zip*

15.7.4.13.2. Head loss measurement of water softening beds

A manometer and two pressure transmitters, placed upstream of the bed and downstream of the floor monitor the pressure value during operation.

This value determines the bed's stability and exchange surface. The higher the head loss, the more stable the bed, but the lower the exchange surface. Compensation then becomes necessary by substantially adding microsand, which will be blocked by the mass of the largest carbonate pellets.

From this information (operating value), it is possible to manage the bed's purge and/or to jointly stop the sand transfer at the top level.

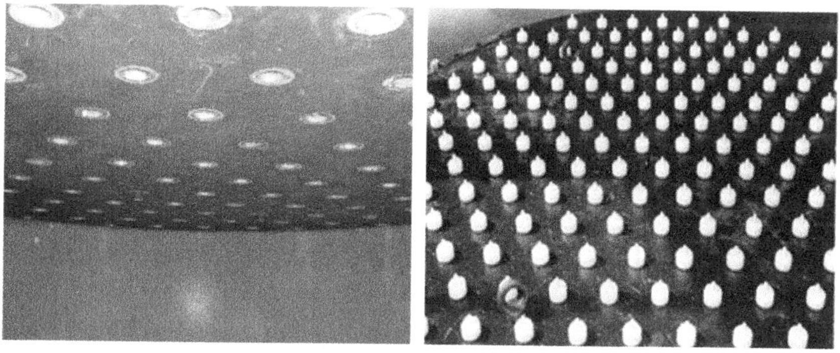

a) b)

Figure 15.36. *Floor strainers with bottom view (a) and top view (b). For a color version of this figure, see www.iste.co.uk/gaid/watertreatment3.zip*

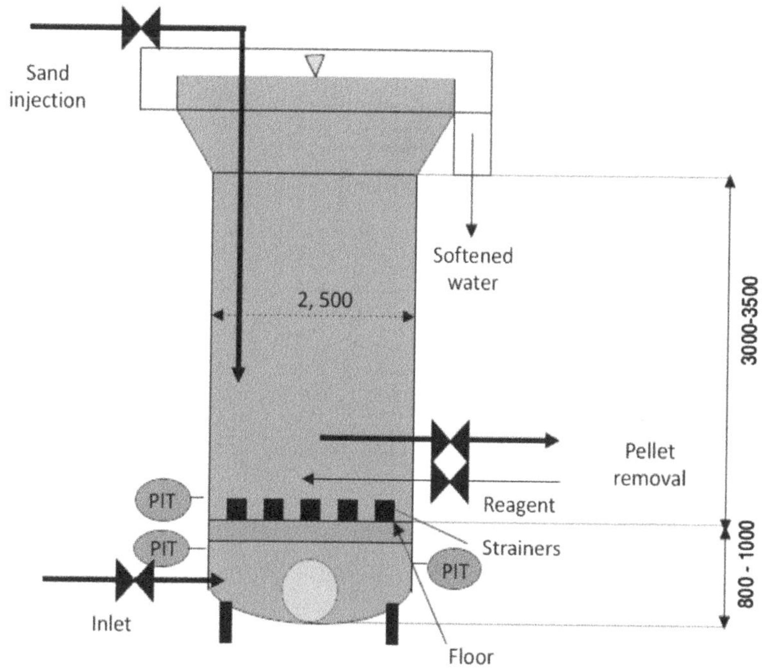

Figure 15.37. *Catalytic Softener (diameter 2.5 m), static sand height: 1.5 m. For a color version of this figure, see www.iste.co.uk/gaid/watertreatment3.zip*

Figure 15.38. *View of an Actina® reactor with a perforated floor and pellet extraction tubes. For a color version of this figure, see www.iste.co.uk/gaid/watertreatment3.zip*

Figure 15.39. *View of a reactor with inlet area and inspection window. For a color version of this figure, see www.iste.co.uk/gaid/watertreatment3.zip*

15.7.4.13.3. Injection nozzles

The density of injection points is at least 2–3 per m^2. The distribution in the reactor is done by pipes equipped with injection nozzles, located at a height of 10–20 cm above the distribution floor in order to ensure a good mixing of the reagent (lime milk or caustic soda) and raw water. The network is calculated for flow velocities of approximately 0.8–1.0 $m \cdot s^{-1}$. The nozzles are fixed on a vertical, retractable support, enabling their examination and unclogging.

15.7.4.13.4. Pellet removal

Pellet removal is done at the level of the distribution floor, since the pellets with the largest diameter are located at the base of the fluidized bed.

Figure 15.40. *Caustic soda injection nozzle. For a color version of this figure, see www.iste.co.uk/gaid/watertreatment3.zip*

Different removal methods can be considered:

– gravity removal (small units only), which requires a pellet transfer tank toward the storage silo by means of a drainage system. Instead of the dripping silo, filter buckets can also be used. Their advantage is that they grant a suitable water removal, enabling the direct evacuation of pellets, without truck transfer;

– by hydro-ejection for small installations;

– by hydrostatic pressure difference, in dilution with pressurized water. The dilution rate to be provided may vary from 10 to 30, depending on the removal network's setting and design;

– by direct pumping. A water feed should be provided for to prevent the clogging of pipes.

The bed level and its composition are very important to obtain the lowest possible hardness on the treated water. Pellet size in the upper and lower sections makes it possible to obtain the ideal hydraulic and chemical conditions for reaching optimal hardness and turbidity performances.

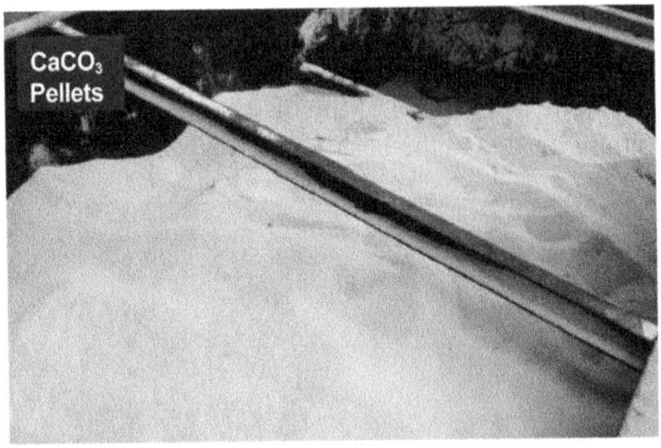

Figure 15.41. *Filtering bucket. For a color version of this figure, see www.iste.co.uk/gaid/watertreatment3.zip*

The volume the bed occupies in the reactor must be as large as possible to enable sufficient contact time and the softening reaction to take place. The more contracted the bed, the faster the reaction to compensate for the lower contact time. The smaller the pellets, the larger their reactive surface. However, on starting up the reactors, it is advisable to avoid too strong hydraulic flows, which could entrain this group of pellets with the treated water.

Option for small diameters (<300 mm)

For small diameters (<300 mm), an option is to install a gravel layer, instead of the perforated floor. In this case, regular monitoring of the gravel must be implemented in order to remove the layer which, eventually, becomes scaled with carbonates. Pellet removal is done manually by gravity flow.

15.7.4.14. Operation and monitoring parameters

All of the equipment making up the process must be checked before start. The operation of the micro-sieving (pre-filter) and its unclogging (if existing, because it is optional), the raw water circuit (raw water pumps and flow meter) will be filled and rinsed before start-up. This is to avoid untimely clogging the perforated floor, the compressed station (for the solenoid valves) with the compressor control and the tightness of the compressed air circuit. Other considerations include the triggering and damper pressure activation values of the pressure switches, measuring the floor's head losses initially without microsand, monitoring the pellet removal circuits as well as the automatic valves and the pumps making up the circuit, watching for the correct operation of softeners and proper distribution of softened water in the dilution circuit, the safety measures of the reagent circuit, the level measurements in the tanks, adjusting the dosing pump flow rates and the dilution circuit, the microsand water softener feed circuit, and adjusting the injection nozzles and counter-nozzles.

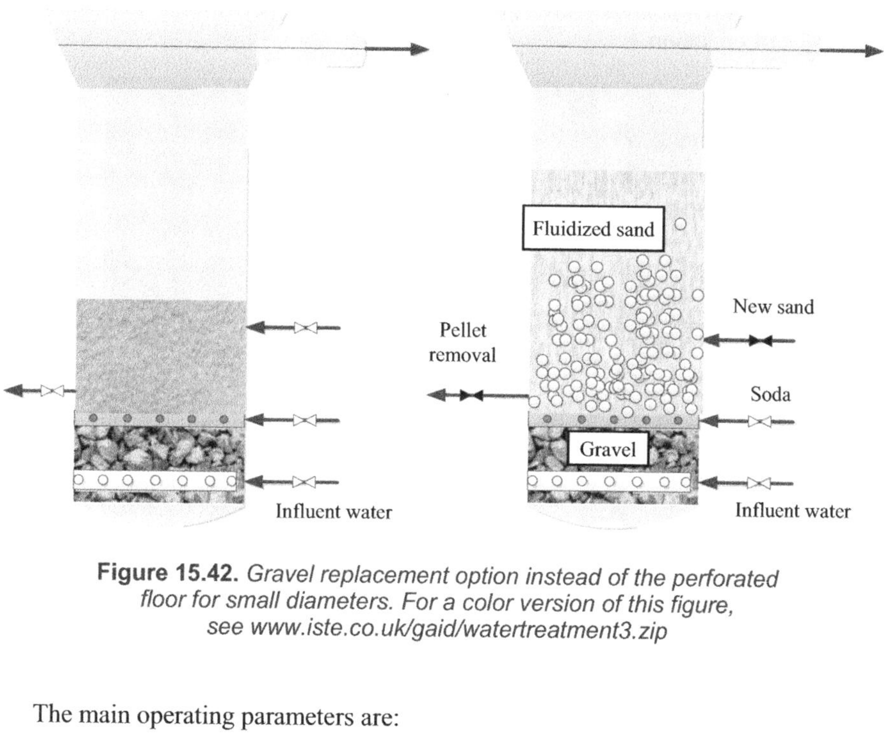

Figure 15.42. *Gravel replacement option instead of the perforated floor for small diameters. For a color version of this figure, see www.iste.co.uk/gaid/watertreatment3.zip*

The main operating parameters are:

– pellet bed static and expanded height;

– particle size of the pellets removed;

– head loss at the perforated floor;

– head loss on the reagent and softened water distribution network (dilution water).

The following parameters measured on raw water determine the reagent rates:

pH, turbidity, color, Alkalinity, total hardness TH, calcium hardness THCa

The treatment rate also depends on the difference between raw water's TH and the desired TH for treated water, the neutralization of raw water's free CO_2, of the target pH and alkalinity values for the calco-carbonic equilibrium of treated water.

15.7.4.14.1. Quality of treated water

In a few hours (3–5 h), the desired quality of treated water is obtained. The system's equilibrium, namely the optimum magnification of pellets, is reached after 200–300 h of operation.

Figure 15.43. *Some Veolia references in France. For a color version of this figure, see www.iste.co.uk/gaid/watertreatment3.zip*

Regular monitoring of the quality of treated water makes it possible to adjust the dose of reagent required to attain the desired hardness.

15.8. Saphira® process

The Saphira® process (Veolia) is a water softening reactor including a mixing settler made in epoxy coated black steel, concrete or high-density polyethylene (HDPE). This reactor comprises three zones:

– an upper section consisting of a lamellar block 0.6–1 m long (drinking water type);

– an intermediate section: a tranquillizing zone which promotes liquid–solid separation;

– a lower section consisting of a Turbomix mixer™, which mixes raw water and alkaline reagents. Water softening reactions take place in this zone.

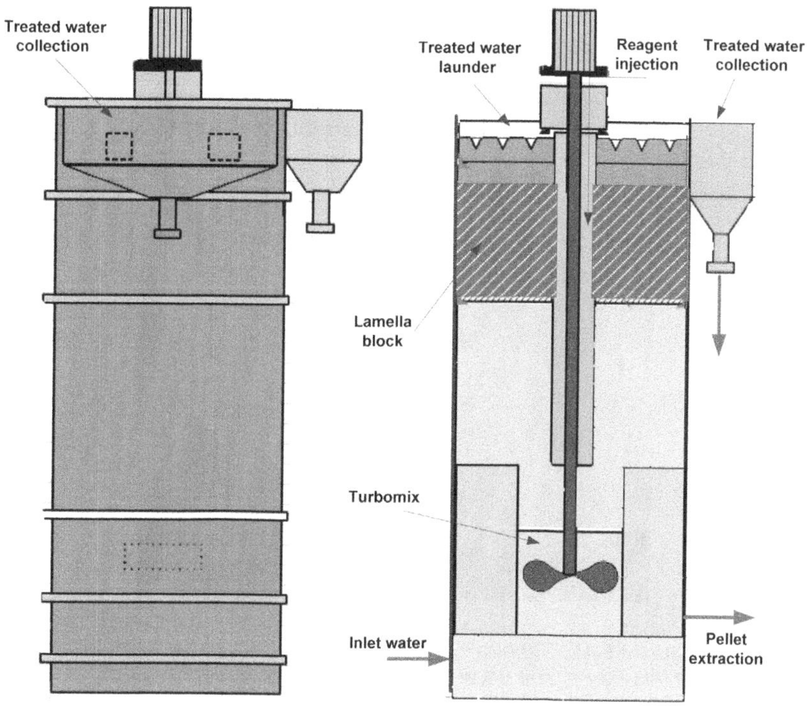

Figure 15.44. *Description of the Saphira® process (Veolia). For a color version of this figure, see www.iste.co.uk/gaid/watertreatment3.zip*

In case of raw water turbidity, a coagulant and a polymer are injected. The coagulant can be injected in line or by means of a coagulation tank. In the case of coagulation–flocculation–settling, Saphira® is inserted between clarification and sand filtration. In this case, neither the coagulant nor the polymer is injected.

Figure 15.45. *Saphira® process (Veolia). For a color version of this figure, see www.iste.co.uk/gaid/watertreatment3.zip*

Mirror velocities are 12–20 $m \cdot h^{-1}$. The contact times are 10–20 min. pH and turbidity sensors are installed at the inlet and in the tranquillizing zone. The water flow is measured using a flow meter. Lime is injected in the form of whitewash. Sodium hydroxide is injected by online dilution.

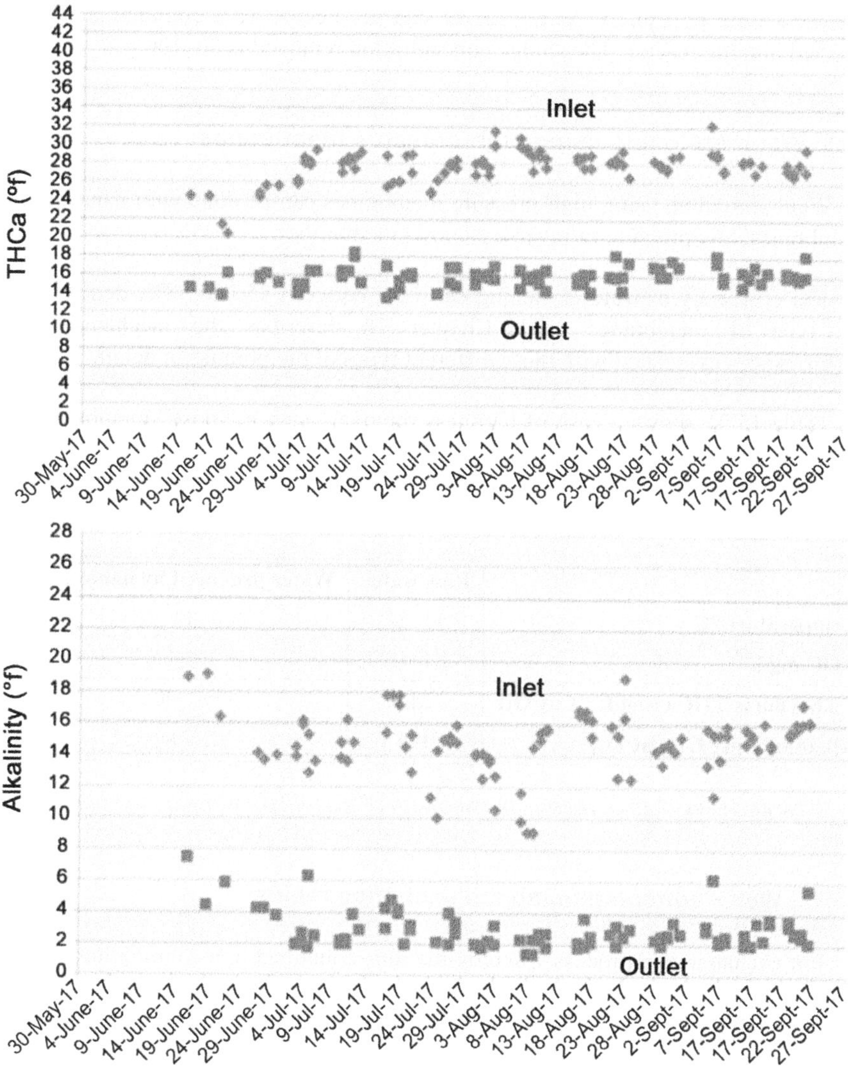

Figure 15.46. *Performances of the Saphira® process in relation to calcium hardness and alkalinity (1°f = 10 mg·L^{-1} $CaCO_3$). For a color version of this figure, see www.iste.co.uk/gaid/watertreatment3.zip*

These results are obtained with a raw water containing a turbidity of 20–50 NTU, a pH around 7.5, and a concentration of Ca^{2+} 280–300 mg·L^{-1} $CaCO_3$ and an alkalinity of 150 and 200 mg·L^{-1} $CaCO_3$. Lime dosages are 120 and 180 mg·L^{-1} (92–94% lime purity). A 20 mg·L^{-1} dosage of $FeCl_3$ (41% commercial) is also injected into the reactor. The sludge concentration removed is 80–100 g·L^{-1}.

15.9. Water softening using high-pressure membranes

Water softening using high-pressure membranes can be done with nanofiltration membranes, low pressure or reverse osmosis membranes.

The efficiency of water softening can vary between 45 and 98%, depending on the type of membranes used. Chapter 18 describes the mechanisms and performances associated with these different types of membranes in detail.

Table 15.21 shows a case of hardness removal using a "loose" membrane. The type of membrane is chosen depending on the treatment goals and inlet concentration. Veolia has numerous drinking water plant references in which partial or high hardness removal is performed.

	Raw water	**Water produced by nanofiltration**
Temperature °C	11–22	11–22
TOC mg·L^{-1}	1.1–1.7	0.5
Ca hardness THCa (mg·L^{-1} $CaCO_3$)	240	90
Alkalinity (mg·L^{-1} $CaCO_3$)	190	80

Table 15.21. *Partial water softening with a "loose" nanofiltration membrane*

15.10. Water softening using ion exchange resins

Ion exchange is a process whereby the ions contained in a solution are removed and replaced by an equivalent quantity of other ions having the same electrical charge (positive charge for cations, negative one for anions). The process takes place in an ion exchanger: a reactor containing the ion exchanger resin, and an environment which performs the exchange function, as well as the associated functions.

The purpose of ion exchange can be:

– the removal of Ca^{2+} and Mg^{2+} ions and their replacement by sodium ions;

– the total demineralization of water (replacement of mineral salts by hydroxyl and hydronium ions).

Resins are presented in the form of pellets with a 0.5–2 mm diameter, and a density slightly higher than water (d = 1.05–1.40). Their color is generally amber-like, although some are intentionally colored in order to identify their saturation.

They are organic compounds obtained by polymerization of a monomer (styrene, acrylate, etc.) whose long chains are "attached" to one another by means of a cross-linker (divinylbenzene [DVB] for polystyrene resins). The "skeleton" previously obtained is treated by grafting functional groups to it. These will define the ions which can be captured during the exchange phase (the phase fixing the ions in the solution).

In this way, the same cross-linked polystyrene skeleton makes it possible to obtain resins capable of fixing all the cations with a sulfone functional group, for example, and all the anions with a quaternary ammonium functional group.

15.10.1. *Resin resistance: use precautions*

Mechanical stability is the ability of a resin to withstand the compression due to head loss (anionic resins are more sensitive than cationic resins), as well as abrasion, due to the friction between the resin pellets. This is also known as "attrition".

Chemical stability is the ability of a resin to tolerate pH conditions, the presence of oxidants (which de-crosslink the resins) and the presence of colloids.

A resin must always be kept moist in order to avoid an external dehydration of the pellets, which would create too high osmotic pressures. The lifespan of a resin is limited in time by the constraints to which it is subjected: around a 3% resin mass loss is estimated for 1,000 cycles.

– *Resistance to organic poisoning*: organic acids (humic acid from surface water, for example) penetrate and remain trapped in strong anionic resins. They "poison" the resin, decreasing its ion exchange possibilities.

– *Bacterial development*: resin particle size sometimes causes them to play the role of a filter blocking organic matter and minerals in suspension. If temperature and pH conditions are met, they can be the seat of bacterial development. A regular disinfection of resins must be carried out when the circuit is sensitive. This disinfection must be carried out using a solution adapted to the type of resin: bleach for sanitary softening, formalin for cationic resins, quaternary ammoniums for anionic resins.

Two types of resins meet the water softening criterion:

– weak cationic resins in the hydrogen cycle: water softening (either total or partial, depending on the respective TH values);

– strong cationic resins in the sodium cycle: total softening.

Calcium and magnesium removal is often done with resins in which the Na^+ ion is exchanged. The removal of bicarbonates is obtained because of a resin in which chlorides replace other anions.

In short, the removal of hardness and alkalinity requires an acid resin for cation exchange and a strongly basic resin for carbonates and bicarbonates. They are arranged in series, or in a mixed bed within a single device.

15.10.2. *Ca^{2+} and Mg^{2+} removal*

Exchange reactions with a resin in the sodium cycle are:

$$NaZ + Ca^{2+} \rightarrow CaZ + 2Na$$

$$NaZ + Mg^{2+} \rightarrow MgZ + 2Na$$

A given volume of resin can bind a limited quantity of ions. When this threshold is reached, the influent water (which enters the exchanger) exits the unit with its characteristics unchanged. The resin column (or bed) is said to be pierced. Regeneration therefore comes down to restoring the resin's exchange capacity by reversing the service phase reaction.

Regeneration is carried out with NaCl such as, for example:

$$(Ca, Mg)\ Z + NaCl \rightarrow NaZ + (Ca, Mg)\ Cl_2$$

Other solutions exist with acid resins (H_2Z).

Resin regeneration can be done by sending the regeneration flow in the same direction as during the service phase (co-current regeneration) or in the opposite direction (countercurrent regeneration). Countercurrent regeneration produces better water quality: the best regenerated layers are the last layers to be crossed during the service phase.

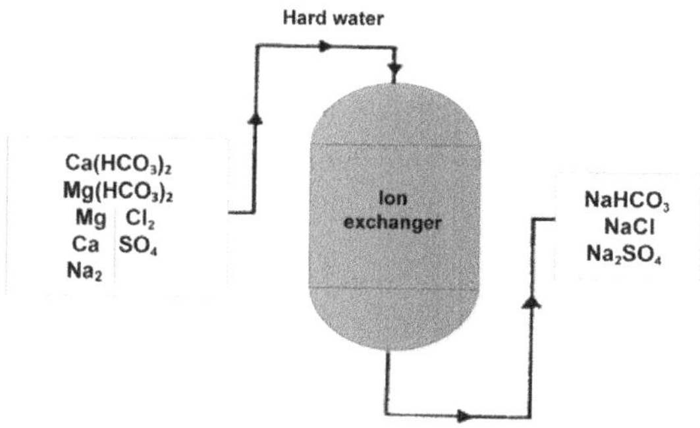

Figure 15.47. *Example of a $Ca^{2+}$and Mg^{2+} exchanger with a Na resin. For a color version of this figure, see www.iste.co.uk/gaid/watertreatment3.zip*

15.10.3. *Bicarbonate removal*

$$HCO_3^- + RCl \rightarrow RHCO_3 + Cl^-$$

Regeneration is achieved by means of a salt solution:

$$RHCO_3 + NaCl \rightarrow RCl + NaHCO_3$$

15.10.3.1. *Ion exchange*

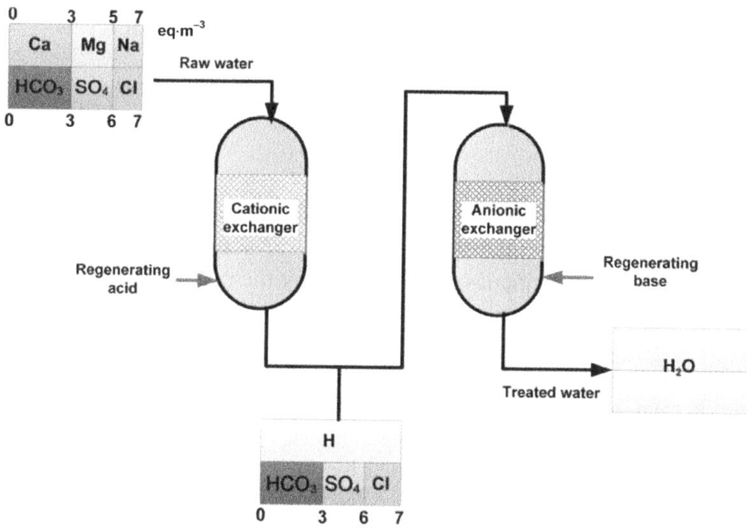

Figure 15.48. *Demineralization using ion exchangers in series. For a color version of this figure, see www.iste.co.uk/gaid/watertreatment3.zip*

15.10.4. *Resin operating capacity*

The capacity of a resin to exchange a quantity of matter is defined by its operating capacity, Cu. This denotes the number of ion sites available for exchange per liter of resin.

The units used are:

– the equivalent per liter of resin: $eq \cdot L^{-1}$;

– 10 $mg \cdot L^{-1}$ $CaCO_3$ per cubic meter and per liter of resin: 10 $mg \cdot L^{-1}$ $CaCO_3 \cdot m^{-3} \cdot L^{-1}$:

$$1 eq \cdot L^{-1} = 50\ mg \cdot L^{-1}\ CaCO_3 \cdot m^{-3} \cdot L^{-1}$$

A volume of 1 L of strong cationic resin in sodium cycle with an operating capacity equal to Cu = 1.2 $eq \cdot L^{-1}$ blocks 1.2 equivalents of calcium, that is to say, 48 g of calcium.

The manufacturers provide charts for determining the operating capacity, which all the same needs to be corrected for factors depending on operating conditions.

15.10.5. *Operating parameters*

The operating parameters of an ion exchange resin water softening system are related to the type of resin, hydraulics, regeneration and cleaning of resins.

The type of resin and height influence the efficiency of water softening, since there are many types of resins with different removal capacities which affect the overall efficiency.

The suitable height for a correct ion exchange (under flow conditions in the service phase) is in the range of 600–800 mm, as a minimum. This quantity is conditioned by the resin's kinetics: that is to say, by the rate of fixation of influential ions. If the height is too low, it leads to a high ion leak for the same volume of resin.

Exchangers are designed to obtain a diameter/bed height ratio lower than 1. The exchanger's total height must allow for the expansion of the resin bed during the rising phase (1.2–2.5 the resin bed's height). Flow rates are between 25 and 50 $m \cdot h^{-1}$. The volumic load is represented by the flow rate circulating in the resin bed in relation to the resin volume: it is expressed as solution volume/resin volume/h. by manufacturers. Its dimension is the inverse of one time: [h].

The salt quantity during regeneration, the brine concentration and brine contact time influence sodium regeneration. If the regeneration is not fully completed, the exchange cycles will be shorter.

As with any other treatment process, water softening by ion exchange requires water quality control in accordance with the following parameters:

– raw water hardness affects cycle times. The higher the calcium and magnesium concentrations, the faster the sodium bound to the resin wears out, implying a regeneration;

– iron and manganese can precipitate onto the resins, causing clogging. It is recommended to strongly reduce these two elements before water softening by resins;

– the sodium present in the water reacts with the resin as it would during regeneration, causing magnesium and calcium to leak into the treated water;

– excessive residual chlorine can damage the resin;

– the eluates must be neutralized before discharge.

Figure 15.49. *Veolia water softening facility (Germany). For a color version of this figure, see www.iste.co.uk/gaid/watertreatment3.zip*

15.11. Comparison between the four water softening solutions discussed

Compared to a water softening by resins, the Actina® process has a number of advantages:

– compactness due to a short retention time in the device. This is because of the fast kinetics which take place therein;

– very fast response time;

– in addition, catalytic water softening with caustic soda:

- reduces alkalinity,

- adds less sodium than resins do.

Compared to conventional water softening, the catalytic water softening process:

– produces easily wringable and transportable pellets, whereas classic lime softening produces a liquid carbonate sludge which must be settled, concentrated and dried. The carbonate pellets thus produced are recoverable, depending on the composition of raw water (soil enrichment, adjuvant for cement works, poultry feed, etc.);

– removes a fraction of iron, manganese and any other heavy metals (Ni, for example) by co-precipitation within the carbonate pellets (direct influence on the choice for recovery purposes).

15.11.1. *Advantages and limitations of the different solutions*

Settler softening	Catalytic softening	Ion-exchange resins	High-pressure membranes
Equipment A settler with cont, coagulation and flocculation tank Sand filtration with backwashing A caustic soda or lime preparation A softened water station Electrical equipment and automation Acidification with sulfuric acid or carbon dioxide	*Equipment* A catalytic reactor Sand filtration with backwashing A microsand preparation: preparation + storage tank A caustic soda or lime preparation A softened water station A final disinfection Electrical equipment and automation	*Equipment* Softener Salt storage Operating building Connection to the wastewater network subject to discharge acceptance by the treatment plant *Advantages* Limited investment Space-saving technology Automatable technique	*Equipment* Pretreatment Membrane train High-pressure pumps Prefilters Outlet pH adjustment Operating building Connection to the wastewater network for the discharge of concentrates subject to acceptance by the treatment plant

Settler softening	Catalytic softening	Ion-exchange resins	High-pressure membranes
A building for protecting all the equipment *Advantages* Accepts raw water turbidity Calco-carbonic equilibrium with a decrease of more than 50% in the calcium bicarbonate content Simple measurement of changes in water characteristics Process control, automatable Technique suitable for large flows Valuable by-products for agriculture *Limitations* Use of reagents (sand, softened water, soda or lime) Limited energy cost Qualified personnel operation Removal of solid by-products	Acidification with sulfuric acid or carbon dioxide A building for protecting all the equipment *Advantages* Very compact solution Calco-carbonic equilibrium with a decrease of more than 50% in the calcium bicarbonate content Simple measurement of changes in water characteristics Process control, automatable Technique suitable for large flows Valuable by-products *Limitations* Only accepts low inlet turbidity (<5 NTU) Use of reagents (sand, softened water, soda or lime) Limited energy cost Qualified personnel operation Removal of solid by-products	*Limitations* Does not accept inlet turbidity (<1 NTU) Presence of sodium in treated water High consumption of salt to regenerate resins 20% reduction of precipitated bicarbonate Connection to the wastewater network to discharge the eluates Qualified personnel operation Limited processing	Electrical equipment and automation *Advantages* Automatable technique Stable performance Removes other elements at the same time (heavy metals, organic matter) *Limitations* Significant investment Does not accept inlet turbidity (<1 NTU) Water loss between 20 and 30% Final pH adjustment Connection to the wastewater network to discharge the concentrates Qualified personnel operation

Table 15.22. *Advantages and limitations*

15.12. References

Aldaco, R., Garea, A., Irabien, A. (2007a). Modeling of particle growth: Application to water treatment in a fluidized bed reactor. *Chem. Eng. J.*, 134, 66–71.

Aldaco, R., Garea, A., Irabien, A. (2007b). A particle growth kinetics of calcium fluoride in a fluidized bed reactor. *Chem. Eng. Sci.*, 62, 2958–2966.

Barkel, R. (1974). The reactivity of calcium oxide towards carbon dioxide and its use for energy storage. *J. Chem. Technol. Biotechnol.*, 24, 221–227.

Caldwell, D.H. and Lawrence, W.B. (1953). Water softening and conditioning problems: solution by chemical equilibrium methods. *Indust. Eng. Chem.*, 45(3), 535–548.

Chen, Y., Fan, R., An, D., Cheng, Y., Tan, H. (2016). Water softening by induced crystallization in fluidized bed. *J. Environ. Sci.*, 50, 109–116.

Ergun, S. (1952). Fluid flow through packed columns. *Chem. Eng .Progr.*, 48(2), 89–94.

Gaid, K. and Sauvignet, P. (2013). Actiflo®Softening : la combinaison optimale de la clarification à grande vitesse et de la décarbonatation. *L'eau, l'industrie, les nuisances*, 359, 75–80.

Gaid, K., Philipot, J.M., Goguel, F. (2012). Le procédé Actina : la décarbonatation catalytique appliquée aux eaux potables et industrielles. *L'eau, l'industrie, les nuisances*, 281, 57–62.

Graveland, A., Van Dijk, J.C., de Moel, P. (1983a). Developments in water softening by means of pellet reactors. *J. Am. Water Works Assoc.*, 75, 619–625.

Graveland, A., Van Dijk, J.C., de Moel, P., Oomen, J.H.C.M. (1983b). Developments in water softening by means of pellet reactors. *J. Am. Water Works Assoc.*, 75(12), 619–625.

Harms, W. and Bruce Robinson, R. (1992). Softening by fluidized bed crystallizers. *J. Environ. Eng.*, 118(4), 513–529.

He, C., Gross, M., Westerhoff, P., Fox, P., Li, K. (2011). Comparing conventional and pelletized lime softening concentrate chemical stabilization. Report, Water Research Foundation, Denver.

Hofman, J., Kramer, O., Hoek, J.P.V.D., Nederlof, M., Groenendijk, M. (2006). Twenty years of experience with central softening in The Netherlands: Water quality, environmental benefits, and costs. In *Proceedings of the International Symposium on Health Aspects of Calcium and Magnesium in Drinking Water*, International Life Sciences Institute, Baltimore.

Hu, R.Z., Huang, T.L., Wen, G., Yang, S. (2017). Modelling particle growth of calcium carbonate in a pilot-scale pellet fluidized bed reactor. *Water Sci. Technol. Water Supply*, 17, 643–651.

Hu, R.Z., Huang, T., Zhi, A., Tang, Z. (2018). Full-scale experimental study of groundwater softening in a circulating pellet fluidized reactor. *Int. J. Environ. Res. Public Health*, 15, 1592–1602.

International Water Supply Association (n.d.). Proceedings of the specialized conference on new technologies in water treatment. Report, International Water Supply Association, Amsterdam.

Jiang, K., Zhou, K.G., Yang, Y.C., Du, H. (2014). Growth kinetics of calcium fluoride at high supersaturation in a fluidized bed reactor. *Environ. Technol.*, 35, 82–88.

Kramer, O.J.I., Jobse, M.A., Baars, E.T., Van der Helm, A.W.C., Colin, M.G., Kors, L.J., Van Vugt, W.H (2015). Model-based prediction of fluid bed state in full-scale drinking water pellet softening reactors. In *Proceedings of the 2nd IWA New Developments in IT & Water Conference*, The Hague.

Maeng, S.K., Carlos, N.A., Rodriguez, S., Choi, J., Kim, S.H., Yu, Y., Kim, B.S., Hong, S.W. (2016). Organic micropollutant removal from groundwater: Comparison of pellet softening and nanofiltration. *J. Water Supply: Res. Technol. – Aqua*, 1–10.

Mahvi, A.H., Shafiee, F., Naddafi, K. (2005). Feasibility study of crystallization process for water softening in a pellet reactor. *Int. J. Environ. Sci. Technol.*, 1, 301–304.

Richardson, J.F. and Zaki, W.N. (1954). Sedimentation and fluidisation: Part I. *Trans. Instit. Chem. Eng.*, 32, 35–53.

Rietveld, L.C., Schagen, K.M.V., Kramer, O.J.I. (2006). Optimal operation of the pellet softening process. In *Proceedings of the AWWA Workshop*, Austin.

Rossum, J.R. and Merrill, D.T. (1983). An evaluation of the calcium carbonate saturation indexes. *J. Am. Water Works Assoc.*, 95–100.

Schagen, K.M.V., Rietveld, L.C., Babuška, R. (2008). Dynamic modelling for optimisation of pellet softening. *J. Water Supply Res. Technol.*, 57, 45–56.

Schetters, M.J.A. (2013). Grinded dutch calcite as seeding material in the pellet softening process. Master's Thesis, Delft University of Technology.

Schetters, M.J.A., Van der Hoek, J.P., Kramer, O.J., Kors, L.J., Palmen, L.J., Hofs, B., Koppers, H. (2015). Circular economy in drinking water treatment: Reuse of ground pellets as seeding material in the pellet softening process. *Water Sci. Technol.*, 71, 479–486.

Slack, J.G. (1980). Calcium carbonate hexahydrate: Its properties and formation in lime soda softening. *Water Res.*, 14(7), 799–810.

Su, C.C., Dulfo, L.D., Dalida, M.L.P., Lu, M.C. (2014). Magnesium phosphate crystallization in a fluidized-bed reactor: Effects of pH, Mg: P molar ratio and seed. *Sep. Purif. Technol.*, 125, 90–96.

Tai, C.Y. (1999). Crystal growth kinetics of two-step growth process in liquid fluidized-bed crystallizers. *J. Cryst. Growth*, 206, 109–118.

Tai, C.Y. and Hsu, H.P. (2001). Crystal growth kinetics of calcite and its comparison with readily soluble salts. *Powder Technol.*, 121, 60–67.

Van Ammers, M., Van Dijk, J.C., Graveland, A. (1986). P.A.N.M. Nühn: State of the art of pellet softening. *Water Supply*, 4, 223–235.

Van der Veen, C. and Graveland, A. (1988). Central softening by crystallization in a fluidized-bed process. *J. Am. Water Works Assoc.*, 80, 51–58.

Van Dijk, J.C. and Wilms, D.A. (1991). Water treatment without waste material-fundamentals and state of the art of pellet softening. *J. Water SRT-Aqua*, 40(5), 263–280.

Van Houwelingen, G., Bond, R., Seacord, T., Fessler, E. (2010). Experiences with pellet reactor softening as pretreatment for inland desalination in the USA. *Water Treat.*, 13(1–3), 259–266.

Van Schagen, K.M., Rietveld, L.C., Babuška, R., Baars, E. (2008a). Control of the fluidised bed in the pellet softening process. *Chem. Eng. Sci.*, 63, 1390–1400.

Van Schagen, K.M., Rietveld, L.C., Babuška, R., Kramer, O.J.I. (2008b). Model-based operational constraints for fluidised bed crystallisation. *Water Res.*, 42(1–2), 327–337.

Van Schagen, K.M., Babuška, R., Rietveld, L.C., Veersma, A.M.J. (2009). Model-based dosing control of a pellet softening reactor. *IFAC Proc.*, 42, 267–272.

Wiechers, H.N.S., Sturrock, P., Marais, G. (1975). Calcium carbonate crystallization kinetics. *Water Res.*, 9, 835–845.

16

Metal Removal

16.1. Iron and manganese removal: general aspects

The water to be treated, be it of underground or surface origin, may contain iron and/or manganese. These two elements can lead to degradation in the quality of drinking water due to a red or black coloring. This is induced by their strong presence and may cause their precipitation at the interior or networks, staining clothing and sanitary facilities. It is therefore crucial to reduce their concentration in order to prevent any inconvenience on the part of the consumer.

16.1.1. *Presence of iron and manganese at the resource level*

Iron and manganese are natural elements present in large quantities in the Earth's crust, mainly in the form of mineral oxides. In deep wells, where the oxygen content is low, water containing iron/manganese is clear and colorless (iron and manganese are dissolved). Tap water may be clear, but when exposed to air, iron and manganese become oxidized and change from colorless dissolved forms into colored solid forms. When iron does not form particles large enough to settle and these remain in suspension (colloidal iron), water becomes tarnished with an orange-reddish color. Manganese is usually dissolved in water, although some shallow wells contain colloidal manganese (black tint). These sediments are responsible for the coloring properties of water containing high concentrations of iron and manganese.

These precipitates or sediments can be quite severe to the point of clogging waterways. Upon contact with water, these minerals can be dissolved and release ferrous Fe^{2+} and manganous Mn^{2+} ions which, in turn, can be oxidized into iron hydroxides ($Fe(OH)_3$) and insoluble manganese oxides (MnO_2), depending on the oxidation–reduction conditions of water.

Leaching from soils rich in iron and manganese in alluvial boreholes and in surface waters could be an additional, more or less specific, underlying cause for this phenomenon.

Figure 16.1. *Water resource containing iron. For a color version of this figure, see www.iste.co.uk/gaid/watertreatment3.zip*

Industrial activities can also be an iron and manganese contamination source through:

– the steel industry (grinding of ores, pickling with acid, dust, alloys, etc.);

– the mining industry (mines, heaps, dust, etc.);

– the ceramics industry (dyes);

– the electric industry (electrodes);

– the chemical industry (particularly fertilizers).

In most cases, the iron and manganese present in the water arise from the aquifer's sediments. The forms in which these two metals appear strongly depend on the hydrodynamic relations between groundwater and river water. Introducing river water – which is generally quite oxygenated – into groundwater can lead to oxidation and retention by an alluvium of ferrous Fe^{2+} ions, which are transformed into iron oxides and hydroxides.

In the case of "stagnant" surface waters (reservoirs, lakes), and more rarely, in groundwater, iron and manganese can be complexed with certain organic compounds (mainly humic and fulvic acids), which make their removal even more difficult.

16.1.2. *Presence of iron and manganese at the production level*

Through the course of the water treatment, the use of chemical reagents may induce an additional supply of iron and/or manganese at some treatment stages.

These are mainly the following stages:

– coagulation with an iron salt, such as ferric chloride, which may contain high levels of manganese (50–600 mg $Mn \cdot kg^{-1}$ $FeCl_3$);

– oxidation against potassium permanganate.

Depending on the treatments carried out downstream of the injection points for these reagents, iron and manganese can be present in soluble (ions) or insoluble (oxides) forms. The soluble form is easily removable during filtration, whereas the dissolved form crosses this barrier and is found in the water leaving the treatment unit.

16.1.2.1. *Presence of iron and manganese at the distribution level*

Water transported in a network may contain suspended mineral and organic matter, be them corrosion products or particles whose removal was not ensured upstream. Slow water transport velocities or stagnant areas in a pipe may induce the inconsistent formation of deposits.

The presence of iron and manganese at the distribution stage is due to the chemical and/or biological corrosion phenomena within the metallic equipment used for distribution circuits (pipes, valves, etc.). Corrosion requires little oxygen: 1 $mg \cdot L^{-1}$ of oxygen creates 3.6 $mg \cdot L^{-1}$ ferrous iron. After dissolved oxygen wears down, the anaerobic corrosion mechanisms come into action, forming sulfides from sulfates. These products are not necessarily corrosive, but adhere poorly to the pipe walls and are likely to cause users to complain about the water color.

16.1.2.2. Effects of these two metals

Iron and manganese are essential trace elements, and drinking water contributes only a small part to our body's needs. The effects of the presence of these two metals are felt mainly at the economic level and at the level of the organoleptic quality of distributed water. Indeed, iron and manganese can affect the flavor and color of food and water. They can react with tannins in coffee, tea and some alcoholic beverages, producing a black deposit which affects both taste and appearance. Manganese is not acceptable in water, even when present in smaller concentrations than iron. Iron causes reddish-brown stains on linen, dishes, utensils, and even glassware. Manganese acts similarly, but causing a brownish black stain. While soaps and detergents do not remove these stains, the use of chlorine bleaches and alkaline products (such as carbonate) can even intensify them. Iron and manganese deposits accumulate in pipelines, pressure tanks, water heaters and water softeners. This reduces the water supply quantity and pressure availability.

16.1.2.3. At the economic level

The presence of iron and manganese at the resource level can lead to the clogging of collection structures (premature aging of wells, boreholes, flow decrease, etc.). At the production level, it can result in excessive clogging of the filtering material (which is sometimes irreversible, as in the case of granular activated carbon), the accelerated degradation of metallic equipment and pipes (corrosion under the deposits), as well as an increase in treatment rates (oxidants), which may in the end have an impact on production costs. The creation of an environment favorable to the development of specific non-pathogenic bacteria (Fe-oxidizing bacteria) causes network corrosion and also a "fishy rotten" taste and smell due to the decomposition of Fe-oxidizing bacteria zoöglœa.

Figure 16.2. *Presence of iron in tap water. For a color version of this figure, see www.iste.co.uk/gaid/watertreatment3.zip*

16.1.3. *Regulatory aspects of iron and manganese*

The French Government Resolution of January 11, 2007 relating to the quality reference values for raw water and potable water specifies that the concentration limit for iron is 0.2 $mg{\cdot}L^{-1}$ and the concentration limit for manganese is 0.05 $mg{\cdot}L^{-1}$.

Concentrations beyond these thresholds cause unpleasant tastes in water facilities and it is not uncommon to treat water containing less than 0.2 $mg{\cdot}L^{-1}$ of iron.

Water treatment companies aim for concentrations in treated water equal to 0.05 $mg{\cdot}L^{-1}$ of iron and 0.02 $mg{\cdot}L^{-1}$ of manganese, because beyond these levels operating problems start to be noticed.

These recommendations are issued for the purpose of user comfort, rather than for attenuating health risks.

16.1.4. *Iron and manganese treatments*

Iron and/or manganese treatments can be:

– physicochemical;

– biological;

– a combination of physicochemical and biological procedures.

	Iron	Manganese
Chemical oxidation	pH > 6 Dissolved oxygen (pH > 7.5) Nitrates rH > 20 eH > 0.25v Formation of iron oxides O_3, Cl_2, ClO^-, ClO_2, $KMnO_4$ (Catalytic) MnO_2	pH > 7.2 Dissolved oxygen Nitrates rH > 23 eH > 0.25v Manganese oxide formation O_3, ClO_2, $KMnO_4$ (Catalytic) MnO_2
Biological oxidation	Oxidation by autotrophic bacteria Dissolved oxygen < 2.5 $mg \cdot L^{-1}$	Oxidation by heterotrophic bacteria Dissolved oxygen > 5 $mg \cdot L^{-1}$
Mixed chemical and biological oxidation	Biological iron removal	Chemical oxidation or controlled biological manganese removal

Table 16.1. *Summary of different iron and manganese processing techniques and operating conditions*

Iron and manganese oxidation and precipitation reactions can be chemical, biochemical or a combination of both. Biochemical reactions are carried out by autotrophic bacteria which act directly, or by heterotrophic bacteria excreting reaction catalyst polymers. Table 16.1 summarizes the factors involved in the chemical and biochemical oxidation or reduction of iron and manganese.

The rH parameter comprises pH and eH. It should be noted that the chemical or biological oxidation of iron is much easier and faster than that of manganese. Conversely, the reduction of manganese is easier than that of iron. When the oxygen content is above 4 $mg \cdot L^{-1}$, the oxidation is mainly chemical. If it is below 2 $mg \cdot L^{-1}$, biochemical oxidation is involved.

Corrective treatments are therefore based on the oxidation (neutralization) of these metals in insoluble form by modifying the water's redox potential and pH.

16.1.4.1. *Iron treatment: oxidation–reduction potential – pH*

In solution, iron can be found in different forms depending on the water's pH and oxidation–reduction conditions (eH potential). The simplified potential/pH diagram (Figure 16.3) swiftly portrays the different preponderant forms likely to be found. Despite this representation not being very accurate, it makes is possible to understand the likely reduction and solubilization phenomena of iron oxides (or Mn found in aquifers).

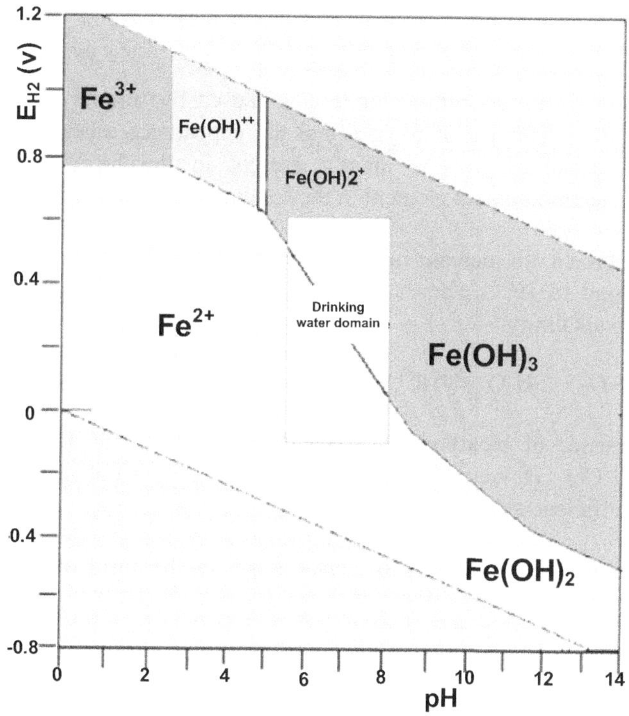

Figure 16.3. *Potential diagram – iron pH. For a color version of this figure, see www.iste.co.uk/gaid/watertreatment3.zip*

The domain of aqueous solutions lies between the two parallels representing the water's oxidation and reduction. The more restricted domain corresponding to feed water can be limited to pH between 5.4 and 8, and to redox potentials between 0 and 0.6 V. In natural waters whose pH > 5–6, iron is mainly present in its dissolved form, Fe^{2+}. It must then be oxidized into the ferric ion Fe^{3+}, which precipitates in the form of fine ferric hydroxide $Fe(OH)_3$ suspensions.

In dissolved oxygen-saturated surface and dam waters with a pH > 7.5, iron is precipitated naturally. Regardless of the water's pH, ferric iron can form stable complexes with certain compounds such as organic matter, silicates, phosphates and sulfates. In these specific cases, it is necessary to adopt special provisions in order to destabilize colloids and destroy the complexes.

16.1.4.1.1. Relationships with other compounds

Bicarbonates

In deep waters, the dissolution of iron is caused by soil fermentation, which results in the disappearance of oxygen and the appearance of carbon dioxide. In these waters, dissolved iron is mainly present in the form of bicarbonates [$Fe(HCO_3)_2$], the predominant element in the solubility of iron and manganese.

During treatment, an increase in pH and/or alkalinity decreases the solubility of this metal bound to HCO_3. For a given pH, the iron concentration is inversely proportional to alkalinity:

$$4\ Fe^{2+} + O_2 + 2\ H_2O + 8\ HCO_3 \rightarrow 4\ Fe(OH)_3 + 8\ CO_2$$

In the presence of bicarbonates, oxygenation leads to the formation of iron hydroxide and CO_2. However, this concentration of CO_2 is not sufficient to affect the oxidation efficiencies of Fe^{2+}.

Near neutrality, the solubility of ferrous iron in the form of bicarbonates is approximately:

$$5\ mg \cdot L^{-1} \text{ of } Fe^{2+} \text{ for } 50\ mg \cdot L^{-1}\ CaCO_3 \text{Alkalinity}$$

and of:

$$1.3\ mg \cdot L^{-1} \text{ of } Fe^{2+} \text{ for } 210\ mg \cdot L^{-1}\ CaCO_3 \text{Alkalinity}$$

Silica and phosphates

These compounds are iron-complexing agents. A concentration greater than 15 $mg \cdot L^{-1}$ of SiO_2 and 2 $mg \cdot L^{-1}$ of phosphates (P_2O_5) complexify iron. These "iron-silica" complexes are sometimes used for preventing the precipitation of ferric iron. The treatment of water loaded with iron-silica complexes can be solved by means of flocculation with a low ferric chloride dose. This is due to the anionic nature of the complex ion previously formed by iron and silica, which then becomes coagulable like a simple colloid. At high concentrations, silicates and phosphates are ferric iron complexing agents, and therefore iron removal inhibitors. Silicates and polyphosphates are sometimes used for preventing the precipitation of ferric iron within pipes.

Silicates are more effective than phosphates (and polyphosphates) for the anti-scaling of ferric iron. During potabilization treatments, they can interfere with iron retention on the filters, leading to poor operating results. The raw waters involved are relatively rare.

Although the presence of silica (over 15 $mg \cdot L^{-1}$ of SiO_2) imposes an additional oxidation and/or flocculation stage, it does not require any significant modification to the process.

Organic matter

The phenomenon of complexation with organic matter mainly takes place in surface waters and, exceptionally, in groundwater. In the case of lakes, organic matter results from their primary productivity, as well as from the leaching of the catchment area.

Humic, fulvic, tannic acids, etc., are then found in solution. The metals, transported in the anionic form in water, are reduced by these acids, producing complexes or chelates in the cationic form. Among others, ferro- and ferri-silicic, ferri-phosphoric and humo-ferric complexes are also found. Organic matter leads to the disappearance of oxygen and nitrates from the water, as well as the reduction of iron and manganese.

A good removal of iron requires a preliminary organic matter (TOC, NOM) treatment. A concentration of organic matter (Ox. $KMnO_4$) beyond 5 $mg \cdot L^{-1}$, and/or a true color greater than 20 $mg \cdot L^{-1}$ Pt-Co lead to a modification in the process because the iron can be in a complexed form.

The different states of iron

The different states in which iron can be found in water are shown in Table 16.2.

Figure 16.4 describes in detail the forms of total iron in water, as well as its other available forms, depending on its ferrous or ferric form.

States of divalent iron Fe^{2+}	States of oxidized iron Fe^{3+}
Free ion in water (rare) Adsorbed ion (frequent) Ion complexed with organic and mineral compounds (frequent) Ion precipitated as iron carbonate and iron sulfide	Crystal oxides Amorphous oxides $Fe(OH)_3$ hydroxides

Table 16.2. *Forms in which iron may be present*

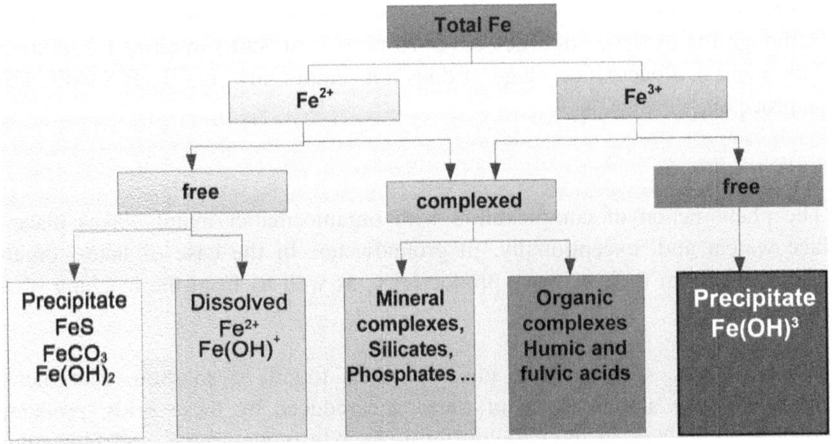

Figure 16.4. *The different existence forms of iron in water. For a color version of this figure, see www.iste.co.uk/gaid/watertreatment3.zip*

16.1.4.1.2. Treatment principles

In natural waters whose pH is always above 4, dissolved iron can only exist in the divalent form (Fe^{2+} or ferrous iron). In principle, it suffices oxidizing it to trivalent iron (Fe^{3+} or ferric iron), which is insoluble in the medium's pH conditions, in order to obtain its precipitation and thus be able to separate it. Corrective treatments are based on iron oxidation, which turn to its insoluble form by modifying its oxidation potential and pH. There is passage from the dissolved form of iron (Fe^{2+}) to a precipitated form [$Fe(OH)_3$]. The oxidized forms precipitate and are retained, either by settling or filtration.

On the other hand, ferric iron can be found at pH > 4 in colloidal form. It can also form stable complexes with organic matter, silica or phosphates. In all these cases, it is necessary to break the complex in order to release the metal and then be able to remove it.

There are also biological treatments relying on specific micro-organisms whose metabolism is closely dependent on rH conditions and, consequently, on pH.

It is essential to properly characterize the water to be treated, knowing:

– its pH;

– the redox potential;

– dissolved oxygen and CO_2;

– total and dissolved iron;

– its alkalinity.

16.1.4.1.3. Physical treatments without reagents

Natural oxidation

Oxidation is generally obtained by simple natural or mechanical aeration, surface runoff or sparging. Pure oxygen also has sufficient oxidizing power because of the diffusion of the gas under pressure in the water.

Cascade aeration or air bubble induced oxidation in a contact tank

Iron oxidation by aeration is limited if iron is in its complexed form. The air oxidation reaction is:

$$4Fe^{2+} + 8H_2O \rightarrow 4Fe(OH)_2 + 8H^+$$

and $4Fe(OH)_2 + O_2 + 2H_2O \rightarrow 4Fe(OH)_3$

that is, $4Fe^{2+} + O_2 + 10H_2O \rightarrow 4Fe(OH)_3 + 8H^+$

The oxidation of 1 mg of Fe^{2+} requires 0.14 mg O_2.

Figure 16.6 illustrates iron oxidation by air injection using oxytubes in a tank. Air ramps can be lifted. A pH probe and an oxygen probe monitor these two parameters.

The stoichiometric dose is 0.14 $g \cdot m^{-3}$ of O_2 per g Fe^{2+}, or in other words, 0.47 L of air (0°C and 760 mm Hg) per gram of Fe^{2+}. Kinetics are favorable starting at pH 7.5, which implies a contact time between 3 and 5 min. Figure 16.7 shows the contact times required for iron oxidation as a function of pH and temperature.

Figure 16.5. *Cascade aeration on surface water (Veolia France Ouest site). For a color version of this figure, see www.iste.co.uk/gaid/watertreatment3.zip*

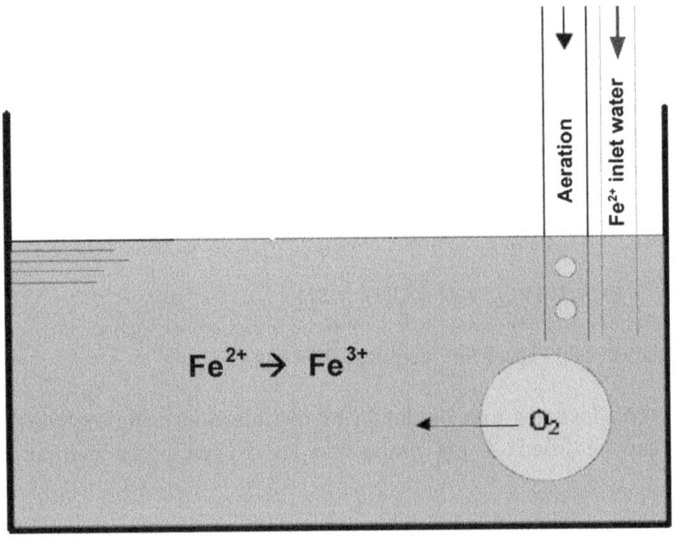

Figure 16.6A. *Air bubble induced oxidation in a tank. For a color version of this figure, see www.iste.co.uk/gaid/watertreatment3.zip*

Figure 16.6B. *Air bubble induced oxidation in a tank. For a color version of this figure, see www.iste.co.uk/gaid/watertreatment3.zip*

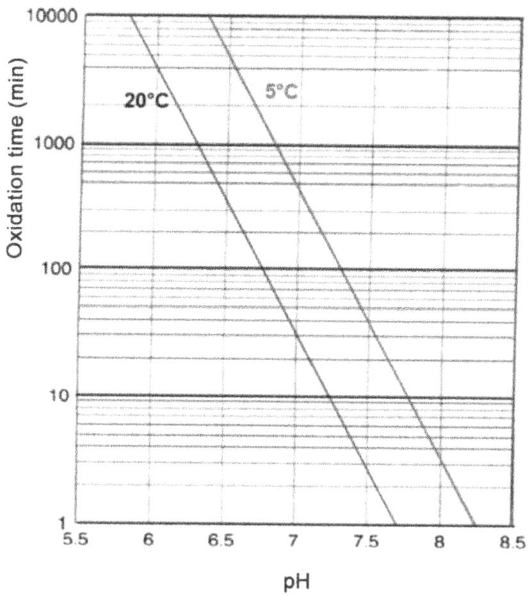

Figure 16.7. *Contact time as a function of pH for air oxidation*

An empirical ratio obtained on various sites shows the oxidation efficiency as a function of pH:

$$Fe^{2+} \text{ oxidation efficiency } (\%) = -54.322\ pH^2 + 897.32\ pH - 3574.8$$

The number of cascade steps required for this oxidation depends on the initial concentration of Fe^{2+}. At pH 7.5, for example, three steps are enough for an inlet concentration of 1 $mg \cdot L^{-1}$ Fe^{2+}.

Water sprinkling oxidation on a support

Water sprinkling systems break water into small droplets, creating oxidation at the air–water interface. The effectiveness of these systems depends on the fall height of droplets, pH, temperature and the iron concentration in raw water. However, after a certain cascade height, the efficiency remains constant because the air–water interface is not renewed. Therefore, when the interface is saturated, gas transfer stops. These systems are used for low concentrations of oxidizable iron (<1 $mg \cdot L^{-1}$).

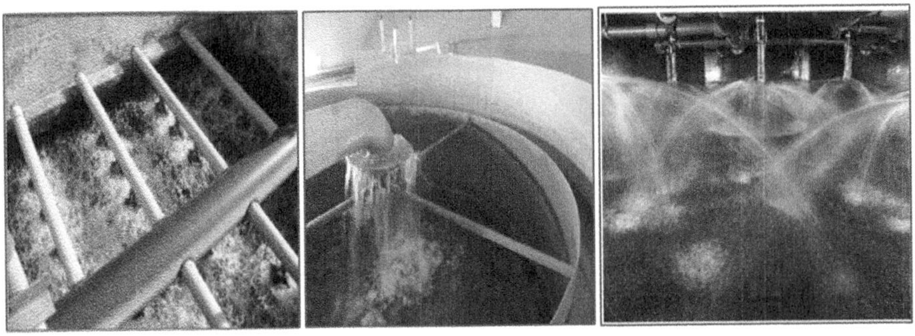

Figure 16.8. *Water sprinkling oxidation. For a color version of this figure, see www.iste.co.uk/gaid/watertreatment3.zip*

Forced air oxidation in oxidation tower

The oxidation tower is a reactor packed with a support which can be made of pozzolan or plastic components, such as Rashig (Techim), Levapac, Pall rings, etc.

The water's flow direction can be ascending, concomitantly with air (co-current system) or descending, against the air current (countercurrent system).

A large contact surface between air and water results from this combination. The air–water interface is permanently renewed with every cascade fall. The air is supplied continuously by a compressor. This stage requires good mixing and a few minutes contact time (depending on temperature and iron content to be oxidized).

The higher the pH, the higher the oxidation velocities. The oxygen supply also serves to strip the hydrogen sulfide, which is frequently found there in trace quantities, and to bring the water almost to saturation with dissolved oxygen, so as to help maintain good organoleptic qualities in distribution and to prevent network corrosion.

Figure 16.9. *Plastic support in an oxidation tower. For a color version of this figure, see www.iste.co.uk/gaid/watertreatment3.zip*

The higher the pH, the higher will be the oxidation velocity. On the contrary, the lower the pH, the longer the oxidation times (contact time required between O_2 and Fe^{2+}).

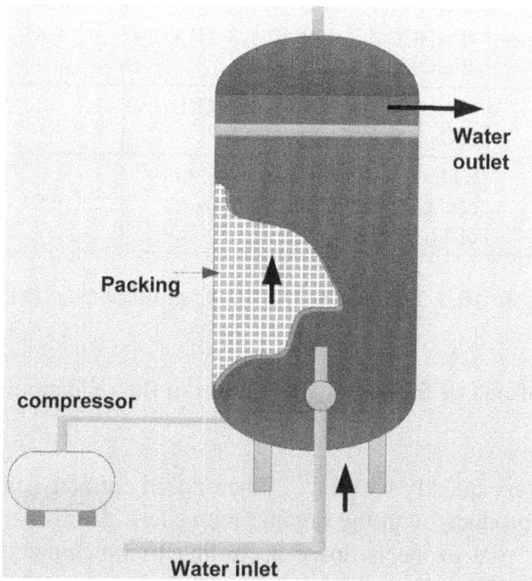

Figure 16.10. *Air–water co-current oxidation tower. For a color version of this figure, see www.iste.co.uk/gaid/watertreatment3.zip*

Depending on the iron concentration present in raw water and the treatment goals, media height varies between 0.8 and 2 m. The flow rate is 40 $m \cdot h^{-1}$ and the contact time can vary between 3 and 6 min.

16.1.4.1.4. Chemical oxidation treatments

Ozone, chlorine, chlorine dioxide, potassium permanganate

Chemical oxidation is obtained by injecting an oxidant such as ozone, chlorine, chlorine dioxide or potassium permanganate.

The chemical reactions and the theoretical oxidant quantities are summarized in Table 16.3.

Oxidants	Chemical reactions	Iron oxidant quantities, mg/mg	Iron precipitates, mg/mg
Ozone (O_3)	$2Fe^{2+} + O_3 + 5H_2O \rightarrow 2Fe(OH)_3 + O_2 + 4H^+$	0.43	1.91
Chlorine (Cl_2)	$2Fe(HCO_3)_2 + Cl_2 + 2H_2O \rightarrow 2Fe(OH)_3 + 2HCl + 4CO_2$	0.64	1.91
Sodium hypochlorite (NaOCl)	$2Fe(HCO_3)_2 + NaOCl + H_2O \rightarrow 2Fe(OH)_3 + NaCl + 4CO_2$	0.45	1.91
Calcium hypochlorite ($(Ca(OCl))_2$)	$4Fe(HCO_3)_2 + Ca(OCl)_2 + 2H_2O \rightarrow 4Fe(OH)_3 + CaCl_2 + 8CO_2$	0.45	1.91
Chlorine dioxide (ClO_2)	$Fe^{2+} + ClO_2 + 3H_2O \rightarrow Fe(OH)_3 + ClO_2^- + 3\,H^+$	1.21	1.91
Potassium permanganate ($KMnO_4$)	$KMnO_4 + 3Fe(HCO_3)_2 + 7H_2O \rightarrow MnO_2 + 3Fe(OH)_3 + KHCO_3 + 5CO_2 + 5H_2O$	0.94	1.91

Table 16.3. *Stoichiometric ratios for various oxidants*

The addition of one of these oxidants results in the oxidation of Fe^{2+} into Fe^{3+}, which precipitates.

Ozone reacts very quickly with Fe^{2+}. When dosed correctly, there is no residual of O_3 to form by-products with the organic molecules. As iron and manganese are often both present, the ozone is dosed according to stoichiometric ratios, so that there will be no residual:

$$2Fe(HCO_3)_2 + O_3 + H_2O \rightarrow 2Fe(OH)_3 + 4CO_2 + O_2$$

Chlorination is a solution used for applications in groundwater, not in surface water. This is because the oxidation kinetics of organic matter and other elements are as fast as those of iron. The chlorine demand may be high and iron oxidation adversely affected.

Chlorination can be obtained with chlorine gas, sodium hypochlorite and calcium hypochlorite.

Sodium hypochlorite is usually the chosen oxidizer because it is relatively inexpensive and readily available:

$$2Fe(HCO_3)_2 + Cl_2 + 2H_2O \rightarrow 2Fe(OH)_3 + 2HCl + 4CO_2$$

0.64 mg Cl_2 per mg Fe^{2+}

$$2Fe(HCO_3)_2 + NaOCl + H_2O \rightarrow 2Fe(OH)_3 + NaCl + 4CO_2$$

$$4Fe(HCO_3)_2 + Ca(OCl)_2 + 2H_2O \rightarrow 4Fe(OH)_3 + CaCl_2 + 8CO_2$$

Potassium permanganate is an alternative to chlorination, with the following reaction:

$$MnO_4^- + 3Fe^{2+} + 7H_2O \rightarrow MnO_2 + 3Fe(OH)_3 + 5H^+$$

Note that 1 mg of iron requires approximately 0.94 ppm $KMnO_4$. In virtue of the acidity formed, the oxidation reaction shows that the alkalinity consumption is 1.49 $mg \cdot L^{-1}$ $CaCO_3$ per mg Fe^{2+} (that is, $(500/2)/(3 \times 56)$) or 1.67 equivalents per mole of Fe^{2+} ($5H^+/3$ Fe). The recommended minimum pH is 6.5. Iron requires about 0.94 ppm of permanganate, whereas H_2S and Mn require 6 ppm and 2 ppm, respectively. Potassium permanganate requires careful dosage control because an excess can tarnish water with a pink to purple color.

Table 16.4 summarizes the iron treatment options using the physicochemical route.

In the case of heavily loaded waters, it is sometimes necessary to resort to an additional coagulation–flocculation–settling stage before filtration in order to prevent the rapid clogging of filters. Regardless of the reaction pH, iron-silica complexes do not make it possible to obtain good quality filtered water. On a similar note, iron complexed with humic substances can be removed using a coagulant (aluminum sulfate, ferric chloride, ferric sulfate).

Iron complexed by humic matter	Uncomplexed iron		
	Iron < 2 mg·L^{-1}	2 < *Iron* < 5 mg·L^{-1}	*Iron* > 5 mg·L^{-1}
	Aeration	Aeration	Aeration
Coagulation/flocculation	-	-	Coagulation/ flocculation
Settling	-	-	Settling
pH control	pH control	pH control	pH control
Sand filter	Sand filter	Dual media filter	Sand or dual media filter
Disinfection	Disinfection	Disinfection	Disinfection

Table 16.4. *Summary of iron treatment characteristics*

16.1.4.1.5. Biological iron removal treatment

Bacteria have different respiratory modes and use dissolved oxygen, nitrates, and then iron oxides as a source of O_2. This is why bacterial biological processes are involved in iron oxidation. The underlying principle is to promote, and then control, bacterial development directly on the filtration support. The developed biomass is made up of bacteria whose metabolism is based on iron oxidation. Most Fe-oxidizing bacteria are autotrophic, meaning that they use carbon dioxide (CO_2) or water bicarbonates (HCO^{3-}) as carbon sources.

There are three main iron-oxidizing bacteria groups:

– filamentous (sheathed) bacteria, which are easily recognizable on microscopic examination, given the fact that typically characteristic species are observed in many cases. For example, the genus *Leptothrix*, which is an aerobic, heterotrophic bacteria (that is to say, which uses organic carbon);

– peduncle bacteria, such as the genus *Gallionella*. It is an autotrophic aerobic bacterium, developing from a biofilm (a grain of sand), with long twisted filamentous peduncles due to the permanent rotation of the bacterium on itself;

– true bacteria or eubacteria (cocci, rods, vibrios, etc.).

Iron (and manganese) bacteria are widespread in nature, not only in groundwater, wells, water treatment stations, raw water or treated water pipes, but also in springs, ponds, on the sediments of lakes, etc. In general, the natural seeding of treatment facilities poses few problems, especially in the case of iron.

It was the observation of physicochemical iron removal stations were functioning correctly when the characteristics of the raw water should have caused serious malfunctions that revealed the presence of a bacterial flora specific to iron.

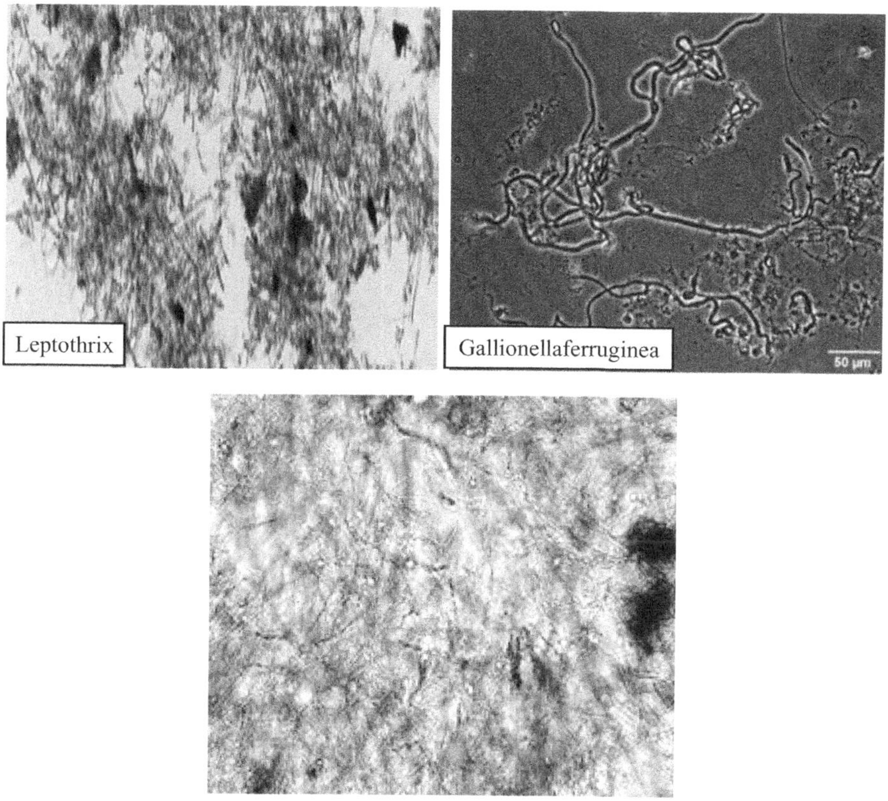

Figure 16.11. *Leptothrix and Gallionella bacteria. For a color version of this figure, see www.iste.co.uk/gaid/watertreatment3.zip*

Fe-oxidizing bacteria oxidize reduced iron (Fe^{2+}) and from this oxidation they draw the energy needed for their metabolism. The quantity of energy thus released is low and bacteria must oxidize a large quantity of iron in order to multiply. These Fe-oxidizing bacteria are aerobic. Two types of biochemical reactions are described depending on the bacterial families:

$$Fe^{2+} + 1/2O_2 + H_3O^+ \rightarrow Fe^{3+} + 2H_2O + 11.3 \text{ kcal}$$

$$2Fe^{2+} + 1/2O_2 + 2H_3O^+ \rightarrow 2Fe^{3+} + 3H_2O + 22.6 \text{ kcal}$$

For some bacteria, the oxidation of this metal is indirect: it is due to the pH increase associated with their growth. This pH increase in the vicinity of the bacteria enables the oxidation of iron by oxygen at a variable rate.

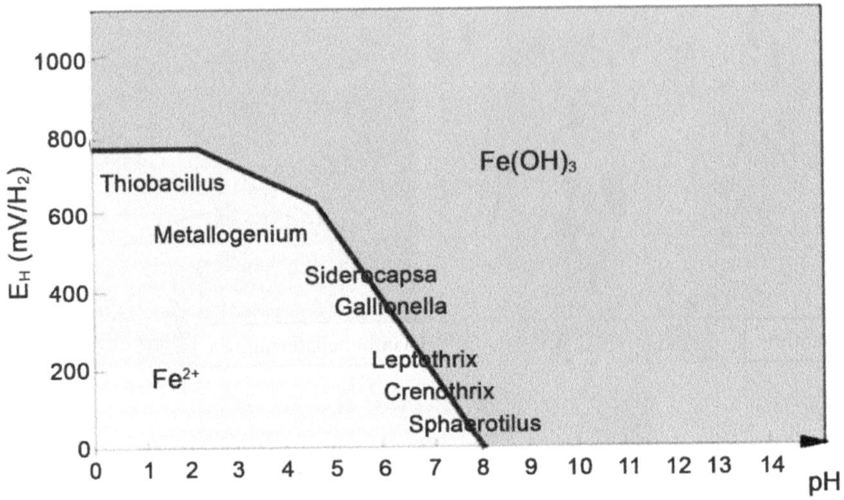

Figure 16.12. *Main bacteria observed as a function of pH and redox potential. For a color version of this figure, see www.iste.co.uk/gaid/watertreatment3.zip*

They oxidize iron either by intracellular enzymatic route, or by extracellular catalytic route under the action of polysaccharide substances excreted by the cell. These two phenomena are extremely rapid.

For metals complexed with organic matter, various heterotrophic bacteria use the organic fraction of the iron/organic matter complex and release the iron, which then precipitates catalytically near their capsules. This return of metals to the mineral state enables autotrophic bacteria to continue their task, triggering oxidation and precipitation. This technique ensures the oxidation of dissolved iron and the retention of insoluble hydroxides within the same filtering material. In all cases, the precipitates formed are more or less complexed ferric oxides, which accumulate around the bacterial bodies. This type of precipitate is much less sticky than the ferric hydroxide formed by physicochemical means. Their growth conditions essentially depend on temperature, pH, dissolved oxygen, redox potential, rH (combined function of pH and redox potential), iron and manganese concentration, salinity and organic matter content.

The potential/pH diagram (Figure 16.13) shows that the favorable conditions for implementing biological iron removal exist in a fairly narrow pH range. If the water characteristics are auspicious for the development of an iron-removing biomass, this seeding takes place by the simple filtration of raw water for a few days.

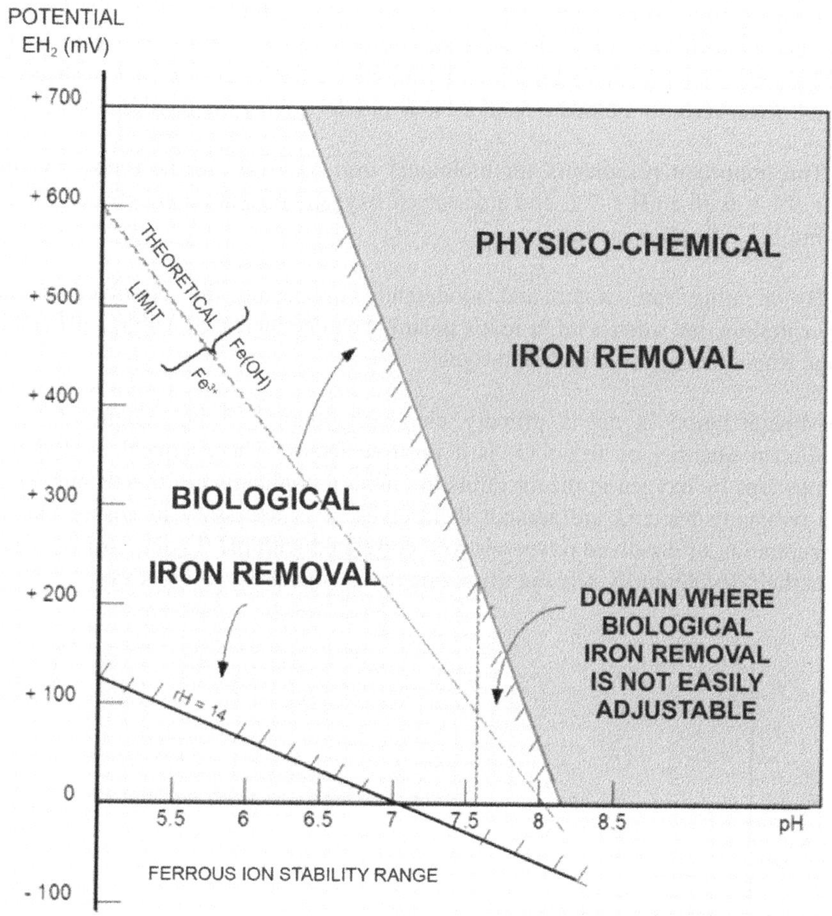

Figure 16.13. *Domains of existence of physicochemical and biological iron removal. For a color version of this figure, see www.iste.co.uk/gaid/watertreatment3.zip*

If the water characteristics are suitable for the development of an iron-removing biomass, this seeding takes place by the simple filtration of raw water for a few days. During the start-up phase—but only during this short period of time—it may be advantageous to wash the filters with raw water.

Once set in, the biomass finds its equilibrium between a development due to the presence of iron (or manganese) in the water to be treated and a partial removal of the biomass during the backwashes of the filter. The unclogging conditions of the filters are similar to conventional air + water backwashing methods. However, it may be necessary to adjust the air scouring conditions, if these prove to be too strong, and cannot maintain a minimum of biomass. The filters are backwashed with treated water free from oxidant residues to avoid destroying the fixed biomass.

The important parameters for biological iron removal are the redox potential eH > 0.1 V with a pH < 7.2, and a dissolved oxygen concentration between 0.2 and 2.5 mg·L^{-1}. The rH must be > 14.

By aerating raw water and moderately increasing the dissolved oxygen concentration, the water's rH or redox potential can be increased within a pH and Eh range where biological oxidation thrives.

Mineralization is not a primary element, although it is recognized that a significant quantity of free CO_2 is a positive factor. The chemical oxidation of ferrous ions by oxygen in the air limits the field of application of this technique. As iron-removing bacteria are aerobic themselves, it is necessary to ensure that the concentration of dissolved oxygen is well correlated with pH (Figure 16.14) so as to grant their development, without triggering the chemical oxidation of iron.

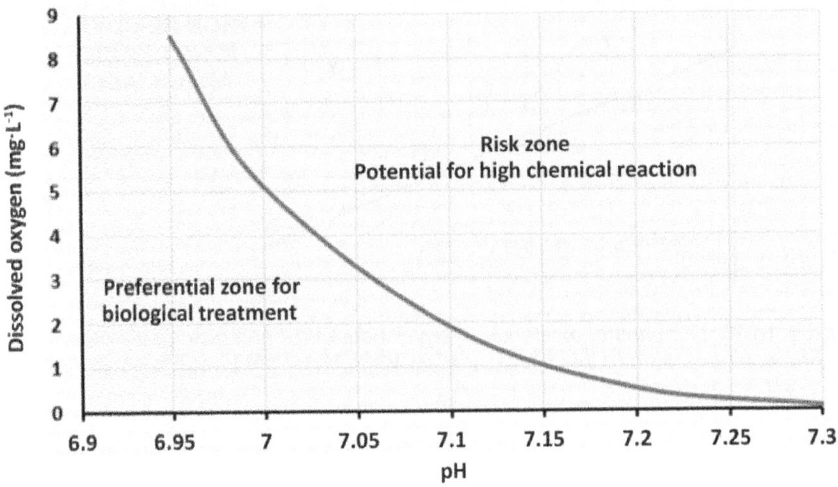

Figure 16.14. *Maximum quantity of O_2 depending on water pH (biological iron removal). For a color version of this figure, see www.iste.co.uk/gaid/watertreatment3.zip*

Oxygen content below ($mg \cdot L^{-1}$)	0.2–2.5
Oxidation–reduction potential (mV)	100–600
rH	14–20
pH	5.5–7.5

Table 16.5. *Optimal conditions for biological iron removal*

The risks of limiting iron-removing bacteria are very rare in natural waters. The only imperative requirement is to place them under pH and redox potential (rH) conditions corresponding to their field of activity.

The biological iron removal domain straddles the theoretical boundary separating Fe(II) and Fe(III):

– if rH < 14, ferrous iron remains in filtered water, the medium is not sufficiently oxidizing;

– if rH > 20, the medium is too oxidizing and iron oxidation is obtained chemically by O_2.

Practical limitations of the biological process

Under standard conditions, treatments can target water with no other potentially harmful effects than those of iron, provided that the content of certain elements does not exceed the values shown in Table 16.6.

Total iron ($mg \cdot L^{-1}$)	<8
Silica ($mg \cdot L^{-1}$)	<15
Manganese ($mg \cdot L^{-1}$)	<0.05
Turbidity (NTU)	<2
Free chlorine ($mg \cdot L^{-1}$)	<0.2
Ammonia ($mg \cdot L^{-1}$)	<2
raw water pH higher than	7.3

Table 16.6. *Limits defining the biological process standard conditions*

Biological iron removal can be implemented aside from the aforementioned limits, in particular when a significant part of the iron is complexed. Dissolved oxygen contents greater than the values expressed by the curve O_2 maxi = f (pH) can be accepted insofar as this complexed iron cannot chemically oxidize in air (for standard quantities of air).

Conditioning the filter is fairly quick and takes less than three weeks if the T °C and pH operating conditions are favorable.

16.1.4.1.6. Treatment stations for iron physicochemical oxidation with air

A simple procedure for treating iron in groundwater comes down to controlling the pH and adjusting it to 7.5 if necessary, then passing the water through a cascade and retaining the formed iron hydroxides on a sand filter. Disinfection is then performed before distribution. Depending on water quality, a balancing or pH adjustment is implemented by adding limewater or caustic soda. This solution is generally proposed for iron concentrations lower than 1 mg·L^{-1}.

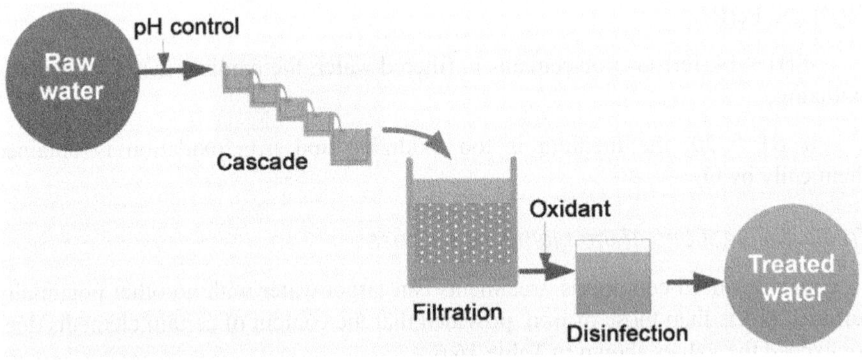

Figure 16.15. *Iron treatment including a cascade. For a color version of this figure, see www.iste.co.uk/gaid/watertreatment3.zip*

This solution has the advantage of operating without the need for energy or chemical reagents. However, its installation on the ground is demanding and a suitable hydraulic profile must be provided. It applies both to surface water and groundwater.

Forced air oxidation with an oxidation tower

When the iron concentration is greater than 1 mg·L^{-1} with peaks reaching up to 5 mg·L^{-1}, an oxidation tower is most often proposed because it makes it possible to achieve the target of a dissolved iron residual <0.2 mg·L^{-1}. A simple procedure for treating iron in groundwater comes down to controlling the pH and adjusting it to 7.5 if necessary, then passing the water through a cascade and retaining the formed iron hydroxides on a sand filter. Disinfection is then performed before distribution.

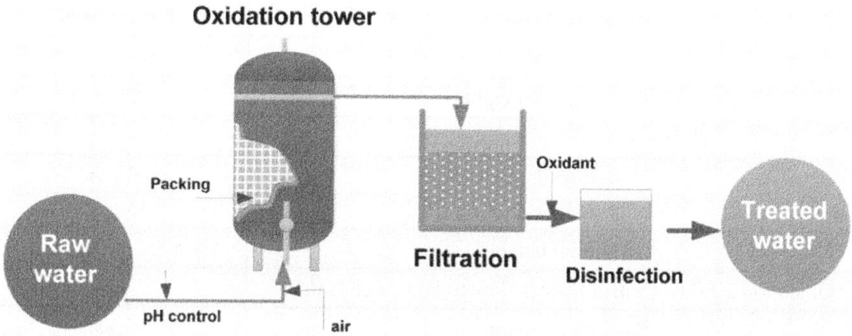

Figure 16.16. *Iron treatment by air oxidation. For a color version of this figure, see www.iste.co.uk/gaid/watertreatment3.zip*

Depending on water quality, a balancing or pH adjustment is implemented by adding limewater or caustic soda. The advantage of this solution is its compactness, in addition to the fact that it can treat high concentrations of dissolved iron. However, it requires an energy requirement, because the air to water ratio has to be of at least 0.1.

Its application requires water with low turbidity (<5 NTU) in order to prevent the packing from being clogged.

Oxidation tower design parameters

The packed oxidation tower includes a Biodagen®-type material or plastic support with a variable layer height, depending on the iron concentration. Pressurized raw water feeds the tower's base. The air is sent to the base of the packed bed. Any excess air is evacuated to the atmosphere through a valve. Water is collected at the top of the tower.

The tower's diameter must allow for a percolation velocity lower than or equal to 40 $m \cdot h^{-1}$. The ferrule height must ensure a useful volume, enabling a contact time $\geq$ 3 min at pH 7.2. As the tower does not have a floor, the raw water inlet is placed under a reversed-dome bottom. Compressed air is injected directly into the tower. Distribution is ensured by a network of perforated pipes embedded in the layer of support materials. The theoretical air flow [in $Nm^3 \cdot h^{-1}$] is 10% of the water flow to be treated.

If the water's pH is lower than 7.2, the standard treatment conditions are no longer met, and the air/water ratio must be increased beyond 10% if one wishes to maintain a limited contact time.

Oxidation tower	Parameters
Flow rate	40 m·h^{-1}
Tower ferrule height	2,000 mm
Pozzolan or plastic support size	30–60 mm
Pozzolan density Plastic support density	0.83 g·m^{-3} 0.35 g·m^{-3}
Material height	1,500 – 2,000 mm
Oxidation air flow (Nm^3/h)/water flow to be treated (m^3/h)	0.1
Operating pressure	>7 bar

Table 16.7. *Oxidation tower design*

When the ferrous iron (Fe^{2+}) content of the water to be treated is between 5 and 10 mg·L^{-1}, the aeration tower is designed for a percolation velocity lower than or equal to 30 m·h^{-1}.

Chemical oxidation

Regardless of the initial dissolved iron concentration, chemical oxidation makes it possible to meet quality requirements. The iron treatment station has a reaction tank (3–5 min) where the chemical oxidant (O_3, Cl_2, ClO^-, ClO_2, $KMnO_4$) is added and pH is controlled. This is followed by sand or dual media filtration (expanded clay/sand or pumice/sand).

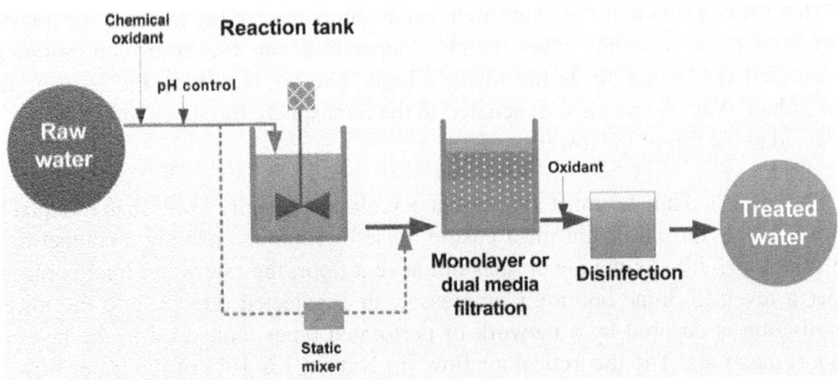

Figure 16.17. *Iron treatment by chemical oxidation. For a color version of this figure, see www.iste.co.uk/gaid/watertreatment3.zip*

When the Fe concentration is lower than 0.8 $mg \cdot L^{-1}$, the chemical oxidant/water mixture can be done with a static mixer. The oxidation reaction takes place in the transport pipeline and on the water volume above the filtering material.

Depending on water quality, a balancing or pH adjustment is implemented by adding limewater or caustic soda, in addition to water disinfection before distribution.

The advantage of this treatment is that it applies to groundwater as well as surface water and that other elements can be oxidized at the same time (arsenic, manganese). It also leads to the partial oxidation of other water constituents (color, algae, etc.), which depend on the type of chemical oxidant.

Coagulants (aluminum salts and iron salts) have an oxidizing power enabling them to oxidize ferrous iron when the levels are not too high ($<0.5\ mg \cdot L^{-1}$). Their efficiency is limited when iron is complexed by organic matter or silica (SiO_2 content greater than 15 $mg \cdot L^{-1}$).

A priori, one may suspect possible iron complexation when the organic matter content (Ox.$KMnO_4$) > 5 $mg \cdot L^{-1}$ O_2, the true color of water > 20 $mg \cdot L^{-1}$ Pt/Co, silica > 15 $mg \cdot L^{-1}$ and phosphates > 2 $mg \cdot L^{-1}$ (P_2O_5).

In this case, an additional treatment is applied between aeration and filtration. It includes a strong oxidant (ozone/permanganate/chlorine dioxide) and/or micro-flocculation (aluminum sulfate or ferric chloride). Flocculant levels are between 3 and 10 $g \cdot m^{-3}$ commercial product, depending on the case. For ozone, treatment rates are between 0.5 and 1.5 g $O_3 \cdot m^{-3}$.

Compressor design parameter for iron oxidation: air compressor capacity

The design of the air compressor used for physicochemical iron removal must take into account not only the quantity of iron to be oxidized, but also the quantity of dissolved oxygen to bring water almost to saturation while maintaining good organoleptic qualities during distribution.

The quantity of oxygen required for the oxidation of iron is given by the following reaction:

$$4Fe^{2+} + O_2 + 10H_2O \rightarrow 4Fe(OH)_3 + 8H^+$$

$$224\ g + 32\ g \rightarrow 0.14\ g \text{ of } O_2 \text{ per g Fe}$$

At 12°C, oxygen saturation is close to 9 $mg \cdot L^{-1}$.

For example, consider the treatment of groundwater (60 $m^3 \cdot h^{-1}$) containing 0.6 $mg \cdot L^{-1}$ of dissolved iron, 4 $mg \cdot L^{-1}$ of dissolved oxygen and a constant temperature of 12°C.

That is:

$$(0.6\ (g \cdot m^{-3}\ \text{iron}) \times 60\ (m^3 \cdot h^{-1})) \times 0.14 = 5.1\ g\ O_2 \cdot h^{-1}$$

The quantity of oxygen to be supplied to approach saturation is then:

$$(9 - 4)\ g \cdot m^{-3} \times 60\ m^3 \cdot h^{-1} = 300\ g\ O_2 \cdot h^{-1}$$

In total, 300 g $O_2 \cdot h^{-1}$ must be dissolved in the water to be treated. Under standard temperature (20°C) and pressure (1 bar) conditions, 1 mole of O_2 (32 g) represents a volume of 24 L. Thus, 300 g O_2 are contained in: 300 g O_2 × 24/32 = 225 L/L of O_2.

Oxygen being present in the air at a proportion of approximately 20% per volume, 1,125 L of air is required to obtain 225 L of O_2. Therefore, 1.125 m^3 would be the volume of air to be injected if the oxygen transfer from air to water was 100% efficient, which is not the case.

This calculated volume should be supplemented by an additional coefficient that globally takes into account other conditions such as injection, mixing and contact times between the air and the water to be treated.

16.1.4.1.7. Filtration design parameters (valid for all oxidation treatments)

The two most common filtration techniques are as follows: monolayer/sand filtration and dual media filtration (expanded clay/sand or pumice/sand).

The filters are open filters or pressure filters. A typical facility (OTV-Veolia) comprises an aeration stage (cascade, water sprinkling system or oxidation tower), followed by filters, which are often vertical.

Sand monolayer filter

Sand has a size of 0.85 mm for a layer height of 1–2 m, depending on the filtration velocity. The filter consists of a floor fitted with long-tailed OTV strainers (40 strainers per m^2 of draining surface). These strainers are fixed to the floor by flanged nuts screwed under the floor. The strainer heads are embedded in a 4/7 pumice layer 0.10 m thick, over which the filtering layer is set. The weight of iron retained per unit of filtering surface is 1,000 g of iron per m^2 of sand for monolayer filters, and 1,800 g of iron per m^2 for dual media filters.

Washing is triggered when:

– the maximum production time, set at start-up, has been reached (setting by clock or counter);

– the maximum head loss, set at start-up, is reached;

– the iron leak exceeds a tolerable concentration (e.g. 50 $\mu g \cdot L^{-1}$).

It has three distinct phases: air, air/water and water alone. Chapter 8 details the conditions for washing monolayer filters. The washwater is admitted through the same tubing as filtered water. Air is admitted through a pipe opening immediately under the floor.

Dual media filtration

In order to increase the length of the filtration cycle between washes, sand filtration can be replaced by expanded clay/sand dual media filtration. The filtration velocities are similar to those recommended for monolayer filtration. Hydro-expanded clay and pumice stone have a size between 1.4 and 2.5 mm for a layer height of 0.50–1 m, depending on the filtration velocity. Sand works as complement in order to obtain a filtering bed height of about 0.5–1 m (sand grain size identical to the monolayer). Dual media filters are washable with air and water, in accordance with the procedure described in Chapter 8.

Filters	**Parameters**
Monolayer filter	
Iron concentration	Fe < 2 $mg \cdot L^{-1}$
Filtration velocity	8–15 $m \cdot h^{-1}$
Sand density	1.65
Sand layer height	800–2,000 mm
Sand grain size	0.8–1.2 mm
Pumice layer height	100 mm
Long-tailed strainer density	40/m^2
Dual media filter	
Iron concentration	2 < Fe < 8 $mg \cdot L^{-1}$
Filtration velocity	10–15 $m \cdot h^{-1}$
Anthracite density	0.7
Pumice stone density	0.7

Anthracite or pumice layer height	500–1,000 mm
Anthracite or pumice grain size	1.4–2.5 mm
Sand layer height	500–1,000 mm
Sand grain size	0.8 mm
Gravel layer height	100 mm
Long-tailed strainer density	45/m^2

Table 16.8. *Filtration design parameters*

16.1.4.1.8. Biological iron treatment

Biological iron treatment involves controlled aeration, followed by sand filtration. The air injection can be carried out in line (directly in the pipes), by inserting a mixing pot with baffles to ensure a good homogenization. Depending on the dissolved oxygen content in raw water, preliminary aeration is applied, followed by filtration and supplemented by additional aeration, to ensure 80–90% O_2 saturation.

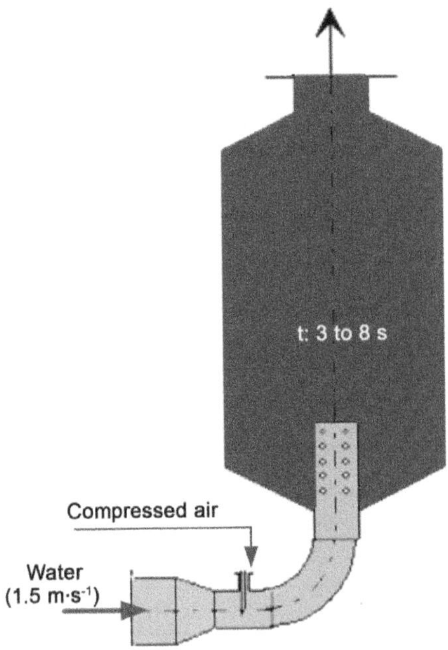

Figure 16.18. *Complementary aeration device. For a color version of this figure, see www.iste.co.uk/gaid/watertreatment3.zip*

This airflow has the role of maintaining the conditions for the dissolved oxygen concentration and redox potential in the pre-aerated raw water, enabling the optimal development of iron bacteria.

In some waters (pH > 7), where the dissolved oxygen content at the filtration inlet must be minutely controlled (maintained at a few tenths of mg·L^{-1}), this solution can be replaced by recycling a small quantity of treated and aerated water. This may require the installation of a small additional pressure-booster.

The filtration medium has a larger particle size than the one used for conventional physicochemical iron removal (1.3–1.5 mm effective size). Biological oxidation is more efficient and faster than chemical oxidation, which makes it possible to treat higher iron values (up to 10 mg·L^{-1}) and at greater filtration velocities (15–35 mh^{-1}), depending on the initial iron concentration. A preliminary seeding phase is always necessary before start-up.

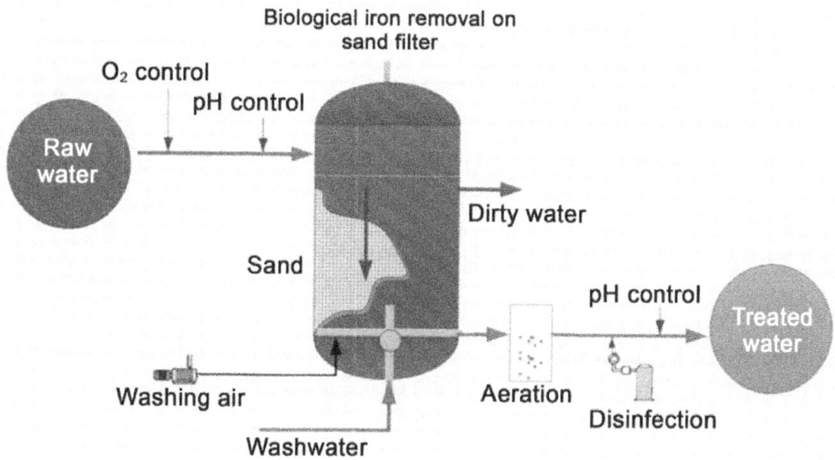

Figure 16.19. *Biological iron removal process with a sand filter. For a color version of this figure, see www.iste.co.uk/gaid/watertreatment3.zip*

A final aeration stage is necessary after a biological iron removal treatment in order to increase the dissolved oxygen content of distributed water.

A simple injection of air suffices, associated with cascades in the recovery basins for the open filters. The filters can be washed with raw water if the turbidity is low, or with chlorine-free treated water.

This is the same type of equipment as the one provided for physicochemical iron removal. Hydraulic loading rates are higher, varying between 15 and 35 $m \cdot h^{-1}$ depending on the iron load to be treated. The effective size of the sand is increased to approximately 1.35 mm.

Neutralization of oxidants (Cl_2, O_3) contained in the washwater: the reagents to be used can be either sodium hyposulfite, sodium bisulfite, or SO_2. The pure product treatment rates to be implemented are shown in Table 16.9.

Washwater consumption is approximately five times lower than for physicochemical iron removal. In addition, it is possible to wash the filters with raw water. On the other hand, it is highly not recommended to wash the filters with chlorinated treated water, which could partially destroy the bacteria population (if $Cl_2 > 0.2$ ppm).

For 1 g of	Cl_2	O_3
Sodium hyposulfite (Na_2SO_3)	4.45	3.3
Sodium bisulfite ($NaHSO_3$)	2.93	2.16
Sulfur dioxide (SO_2)	1.11	1.4

Table 16.9. *Chlorine and ozone neutralization by reducers*

Complementary treatments and equipment

Turbidity

If raw water has a turbidity greater than 1 NTU and this turbidity is not due to the precipitation of iron, an in-line addition of coagulant should be provided.

Ammonia

Ammonia does not significantly disturb the biological processes for removing iron and/or manganese. In other words, this parameter is not limiting. Since its oxidation by chlorine upstream of the biological reactor is not possible, it must be done downstream. A contact time of 20–30 min is required to ensure this oxidation.

Manganese

When the water simultaneously contains iron and manganese (>0.5 $mg \cdot L^{-1}$), the oxidation of these two elements by biological means requires different rH conditions and two separate treatment stages (see biological manganese removal).

Biological filtration design parameters

The filters are open or closed canister under pressure.

Monolayer filter	**Parameters**
Filtration velocity for [Fe] < 2 $mg \cdot L^{-1}$	30–35 $m \cdot h^{-1}$
Filtration velocity for [Fe]: 2–5 $mg \cdot L^{-1}$	20–30 $m \cdot h^{-1}$
Filtration velocity for [Fe] > 5 $mg \cdot L^{-1}$	15–20 $m \cdot h^{-1}$
Sand density	1.65
Sand layer height	1,500 mm
Sand grain size	1.3–1.5 mm
Gravel layer height	100 mm
Long-tailed strainer density	40–45 per m^2

Table 16.10. *Design parameters for biological iron removal on a sand filter*

Advantages of the biological iron treatment

As a general rule, when applicable, biological iron removal should be preferred over the physicochemical method. Its main advantages are listed below:

– rapid oxidation: the oxidation tower is unnecessary. An in-line pressurized air injection is sufficient;

– reagents are not needed for pH correction or flocculation;

– high retention capacity. Iron is retained in a very compact form. This results in a very high iron retention capacity. As for physicochemical iron removal, this capacity depends on raw water's initial iron content, but is about seven times greater than in the case of physicochemical iron removal (up to 5 kg of iron$\cdot m^{-2}$, instead of 1 $kg \cdot m^{-2}$). This results in an appreciable saving on washwater;

– high filtration velocity: because of the nature of the floc, filtration velocities are generally 2.5–5 times faster;

– increase in sand's effective size: high filtration velocities are favored by the possibility of choosing thicker sand, due to the nature of the precipitates formed and bacteria's morphology. In practice, sand with an effective size of 1.35 mm is chosen. This means that the head loss in the material remains practically identical, despite the increase in velocity by two to three times;

– improvement in the operation of old facilities: depending on the case, the station can be renovated by adopting one or more of the following measures:

– changes in aeration conditions;

– changing the filter medium;

– reagents removal, especially strong oxidants;

– change in pH.

16.1.4.2. *Manganese treatment*

In solution, manganese can be found in different forms depending on the water's pH and oxidation–reduction conditions. The manganese potential/pH diagram (Figure 16.20) quickly offers a glimpse of the preponderant forms likely to be found. The shaded area represents the stability zone for divalent manganese Mn^{2+} (pH 6–8.5 and EH = 0.2–0.5 V).

Figure 16.20 shows that manganese can be found mainly in dissolved reduced form Mn^{2+} and the insoluble oxidized forms are MnO_2 and Mn_2O_3. The $MnCO_3$ form, for very high pH/very low potential couples, corresponds to extreme conditions, which are rarely found in drinking water.

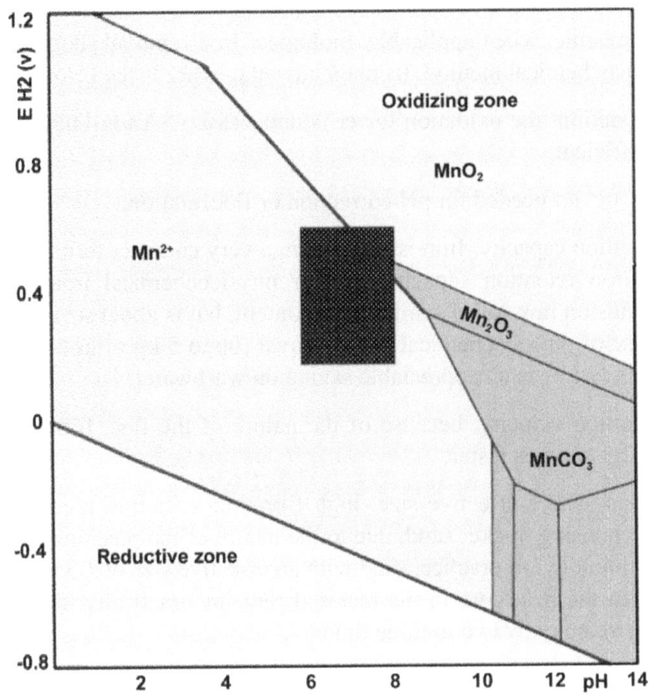

Figure 16.20. *Potential diagram – manganese pH. For a color version of this figure, see www.iste.co.uk/gaid/watertreatment3.zip*

It is more common to encounter the form Mn^{2+} in water that the ferrous iron Fe^{2+} form because the solubility of manganese is higher. Manganese can also form complexes with organic matter (humic and fulvic acids) or be adsorbed onto iron oxides, rendering it more difficult to oxidize.

Under the pH and potential conditions of natural waters, most of the manganese appears in the dissolved divalent state, whereas iron is often present in the precipitated trivalent form.

16.1.4.2.1. Aeration treatments

The kinetics of oxidation by oxygen is very slow under the typical conditions in water treatment stations and requires a long contact time (several hours). The estimated half-life of Mn^{2+} in the presence of O_2 is very long and not acceptable for a design of a drinking water plant. It only begins to be effective from pH 9.5. This removes the solution at the level of a drinking water station:

$$2Mn^{2+} + O_2 + 2H_2O \rightarrow 2MnO_2 + 4H^+$$

Therefore, aeration cannot be used as a pretreatment for oxidizing manganese. This is the reason why manganese is chemically oxidized to be removed.

16.1.4.2.2. Chemical oxidation treatments

Physicochemical manganese removal

Manganese removal requires at least two distinct and specific stages: the oxidation of manganese to form hydroxides and the subsequent retention/separation of the precipitate. Dissolved oxygen and free chlorine readily oxidize Fe^{2+} in the pH range of approximately 6 to 8.5 for which Mn^{2+} oxidation by those oxidants is very slow. This allows for effective use of strategies involving use of the weaker oxidants for conversion of Fe^{2+} to ferric iron particles followed by use of a stronger oxidant for Mn^{2+} oxidation.

1) The oxidation of dissolved manganese is more difficult than that of iron because, as shown by the redox-pH potential diagram (Figure 16.20), it requires higher eH-pH couples.

The reaction rate of chlorine is too slow, and it is only effective from pH 9. For example, the chlorine levels to be implemented are very high ($>2.5\ mg \cdot L^{-1}$) in order to oxidize $0.2\ mg \cdot L^{-1}$ of manganese after more than 1 h 30 min of reaction. Unlike iron, coagulants do not have sufficient oxidizing power.

Three oxidants can be considered to transform dissolved manganese Mn^{2+} into manganese dioxide MnO_2: potassium permanganate, chlorine dioxide and ozone.

2) Manganese oxides are then physically retained by filtration.

Oxidation reactions

Chemical oxidation is obtained by injecting an oxidant such as ozone, chlorine dioxide or potassium permanganate. The addition of one of these oxidants results in the oxidation of Mn^{2+} into Mn^{4+}, which precipitates. The essential condition is for pH to be ≥7.2. Chlorine is considered as an indicator because its kinetics are too slow (at the pH of natural waters) to be implemented at a station.

The chemical reactions and the theoretical oxidant quantities are summarized in Table 16.11.

Ozone reacts very quickly with Mn^{2+}. When dosed correctly, there is no residual of O_3 to form by-products with the organic molecules. As iron and manganese are often combined, the ozone is dosed according to stoichiometric ratios, so that there is no residual.

The limitations of the ozone treatment are related to its very high oxidizing power, even higher than that of oxidants such as ClO_2 and $KMnO_4$. Ozone is highly reactive in relation to the water matrix, be it organic or mineral. The contact time varies between 2 and 3 min. As the O_3 is introduced as a pretreatment for the oxidation of Mn^{2+}, competing reactions do not interfere with its effectiveness, because these are very rapid and identical to those of iron.

Oxidants	**Chemical reactions**	**Mn oxidantquantities, mg/mg**	**Mn precipitates, mg/mg**
Ozone (O_3)	$Mn^{2+} + O_3 + H_2O \rightarrow MnO_2 + O_2 + 2H^+$	0.87	1.58
Chlorine (Cl_2)	$Mn^{2+} + Cl_2 + 2H_2O \rightarrow MnO_2 + 2Cl^- + 4H^+$	1.26	1.58
Chlorine dioxide (ClO_2)	$Mn^{2+} + 2ClO_2 + 2H_2O \rightarrow MnO_2 + 2ClO_2^- + 4H^+$	2.41	1.58
Potassium permanganate ($KMnO_4$)	$3Mn^{2+} + 2KMnO_4 + 2H_2O \rightarrow 5MnO_2 + 2K^+ + 4H^+$	1.90	2.62

Table 16.11. *Stoichiometric ratios for various oxidants*

In case of excessive ozone, it is possible to produce potassium permanganate with a purplish color.

The stoichiometric ratio is 0.87 mg O_3 per mg Mn^{2+}. The reaction rate between O_3 and Mn^{2+} is fast for a rate constant of 2×10^6 $M^{-1} \cdot s^{-1}$ at pH 7.2. The reaction times are very short (<60 s). Experience shows that during preozonation, oxidation kinetics are well controlled, even in the presence of organic matter.

With chlorine dioxide, the direct reaction on the manganous ion does not generate chlorites. On the other hand, the rapid decomposition of the dioxide after its injection can generate chlorite ions.

For chlorine dioxide, the oxidation of manganese constitutes one of the major applications of this reagent in drinking water pretreatment because its use generates no haloforms and only a few organochlorine compounds (even when directly applied to raw water). From a bactericidal and virucidal perspective, this oxidation has an advantage in waters with a low organic matter content.

The efficiency of chlorine dioxide is far less affected than that of potassium permanganate by the complexation of manganese or by the reaction pH. The recommendation is to operate at a pH of 7.2.

For determining the treatment rate, it is necessary to take into account the iron and hydrogen sulfide contents (or other reduced forms of sulfur), dioxide is highly reactive with these compounds.

Mn^{2+} oxidation responds to a stoichiometry of 2.41 mg ClO_2 per mg of Mn^{2+} and proceeds by a fast second-order reaction, with a k2 of $1 \times 10^4\ M^{-1}\cdot s^{-1}$ at pH 7.2.

Contact time is 1–2 min, depending on the ClO_2 dose and the initial concentration of dissolved Mn^{2+}. However, in the presence of Mn^{2+}, chlorine dioxide is not completely reduced into chloride ions (Cl^-), but only into chlorite (ClO_2^-). As this is difficult to remove throughout the treatment, it reacts during the disinfection stage with free chlorine, forming a chlorate ion (ClO_3^-). When a chlorate ion is present in water, it is very persistent and difficult to remove. However, to be able to meet the standards for chlorites and chlorates in the water produced, chlorine dioxide must not exceed a dose of 1.2 $mg\cdot L^{-1}$.

All things considered, in the presence of high concentrations of Mn^{2+}, chlorine dioxide may not be the right solution, and in any case, it should be monitored.

Potassium permanganate is an alternative to O_3 and ClO_2.

$$3Mn^{2+} + 2KMnO_4 + 2H_2O \rightarrow 5MnO_2 + 2K^+ + 4H^+$$

Manganese requires approximately 1.90 ppm $KMnO_4$. This oxidation reaction takes place swiftly, with a reaction rate constant of $1 \times 10^5\ M^{-1}\cdot s^{-1}$ at pH 7.2.

Considering the acidity formed, the oxidation reaction shows that the alkalinity consumption is 1.20 mg·L^{-1} $CaCO_3$ per mg Mn^{2+} (that is, (400/2)/(3 × 56)) or 1.33 equivalents per mole of Mn^{2+}.

This reagent requires careful dosage control because an excess can tarnish water with a pink to purple color. For pH values around 7.2–7.3, the contact time is around 5 min. The redox/pH potential couple seems to be of great importance. When the reaction is efficient, it is generally fast, it requires only a few reagents and does not generate tastes. Injections of permanganate can be applied for the physicochemical removal of iron/manganese from borehole water, just before filtration.

On water containing organic matter, the permanganate injection is preferably done ahead of the treatment because it makes it possible to flocculate the manganese complexed with organic matter. The contact time can reach 20 min. $KMnO_4$ doses are much higher due to the oxidant demand by organic matter. Generally, we consider twice to three times the stoichiometric dose.

Table 16.12 summarizes the manganese treatment options using the physicochemical route.

Manganese complexed by humic matter	**Non complexed Mn^{2+}**		
	Mn^{2+} < 2 mg·L^{-1}	**2 < Mn^{2+} < 5 mg·L^{-1}**	**Mn^{2+} > 5 mg·L^{-1}**
Coagulation/flocculation	-	-	Coagulation/flocculation
Settling	-	-	Settling
pH control	pH control	pH control	pH control
Sand filter	Sand filter	Dual media filter	Sand or dual media filter
Disinfection	Disinfection	Disinfection	Disinfection

Table 16.12. *Summary of manganese treatment characteristics*

In the case of heavily loaded waters, it is sometimes necessary to resort to an additional coagulation–flocculation–settling stage before filtration in order to prevent the rapid clogging of filters.

The oxidation of manganese using chemical oxidants requires controlling the reagent injections because an excessive dosage tarnishes water with a pink-violet color.

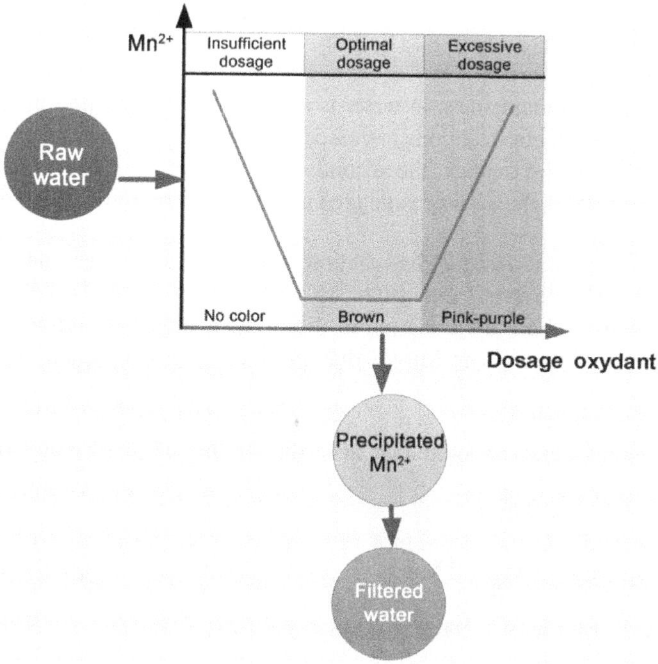

Figure 16.21. *Need to control the chemical oxidant dosage with Mn^{2+}. For a color version of this figure, see www.iste.co.uk/gaid/watertreatment3.zip*

16.1.4.2.3. Catalytic manganese removal (Mangaflo® process)

This technique is based on a filtration support reaction, entirely or partly made up of manganese oxide. When in contact with the water to be treated, it achieves the oxidation of dissolved manganese and the retention of formed oxides in a single step.

In stations treating manganese physicochemically (oxidation at high pH followed by filtration), a black deposit of manganese dioxide MnO_2 is observed on the filtering sand grains, in a more or less pronounced way. The sand, which is coated with MnO_2, is called “manganized sand”.

This deposit acts as a chemical catalyst during the manganese oxidation reaction, promoting and accelerating it.

Manganese dioxide is a natural material, crushed and sifted, borrowed from the mining industry. There are several varieties, its most common one being pyrolusite.

The materials on the market have a 70–85% content of MnO_2 (the rest essentially comprising other oxides: MnO, Fe_2O_3, SiO_2, etc.).

The oxidation of manganese in water is an autocatalytic reaction in three stages, to which different kinetic reactions are associated. The first stage is the oxidation of Mn^{2+} in solution by the oxidant, the second is the adsorption of Mn^{2+} onto MnOx(s) solids, and the third is the surface catalyzed oxidation of the sorbed Mn^{2+} ion.

The overall oxidation–reduction reaction takes place at the material's surface (at the solid–liquid interface) and leads to the formation of (solid) manganese sesquioxide Mn_2O_3, both by oxidation of dissolved manganese and by reduction of solid manganese dioxide. The Mn_2O_3 thus produced gradually coats the material's grains.

The reaction kinetics are quite fast, according to the following reactions:

$$Mn^{2+} + O_2 \rightarrow MnO_2$$

$$Mn^{2+} + MnO_2 + O_2 \rightarrow 2MnO_2$$

$$2Mn^{2+} + MnO_2\,(s) + 2H_2O \rightarrow Mn_3O_4(s) + 4H^+$$

The expression for the catalytic oxidation rate is:

$$-d\,[Mn^{2+}]/dt = k\,[Mn^{2+}] + k1\,[Mn^{2+}]\,[Mn^{4+}]$$

The rate constant is: k cat = k + k1 $[Mn^{2+}]$.

The second stage is important in filtration processes because the layers of MnOx(s) present on the filtering material rapidly adsorb the dissolved Mn^{2+}, which is then oxidized. While the oxidation of manganese by oxygen alone is slow, the reaction Mn^{2+}/O_2 is catalyzed by the presence of oxidized manganese.

Manganese dioxide acts both as a catalyst and as an oxidant. Its modus operandi is:

1) catalytic action: the principle is the same as the catalytic effect obtained with manganized sand, the material serving as a support for fixing the dissolved manganese;

2) oxidizing action: the manganese dioxide acts as an oxidant for the dissolved manganese present in the water to be treated.

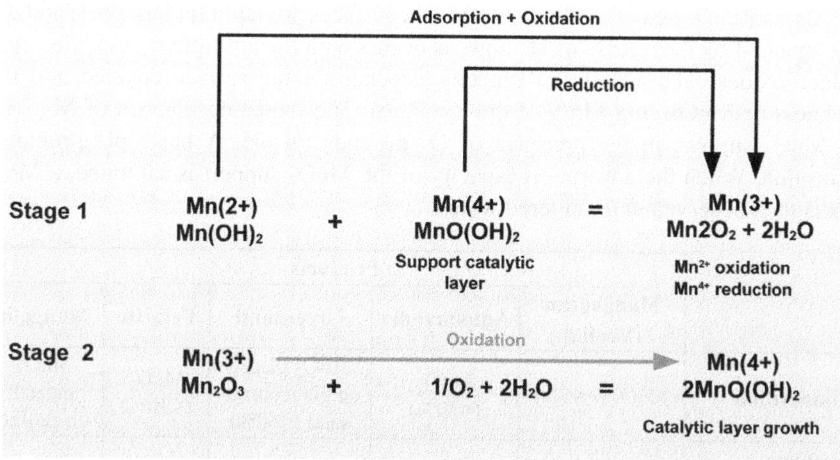

Figure 16.22. *Mn^{2+} oxidation mechanisms on manganese dioxide. For a color version of this figure, see www.iste.co.uk/gaid/watertreatment3.zip*

MnO_2 is not selective for manganese and also oxidizes ferrous iron. Fe^{2+} ions are also oxidized by MnO_2 and are deposited on the grains' surface of the filtering medium:

$$2Fe^{2+} + 3MnO_2\,(s) + 4H_2O \;\rightarrow\; 2Fe_2O_3 + Mn_3O_4\,(s) + 8H^+$$

When all the dioxide is converted into sesquioxide, it is necessary to regenerate the material. Two types of regeneration are possible: mechanical and chemical.

Mechanical regeneration takes place during conventional washing, during the air + water phase, which is carried out for several minutes.

Chemical regeneration involves injecting an oxidant into the material (permanganate, free chlorine), which encourages the reoxidation of the sesquioxide into active dioxide.

The lifespan of manganese dioxide is a minimum of 10 years. Manganese dioxide is composed almost exclusively of manganese oxide in grains whose effective size is 0.8–1 mm, with a high density close to 4. The Mn oxidic support used for the oxidation of Mn^{2+} into Mn^{3+} and Mn^{4+} is often referred to as MnOx(s), "x" being between 1.5 and 2.0. A naturally occurring ion exchange mineral, such as green-colored glauconite (called greensand) which does not contain manganese, is a material which has been surface-coated with an Mn oxide in order to use it as an oxidation support.

The fixation kinetics of Mn^{2+} onto MnOx surfaces are rapid (at basic pH) and are accompanied by the release of H^+ ions, as occurs with the adsorption of cations onto surface oxides. The adsorption capacity depends on the surface covered and the thickness of MnOx (mg Mn/g of dry medium). The oxidation kinetics of Mn^{2+} on the MnO_2 support in the presence of O_2 depends on pH. A basic pH promotes adsorption. When the adsorption capacity of the MnO_2 support is saturated, a Mn^{2+} leak is then observed in the filtered water.

	Products				
	Mangagran (Veolia)	**Aquamandix**	**Greensand**	**Polarite**	**Mangalit**
Composition	MnO_2 (>85%)	MnO_2 (<70%)	MnO_2 placed on glauconized sand (<60%)	MnO_2 (<30%)	MnO_2 placed on zeolite
Hardness (mhos)	6–7	<6	<5	<5	<5
Density (kg·m^{-3})	2,200–2,400	2,000	1,426	1,900	1,000
Effective size (mm)	0.93	0.5–1	0.30–0.35	0.41	0.80
Uniformity factor	<2.5	1.66	1.66	1.35	1.75
Optimum pH	>7.2	>6.5	6.5–8.5	6.8–8.5	7.1–8.5
Washing velocity (m·h^{-1})	30 (MnO_2 alone) 40 in dual media sand-MnO_2	30	30	<25	–
Contact time (min)	3–6	–	15–20	6	–
Implementation	Monolayer Dual media Three-media	Bed mixed with sand (80%) and MnO_2 (20%)	Fixed bed	Fixed bed	Fixed bed
Company	Sibelco	Eurochlore	Inversand	CAS	Zeochem

Table 16.13. *Characteristics of various MnO_2 supports available in the market*

Regeneration is reached by backwashing using water containing an oxidant such as chlorine in order to reactivate the sites. During washing, most of the MnO_2 particles present on the support are removed.

Description of the Mangaflo® process

The Mangaflo® process is the Veolia process for catalytic manganese removal. The commercial product is called Mangagran (Sibelco). Mangagran is a natural manganese granulated dioxide. Its chemical composition and particle size are standardized. Until present, particle size is unique for the drinking water applications.

The specificity of this material makes it possible to maintain a uniform and stable layer of manganese dioxide, contrary to the mixed bed theory. The hardness of this material (much higher than that of other products on the market) preserves the initial grain size and adsorption capacity. The material is not a consumable in the same way as other materials such as limestone, activated carbon, etc. The loss per year should be compared to that of a sand (~2%). Under the design and operating conditions recommended by Veolia, its life can span 10 years.

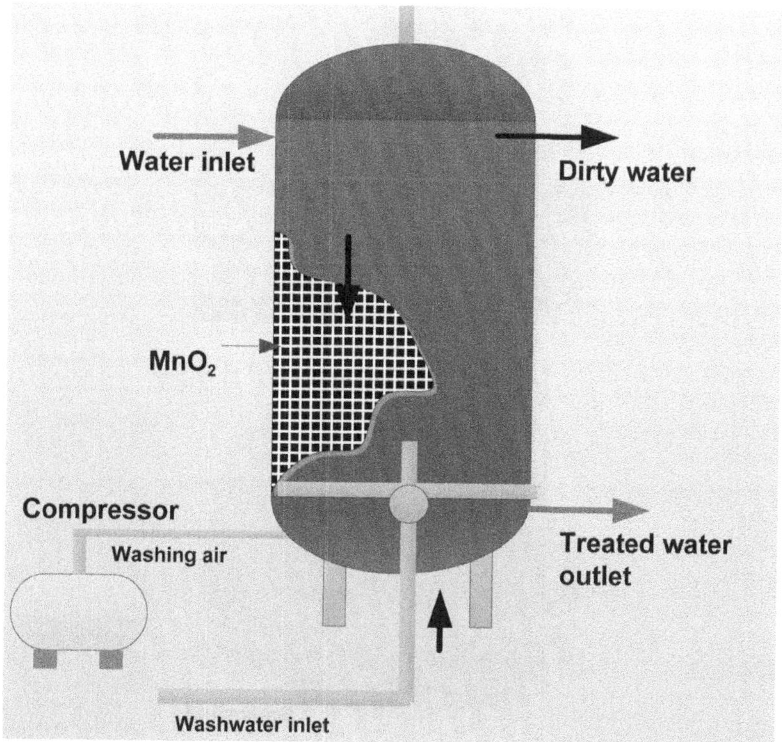

Figure 16.23. *Catalytic manganese removal filter (Mangaflo® process). For a color version of this figure, see www.iste.co.uk/gaid/watertreatment3.zip*

The specifications for Mangagran are summarized in Table 16.14.

Bulk density	2–2.2 g·cm^{-3}
Packed density (dry appearance): dry solid weight	2.3–2.4 g·cm^{-3}
Solid's absolute density	4.2 g·m^{-3}
Porosity	0.44
Void ratio (e)	0.786
Mohs scale hardness	6–7
Effective size	0.85–1 mm
Uniformity coefficient	<2.35

Table 16.14. *Mangagran technical specifications*

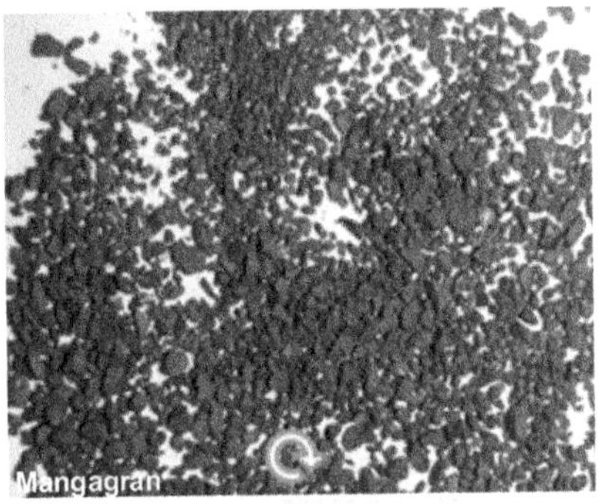

a)

b)

Figure 16.24. *(a) Manganese dioxide (MnO_2) and (b) Greensand, a glauconized sand coated with MnO_2*

Implementation in a filter

Manganese dioxide is considered both as a filtration material (under certain conditions) and as an adsorbent material. This is why it can be used on its own (for manganese removal exclusively) or in combination with another filtering material as part of dual media filtration: most often sand, expanded clay or granular activated carbon (in the presence of micropollutants).

The dual media filter will enable the treatment of both iron and manganese. Iron is oxidized by aeration and retained on sand. The bottom layer of manganese dioxide treats dissolved manganese. Sand height is between 0.5 and 0.8 m. The height of manganese dioxide depends on the retention time and the quantity of manganese to be removed.

The design parameter with MnO_2 is the contact time, not the filtration velocity. Like activated carbon, the filter is considered a reactor. Considering the contact time (approximately 3–8 min), the layer height of the mineral and the filtration velocity are designed based on the contact time.

The dioxide/sand combination guarantees both particle retention and Mn fixation (in the case of first stage filtration).

The removal of Mn can be obtained with the three-media filter:

– anthracite or pumice–sand–MnO_2: after coagulation–flocculation–settling, if the raw water presents an average or high turbidity or if the raw water presents a low turbidity (<15 NTU), in-line coagulant is added upstream the three media filter;

– granular activated carbon–sand–MnO_2: after coagulation–flocculation–settling if the raw water presents an average or high turbidity or if the raw water presents low turbidity (<2 NTU) and micropollutants.

In all applications, the removal of the excess of coating formed by oxidation is performed by a simple mechanical washing because of the air/water combination, without adding any reagent.

When Fe is present in the raw water (>50 $\mu g \cdot L^{-1}$), it is recommended to oxidize it first (chemical oxidation, aeration, etc.). If not, there will be a competition between Fe and Mn on MnO_2 media, which can reduce the performance of the MnO_2.

The filter's design and its position in the treatment chain are summarized in Table 16.15.

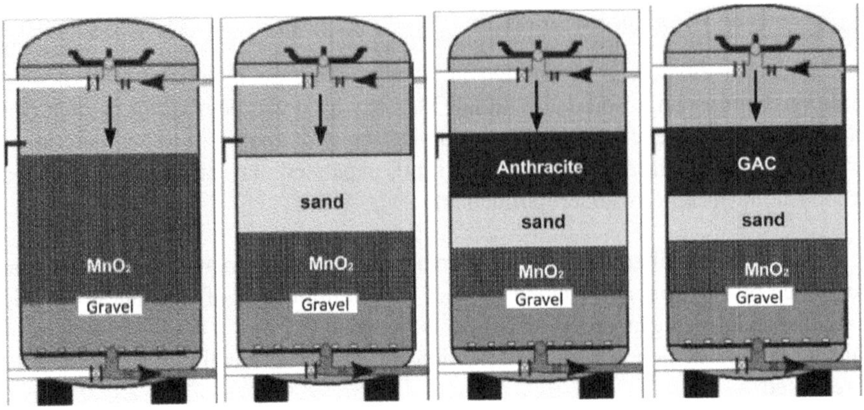

Figure 16.25. *Different MnO_2 implementation solutions (Veolia processes)*

Groundwater	Surface water		
Water < 1 NTU	Raw water < 5 NTU DOC< 2 $mg \cdot L^{-1}$ Pre-oxidation	After coagulation-flocculation-settling water < 3 NTU Fe removed*	After 1st filtration stage (water < 1 NTU) Fe removed*
Fe < 0.2 $mg \cdot L^{-1}$ MnO_2 filter	For all [Fe] In-line coagulant Dual media filter Sand/MnO_2	Dual media filter Sand/MnO_2	MnO_2 filter
Fe EB > 0.2 $mg \cdot L^{-1}$ Pre-oxidation Dual media filter Sand/MnO_2	For all [Fe] Three-layer filter If turbidity peaks, raw water at 10 NTU In-line coagulant Anthracite/sand/MnO_2	Three-layer filter Possible in-line coagulant Anthracite/sand/MnO_2	Dual media filter at the second stage GAC/MnO_2 Presence of micropollutants Taste and odors
Fe EB > 0.2 $mg \cdot L^{-1}$ Pre-oxidation Three-media filter If pesticides, taste and odors GAC/Sand/MnO_2	For all [Fe] Three-layer filter If pesticides, taste and odors GAC/sand/MnO_2	Three-layer filter If pesticides, taste and odors GAC/sand/MnO_2	

*Iron being removed during settling.

Table 16.15. *Possible solutions integrating MnO_2 depending on the quality of the water (Fe and Mn) to be treated*

Design parameters

The volume of MnO_2 is based on a contact time of 3–8 min inside the dioxide layer (layer height = retention time/filtration velocity), reaching values below the 50 $\mu g \cdot L^{-1}$ standard in treated water:

– contact time: 3–8 min;

– variation of raw water Mn: < 2,000 $\mu g \cdot L^{-1}$;

– residual: < 20 $\mu g \cdot L^{-1}$ (guaranteed < 50 $\mu g \cdot L^{-1}$).

In addition to sufficient contact time for that design, material height must be at least equal to 25 cm, even for Mn^{2+} concentrations < 500 $\mu g \cdot L^{-1}$.

Filtration velocities are between 5 and 12 $m \cdot h^{-1}$. The adsorption capacity is around 50 mg Mn^{2+}/kg MnO_2/filter cycle.

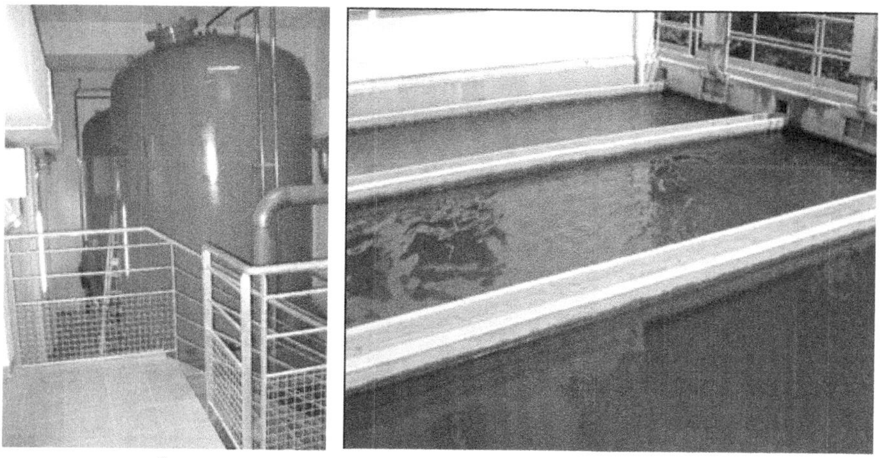

a) b)

Figure 16.26. *Pressure filter (a) and open filter comprising MnO_2 (b) (Veolia sites). For a color version of this figure, see www.iste.co.uk/gaid/watertreatment3.zip*

Performances

The adsorption capacity is optimal if the following conditions are met prior to filtration: pH ≥ 7.2, potential > 200 mV Ag/AgCl and dissolved oxygen > 6 $mg \cdot L^{-1}$.

The oxidation reaction kinetics of Mn^{2+} by manganese dioxide is effective from pH 7.2 and higher. If the pH is not optimum, the adsorption is degraded in a few hours, letting residual escape, whose content increases exponentially in the following hours. To prevent this, it is advisable to provide a pH adjustment station with caustic soda or lime.

Contact time	3 min	6 min	8 min
Raw water in $\mu g{\cdot}L^{-1}$ **Mn**	<500	<500	500–1,000
Treated water in $\mu g{\cdot}L^{-1}$ **Mn**	<50	<5	<50

Table 16.16. *Performances as a function of Mn^{2+} concentration and pH > 7.2*

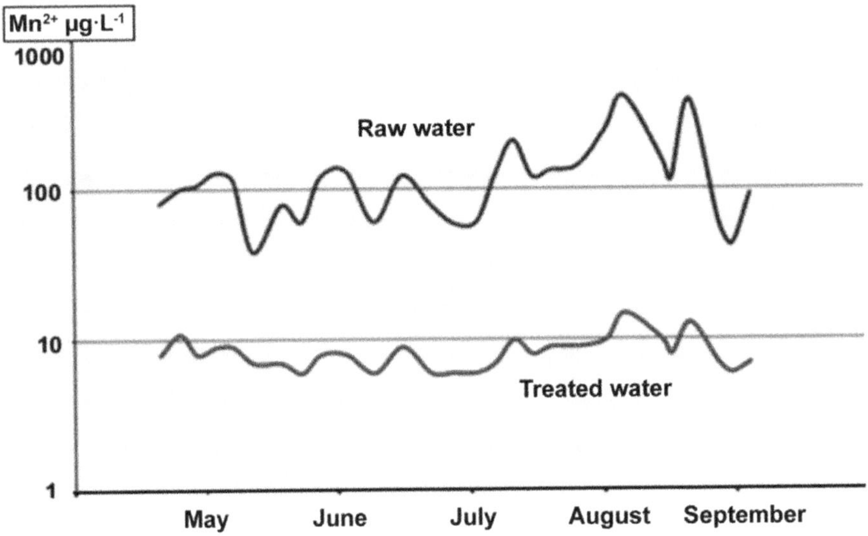

Figure 16.27. *Mn^{2+} removal on a Mangaflo® filter (Veolia). For a color version of this figure, see www.iste.co.uk/gaid/watertreatment3.zip*

Figure 16.27 shows the monitoring of manganese in groundwater at pH 7.3, with a contact time of 3 min. The mean outlet Mn^{2+} concentration is lower than 10 $\mu g{\cdot}L^{-1}$ for a variable inlet between 60 and 300 $\mu g{\cdot}L^{-1}$.

The calcium hardness and the alkalinity of the water are not considered for the design.

The duration of the filtration cycle must correspond to the retention/adsorption dynamic by ensuring that the turbidity and Mn contents in treated water are in accordance with the treatment goals. In the case of a dual media filter, the filtration time between two washes is defined by the retention capacity of the filtering material's particles (pumice, anthracite, expanded clay or sand) in $g \cdot m^{-2}$. After a few hours of filtration (depending on the quality of the water leaving the settler), manganese retention may decrease following the blocking of the dioxide's active sites by the suspended solids accumulated inside the filter.

In the case of a dual media filter, care is taken to respect the washing cycles (24–48 h) estimated when the filter was designed. In the case of a monolayer MnO_2 filter set at the first stage (underground water), or at the second stage (after a sand filter), the filtration cycle duration is directly related to the manganese content of the inlet water.

Manganese dioxide is a much heavier material than sand. The design of the washing equipment is made taking into account the significant head loss resulting from this density. This remark is particularly important in the context of a rehabilitation, considering that the equipment is generally reused by default.

If the total filtering layer (including the MnO_2) does not exceed 1 m, the filter can be washed according to the classic procedure for washing a sand filter, using air and water, with a water rinsing speed of around 25 $m \cdot h^{-1}$. When the total height of materials exceeds 1 m, the reclassification of a MnO_2/sand dual media filter requires a sand expansion of around 10–15%. A water rinsing speed of around 30–40 $m \cdot h^{-1}$ is necessary. The production of sludge is estimated at 70–80% of the manganese removed. It is necessary to provide a wastewater tank in order to recover the settled MnO_2 and to avoid discharging MnO_2-loaded water into the river.

The slightly blackish washwaters contain a part of solid MnO_2 corresponding to the layer removed by attrition during washing.

Regeneration

Regeneration is carried out using an oxidant which makes it possible for the filtering grain surface to be re-oxidized with active MnO_2. This regeneration can be carried out on an ad hoc basis when MnO_2 begins to be "saturated" (or more exactly converted into Mn_3O_4), which results in a shift of the manganese content of filtered water. The appearance of this phase can take place at around the eighth year of operation for Mn^{2+} concentrations in raw water of around 0.1–0.3 $mg \cdot L^{-1}$.

Arguments

– Drinking water certification;

– very low attrition/no consumption;

– adaptation to all types of filters (new station or rehabilitation);

– high adsorption capacity;

– constant refining;

– redox buffer role (bypassed MnO_2 filter depending on raw water content);

– release of oxidized manganese by backwashing;

– no added oxidant;

– low operating cost;

– no dosage concerns for the operator.

Treatment in the presence of ammonia

In groundwater, it is not uncommon to find the presence of ammonia together with the presence of iron and manganese.

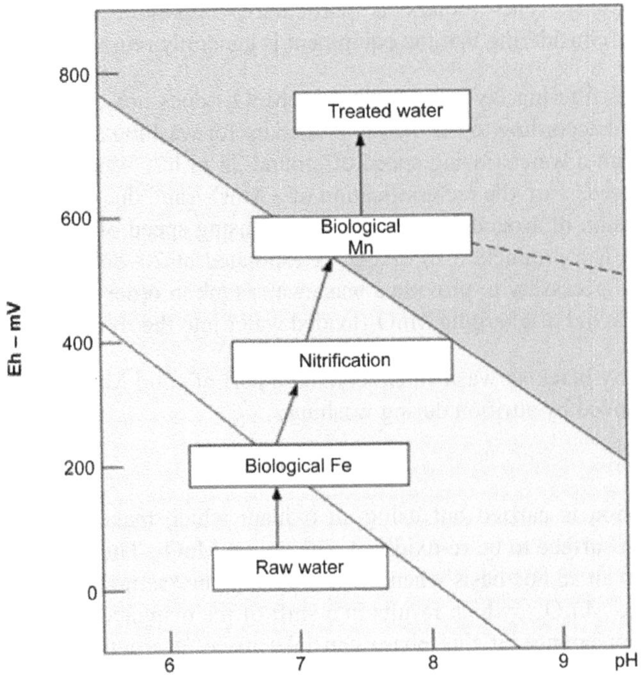

Figure 16.28. *Redox potential conditions for the removal of iron, NH_4^+ and biological Mn. For a color version of this figure, see www.iste.co.uk/gaid/watertreatment3.zip*

Two solutions are possible:

– breakpoint chlorination (see Chapter 21);

– biological nitrification.

The redox potential/pH diagram (Figure 16.28) shows that the redox potential required for the oxidation of ammonia is above iron's potential, but below that of manganese.

The removal of these elements is done in the following order:

1) iron;

2) ammonia;

3) manganese.

It will be necessary to ensure that the quantity of oxygen available will be sufficient to enable the removal of iron and ammonia.

16.1.4.2.4. Biological manganese removal

Within the same filtering material, this technique ensures the oxidation of dissolved manganese and the retention of insoluble oxides. The advantage of the biological treatment is that it generally does not require the addition of chemicals.

The biological mechanisms for manganese removal are:

– direct intracellular oxidation of Mn^{2+} by enzymatic action. Manganese-oxidizing micro-organisms use Mn^{2+} as an electron donor and oxygen as an electron acceptor. Bacteria performing this oxidation often include *Leptothrix, Crenothrix, Hyphomicrobium, Siderocapsa, Siderocystic, Metallogenium, Pseudomonas*;

– adsorption onto the surface of the cytoplasmic membrane: Mn^{2+} can be adsorbed onto negatively charged extracellular polymeric substances (EPS). Oxides such as γ-Mn_3O_4, which are generated by bacterial catalytic reactions, can also adsorb Mn^{2+};

– the catalytic oxidation of Mn^{2+} by biopolymers generated by micro-organisms.

The underlying principle is to promote and to manage the development of a biomass made up of bacteria whose metabolism is based on the manganese oxidation. The adsorption onto MnOx(s) particles which accumulate in the filter during the oxidation process also contributes significantly to the removal of Mn^{2+} during the biofiltration process.

The implementation conditions are substantially different from those for biological iron removal:

– since the redox potential is higher for manganese than for iron, the quantity of air to be injected is larger. Manganese removal requires a much higher quantity of dissolved oxygen (>5 mg $O_2 \cdot L^{-1}$) than the one used for iron removal;

– the oxidation kinetics of manganese is slower than for iron, and requires slower filtration velocities;

– a preliminary seeding phase is always necessary before start-up. The filters' seeding time is far slower (it can take 10–15 weeks). Once set in, the biomass finds its equilibrium between the development due to the presence of Mn, and partial removal due to filter unclogging;

– the filtration support may have a larger particle size than the one used in the context for physicochemical manganese removal. The contact time within the biological filters is between 15 and 25 min;

– growth conditions are essentially dependent on factors such as temperature, pH, dissolved oxygen or redox potential, manganese concentration, salinity, organic matter content.

The risks of not developing manganese bacteria are very rare in natural waterways. The only imperative requirement is to place them under pH and redox potential conditions corresponding to their optimal field of activity.

The biological removal of manganese is generally carried out at pH 6.5–7.2 in order to differentiate it from the physicochemical removal. This is because at higher pH values, it is not easy to distinguish between the precipitation of $MnO_2(s)$ and the biological effect. A stoichiometric nutrient ratio of 100:5:1 for C:N:P is recommended to support micro-organism growth.

	Iron removal	**Manganese removal**
Dissolved oxygen content	0.2–2 $mg \cdot L^{-1}$	≥5 $mg \cdot L^{-1}$
Oxidation–reduction	40–100 mV	300–700 mV
rH	14–20	22 and beyond
pH	5.5–7.5	7.5–9

Table 16.17. *rH and pH conditions for biological iron and biological manganese removal*

The domain of biological manganese removal is above that of iron in terms of redox potential. This implies that biological iron removal must take place upstream of manganese removal. In practice, this means that in the presence of iron and manganese, two successive biological filtration stages are necessary for their removal.

The following schedule can be adopted:

– controlled aeration for biological iron removal, followed by a first filtration: biological iron removal;

– second aeration and pH raising, followed by a second filtration: biological manganese removal.

The advantage of the biological oxidation of manganese results in hydrated forms of manganese oxides y-MnOOH, whereas chemical oxidation creates more amorphous forms of manganese oxides. These oxides tend to accumulate around bacterial cells, sheaths and filaments. As a result, filter clogging is less likely with the biological treatment, which results in longer cycle times and more widely spaced wash frequencies.

16.1.4.2.5. Treatment files for the simultaneous removal of iron and manganese

Treatment files for the simultaneous removal of iron and manganese are many and varied. The choice of the most suitable technique depends on the origin and overall quality of raw water, the concentration and variability of metals in the resource, their dissolved, soluble or complexed form with organic matter, the existent treatment station or the projected one, and the goals to be achieved (quality of treated water, costs, etc.).

Water treatment stations can be with and without reagents, physicochemical, biological or hybrid.

They apply to groundwater, for which the Fe^{2+} and Mn^{2+} concentrations are the most frequent, and also to surface waters.

For the latter, a part of the suggested solutions are valid when turbidity is lower than 15 NTU (columns two and three of Table 16.15).

If turbidity is higher and the presence of organic matter is noticed, iron and manganese are chemically oxidized in pre-treatment (the most frequent case). Mn^{2+} refining can then be carried out on a sand/MnO_2 dual media filter. In the presence of micropollutants whose concentrations are not very high, GAC/MnO_2 filtration can be considered after coagulation–flocculation–settling of the raw water.

Groundwater has a more stable quality than surface water. On the other hand, they are often poor in dissolved oxygen, which gives it a "reductive" character contributing to iron and manganese levels above standard. Iron and manganese are rarely complexed in groundwater whose organic matrix is generally very weak. But iron can be complexed with silica when the latter is present at high levels ($>20\ mg \cdot L^{-1}$). The least protected aquifers (subject to sudden variations in turbidity during rainy periods, for example) are often sufficiently oxygenated, and do not present excessive dissolved iron or manganese contents.

Physicochemical treatments without reagents

This is the simplest case for the implementation of iron air oxidation through a cascade or an oxidation tower, followed by dual filtration with sand (iron hydroxides retention) and MnO_2 (manganese retention).

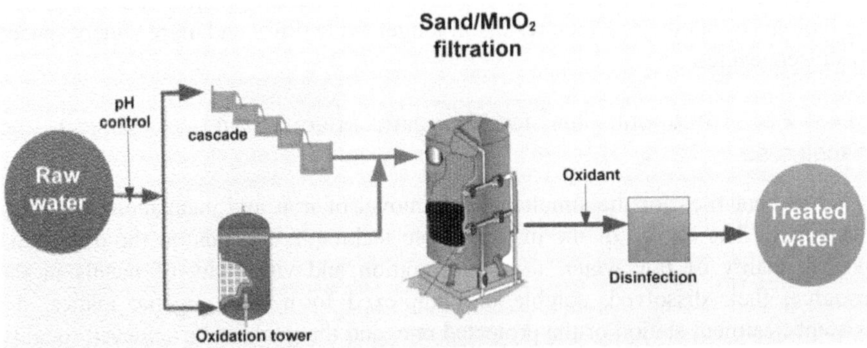

Figure 16.29. *Case of groundwater without chemical reagents. For a color version of this figure, see www.iste.co.uk/gaid/watertreatment3.zip*

The pH must be controlled in order to satisfy the oxidation conditions for the two elements. The addition of caustic soda into raw water may be required. Supplementary balancing may be necessary at the end of the treatment. The cascade requires a surface area, which is not always available. The oxidation tower requires energy for forced aeration and for annual chemical clean-ups, in the event of carbonate precipitation.

Physicochemical treatments with reagents

This solution has the advantage of saving reagents because as the iron oxidizes with air, then it is only Mn^{2+} that requires chemical oxidation.

The case of groundwater with a chemical reagent for Mn^{2+}

The oxidant choice is based on economic considerations as well as on conditions of ease of operation and the initial Mn^{2+} concentration. The same constraints mentioned regarding the cascade and the oxidation tower are applicable.

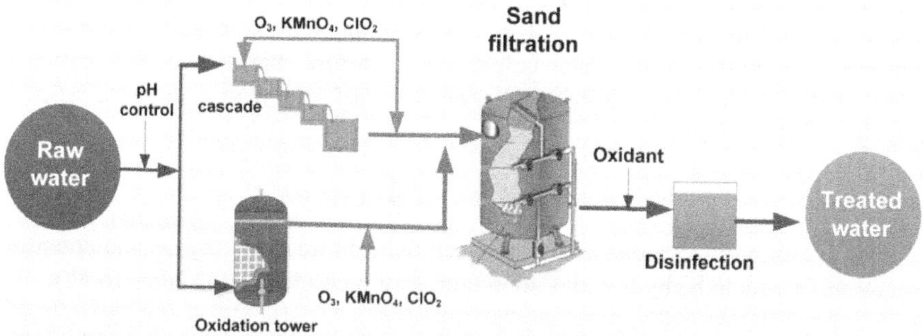

Figure 16.30. *The case of groundwater with a chemical reagent for Mn^{2+}. For a color version of this figure, see www.iste.co.uk/gaid/watertreatment3.zip*

The case of groundwater with a chemical reagent for Fe^{2+} and Mn^{2+}

This solution chemically treats both iron and manganese. The choice of oxidant depends on the same considerations as the ones discussed for the previous solution (Figure 16.30).

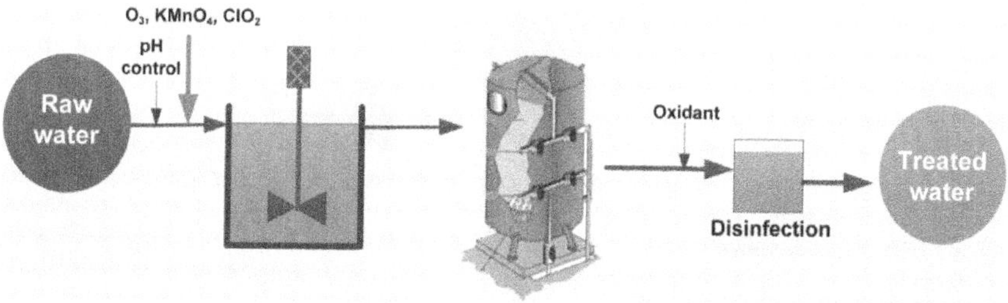

Figure 16.31. *The case of groundwater with a chemical reagent for Fe^{2+} and Mn^{2+}. For a color version of this figure, see www.iste.co.uk/gaid/watertreatment3.zip*

The presence of iron, which often accompanies manganese, may have an effect on the efficiency of manganese oxidation. Since iron oxidizes more quickly than manganese at any pH, the iron oxidant demand must be satisfied before significant manganese oxidation can occur (Figure 16.31). In the case of high Fe^{2+} and Mn^{2+}

concentrations, an anthracite, expanded clay or pumice/sand dual media filter replaces the monolayer sand filter.

In practice, the dosage of permanganate for oxidizing Mn^{2+} must be precisely optimized in order to fully oxidize Mn^{2+}, without leaving excess permanganate in treated water. Excess permanganate can cause water to turn pink, or produce a precipitate of MnOx(s) in the distribution system, and increase the number of consumer complaints about water color. It is therefore essential to pay particular attention to the dose of permanganate used. In most treatment stations, a reaction time of 3–5 min is sufficient.

The case of groundwater with various solutions for Fe^{2+}

This solution draws a distinction between the iron treatment by air and chemical oxidation, in which the hydroxides are retained on the sand filter (Figure 16.32).

The low organic matter content of raw water and the reasonable chlorination rates applied do not carry any risks of THM formation in excessive quantities. Chlorine gas or hypochlorite can be injected for iron oxidation.

Manganese is then retained on the MnO_2 filter. This means there are two filters operating in series. With pressure filters, filtered water from first stage goes directly to the second stage. In the presence of open filters, intermediate pumping and covering are necessary. The advantage is that if one of the filters is momentarily unavailable, production can continue by operating with a single filter. If it is the MnO_2 filter which is unavailable, Mn^{2+} can be chemically oxidized.

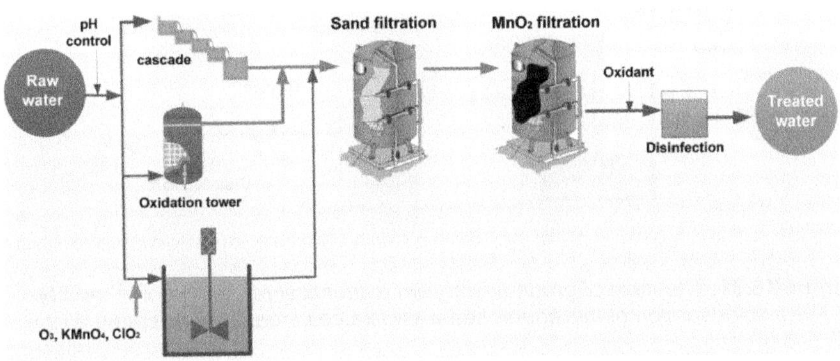

Figure 16.32. *The case of groundwater with various solutions for Fe^{2+}. For a color version of this figure, see www.iste.co.uk/gaid/watertreatment3.zip*

The case of moderate turbid water (<15 NTU), containing Fe^{2+}/Mn^{2+}

When treating surface water or groundwater with variable turbidity, but always lower than 15 NTU and not containing organic matter (<2 mg·L^{-1}), a chemical oxidation of Fe^{2+} and Mn^{2+} is performed, followed by in-line injection of a coagulant (Al or Fe) in order to treat turbidity. The coagulant dosages are related to the inlet turbidity (Figure 16.33).

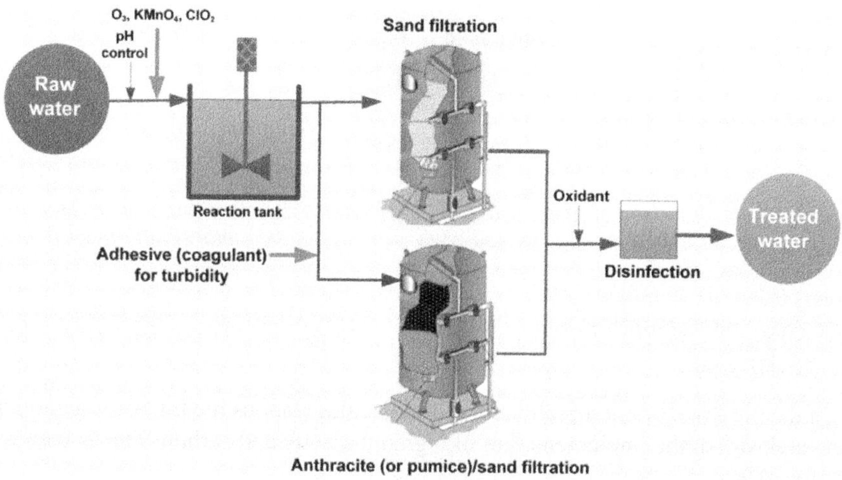

Figure 16.33. *Case of water with turbidity (<15 NTU). For a color version of this figure, see www.iste.co.uk/gaid/watertreatment3.zip*

Most of the time, surface waters are sufficiently oxygenated. This greatly limits the risk of having high dissolved iron contents. Iron oxidation is generally well ensured during the coagulation stage. The iron hydroxides formed are then retained during filtration.

If iron is complexed with organic matter or silica, the use of a more powerful oxidant may be necessary. The oxidant choice is made taking into account the other reagents already implemented at the station, the other quality parameters of raw water and the existing equipment.

This is a variant of the previous case with possible air oxidation of iron, turbidity removal by adhesion in a dual media filter (expanded clay/sand) and MnO_2 removal in a second-stage filter. A second variant is to operate with an anthracite (or expanded clay or pumice)/sand/MnO_2 three-media filter (Figure 16.34) which removes the need for a filter (MnO_2) and intermediate pumping (in the case of open filters). This solution is already proven successful in France.

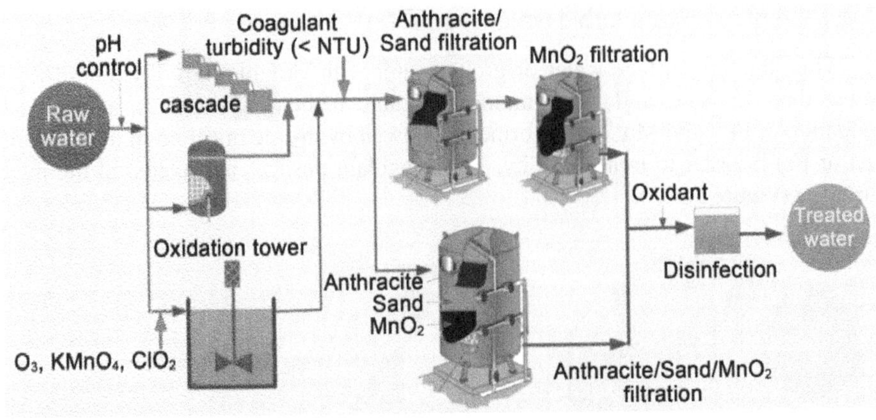

Figure 16.34. *The case of water with turbidity (<15 NTU) with a three-layer filter. For a color version of this figure, see www.iste.co.uk/gaid/watertreatment3.zip*

The case of slightly turbid water (<5 NTU), containing Fe^{2+}/Mn^{2+} and micropollutants

In the presence of micropollutants and molecules responsible for tastes and odors (Fe^{2+} and Mn^{2+}), the implementation of a granular activated carbon filter is required (Figure 16.35).

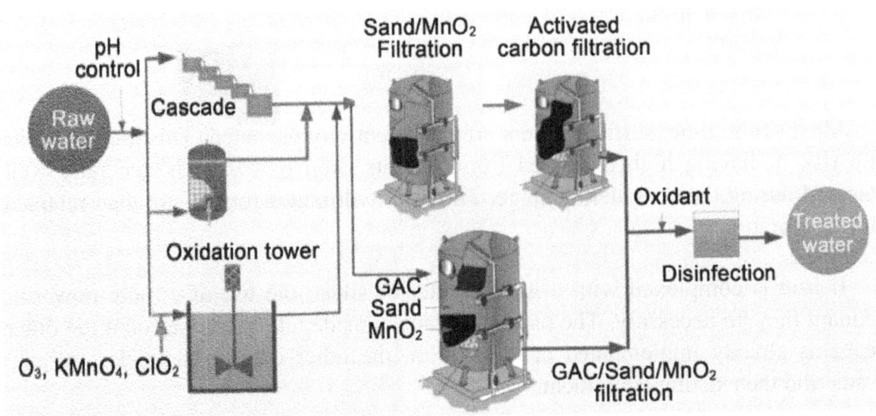

Figure 16.35. *The case of groundwater with Fe^{2+}, Mn^{2+} and micropollutants. For a color version of this figure, see www.iste.co.uk/gaid/watertreatment3.zip*

This filter can be placed as a second stage after iron and manganese removal on a sand/MnO_2 dual media filter. An economically interesting variant is to have a

GAC/sand/MnO_2 three-media filter. The oxidation of iron is preferably carried out by air oxidation.

Biological treatment of iron and manganese

The biological treatment involves two distinct phases, due to the different operating conditions for Fe^{2+} and Mn^{2+}. Biological iron removal works at high velocity (20–35 $m \cdot h^{-1}$) but requires little oxygen, whereas biological manganese removal works at a lower velocity (<15 $m \cdot h^{-1}$), but requires a high concentration of O_2. This is why intermediate aeration is recommended. In short, it is not possible to implement the biological removal of iron and manganese in the same filter (Figure 16.36).

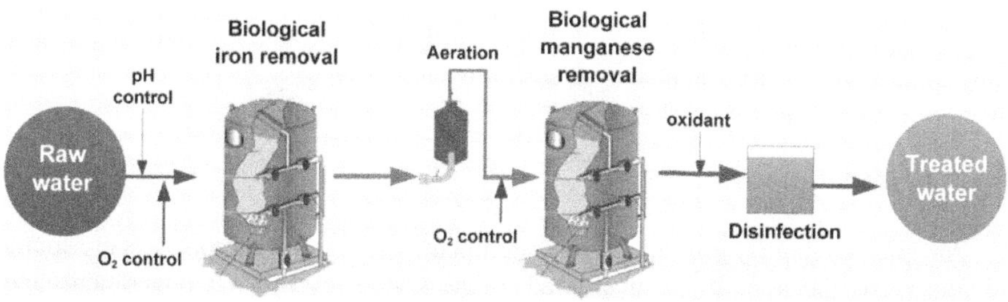

Figure 16.36. *Biological treatment of Fe^{2+} and Mn^{2+} in groundwater. For a color version of this figure, see www.iste.co.uk/gaid/watertreatment3.zip*

It is necessary to strictly control the operating conditions for biological manganese removal because process failure results in a 2–3 month downtime due to the long seeding period required for bacteria growth. Since the quality of surface water is much less stable than that of groundwater, biological iron and manganese removal techniques are not recommended. In fact, the adequate conditions are rarely satisfied.

Hybrid (physicochemical and biological) treatment

While biological iron removal is proposed whenever the qualities of raw water are suitable for it, physicochemical references are much more numerous, due to their ease of use. On the other hand, the constraints of biological manganese removal encourage operators to prioritize a physicochemical solution by setting up a MnO_2 filter at the second stage of biological filtration (Figure 16.37).

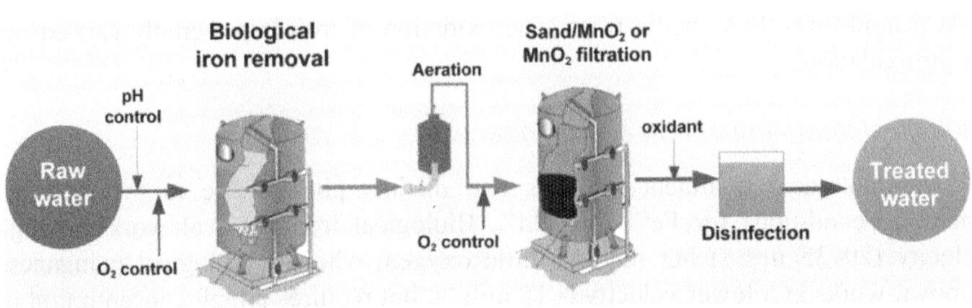

Figure 16.37. *Biological treatment of Fe^{2+} and physicochemical treatment of Mn^{2+}. For a color version of this figure, see www.iste.co.uk/gaid/watertreatment3.zip*

Veolia has many references with this hybrid solution, which is satisfactory not only in economic terms but also in its operating conditions.

16.2. Arsenic removal

Arsenic is a natural groundwater contaminant whose trace components are found in rocks and the soil. It is transported to the waters by erosion phenomena and dissolution mechanisms. It is present in natural waters in organic and inorganic forms. However, it is inorganic arsenic which predominates in water, and often in worrying concentrations for health (Figure 16.38).

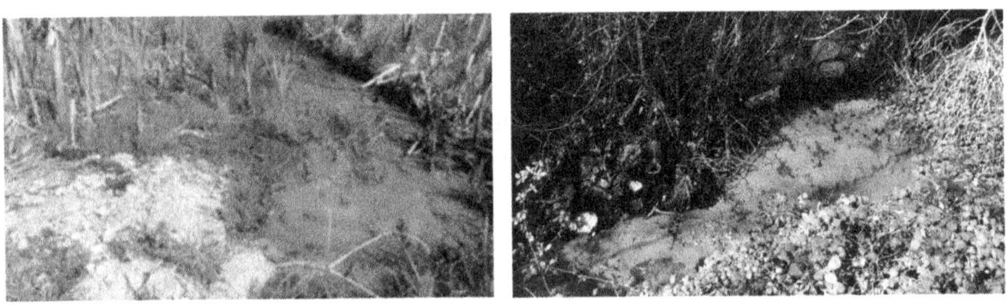

Figure 16.38. *Surface water contamination with arsenic. For a color version of this figure, see www.iste.co.uk/gaid/watertreatment3.zip*

Arsenic is known as an acutely toxic poison, with proven carcinogenic effects in the long term. Trivalent arsenic (As^{3+}) is more toxic than pentavalent arsenic (As^{5+}), and its inorganic forms are more toxic than organic forms. In humans, the lethal dose is between 70 and 180 mg.

As^{3+} combines with the sulfhydric groups from proteins, blocking the activity of certain enzymes, whereas As^{5+} replaces the phosphate ion, causing disturbances in carbohydrate metabolism. Chronic arsenic poisoning is caused by the regular ingestion of contaminated food or drink. In the event of heavy environmental contamination, drinking water generally represents the main contribution of arsenic to the body.

Arsenic is the only mineral substance classified as class A (known to be carcinogenic to humans), without being classified as carcinogenic to animals. The European directive (1998) limited the concentration of arsenic in drinking water to $10\ \mu g \cdot L^{-1}$.

16.2.1. *Arsenic chemistry*

Arsenic exists in the −3, 0, +3, and +5 oxidation states. Environmental forms include arsenous acids (H_3AsO_3, $H_2AsO_3^-$, $HAsO_3^{2-}$), arsenate acids (H_3AsO_4, $H_2AsO_4^-$, $HAsO_4^{2-}$), methylarsonic acid, dimethylarsinic acid, methyl arsonate (MA), dimethyl arsonate (DMA), trimethyl arsine oxide (TMAO), etc. Arsenic (As^{3+}) preferentially forms complexes with oxides and nitrogen. Arsenic (As^{5+}) forms complexes with sulfides.

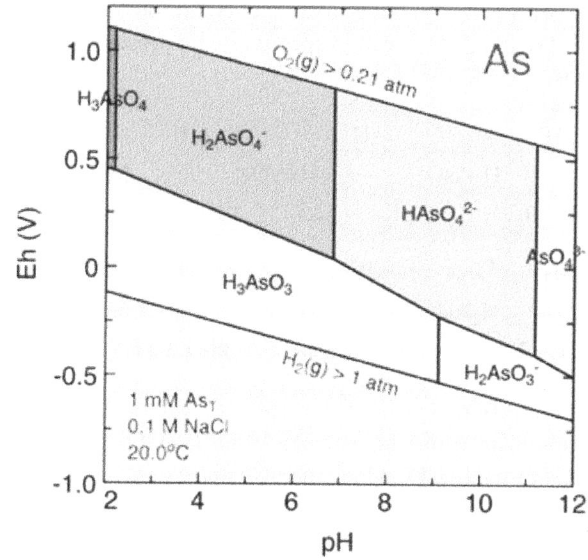

Figure 16.39. *Diagram of the various forms of arsenic. For a color version of this figure, see www.iste.co.uk/gaid/watertreatment3.zip*

In water, the common forms are arsenite (AsO_3^{3-}) and arsenate (AsO_4^{3-}), corresponding to trivalent arsenic As^{3+} and pentavalent arsenic As^{5+}, respectively. Arsenate forms (+5) are AsO_4^{3-}, $HAsO_4^{2-}$, $H_2AsO_4^{-}$, whereas arsenite (+3) forms include $As(OH)_3$, $As(OH)_4^{-}$, $H_2AsO_3^{-}$ and AsO_3^{3-}. Pentavalent forms are predominant and stable in oxygen-rich surrounding environments. Trivalent arsenites are more common in less O_2-concentrated environments, such as groundwater.

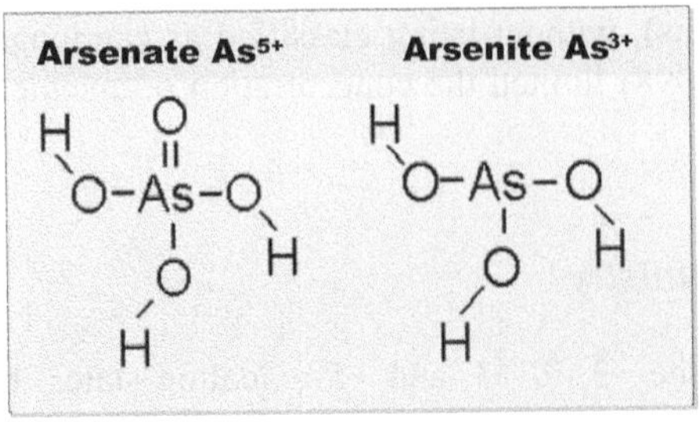

Figure 16.40. *Arsenite and arsenate developed formulas. For a color version of this figure, see www.iste.co.uk/gaid/watertreatment3.zip*

The redox potential (Eh) and pH control arsenic's speciation. $H_2AsO_4^{-}$ dominates at low pH (pH < 6.9) under oxidizing conditions. At higher pH, $HAsO_4^{2-}$ is dominant. H_3AsO_4 and AsO_4^{3-} can, respectively, be present under strongly acidic or basic conditions.

$$H_3AsO_3 \xrightarrow[pK1 = 9.2]{-H^+} H_2AsO_3^{-} \xrightarrow[pK2 = 12.3]{-H^+} HAsO_3^{2-} \xrightarrow[pK3 = 13.4]{-H^+} AsO_3^{2-}$$

$$H_3AsO_4 \xrightarrow[pK1 = 2.1]{-H^+} H_2AsO_4^{-} \xrightarrow[pK2 = 7.08]{-H^+} HAsO_4^{2-} \xrightarrow[pK3 = 11.5]{-H^+} AsO_4^{2-}$$

The pK values of different species show that, in waters whose pH is set between 6 and 9, the predominant forms for valence V are $H_2AsO_4^{-}$ and $HAsO_4^{2-}$, whereas for valence III, it is form H_3AsO_3 which mainly exists.

Forms	Formulas	Names
Arsenic As^{3+}	H_3AsO_3, $HAsO_2$, AsO_3^{3-}	Arsenite ions
Arsenic As^{5+}	$H_2AsO_4^{-}$, $HAsO_4^{2-}$, AsO_4^{3-}	Arsenate ions

Table 16.18. *Main forms of arsenic*

As^{5+} is predominant in oxygenated waters (surface water), whereas As^{3+} becomes predominant when conditions are reductive. The arsenite-arsenate couple is rarely in thermodynamic equilibrium. On the one hand, this can be explained by the very slow kinetics of the natural oxidation–reduction reactions and, on the other hand, by the bacterial populations which modify the arsenite–arsenate ratio.

Thus, in the aerobic zone (Table 16.18), arsenic is found essentially at valence +5. It is then bound to iron oxides and manganese.

pH	0 - 9	10 - 12	13	14
As(III)	H_3AsO_3	$H_2AsO_3^-$	$HAsO_3^{2-}$	AsO_3^{3-}

Figure 16.41. *pH range with arsenite predominance. For a color version of this figure, see www.iste.co.uk/gaid/watertreatment3.zip*

In the anaerobic/anoxia zone, there is a redissolution of iron and manganese with a reduction of arsenic V.

pH	0 - 2	3 - 6	7 -11	12-14
As(V)	H_3AsO_4	$H_2AsO_4^-$	$HAsO_4^{2-}$	AsO_4^{3-}

Figure 16.42. *pH range with arsenate predominance. For a color version of this figure, see www.iste.co.uk/gaid/watertreatment3.zip*

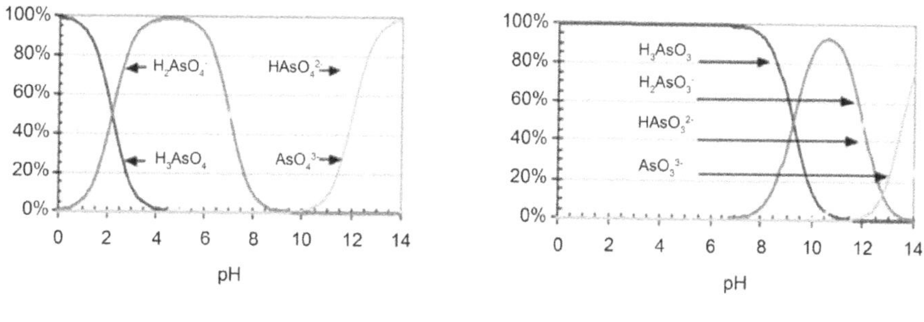

Figure 16.43. *Speciation of As^{5+} as a function of pH (left) and as a function of pH (right). For a color version of this figure, see www.iste.co.uk/gaid/watertreatment3.zip*

Chemical speciation is an essential element when choosing the process to be implemented. Negative charges facilitate removal by adsorption, ion exchange and the co-precipitation process. Since the arsenite [As^{3+}] charge is neutral at natural pH values (6–9), this form is not easily removed. However, the arsenate charge [As^{5+}] is negative (–1 or –2) in the water pH range, so it should be efficiently removed. The conversion of [As^{3+}] into [As^{5+}] becomes an essential element for most treatments.

Environments	Chemical reactions of As
Aerobic zone	Oxidation of As^{3+} into As^{5+} Adsorption and co-precipitation of As into oxides $Fe^{2+} \rightarrow Fe^{3+} \rightarrow Fe(OH)_3, H_2AsO_4$ $Mn^{2+} \rightarrow Mn^{4+} \rightarrow MnO_2$, H_2AsO_4 $HAsO_2 \rightarrow HAsO_4^{2-}$
Anoxic/anaerobic zone	Reduction of As^{5+} into As^{3+} Reduction of Fe/Mn oxides Co-precipitation with FeS $Fe(OH)_3, H_2AsO_4 \rightarrow Fe^{2+} + HAsO_4$ MnO_2, $H_2AsO_4 \rightarrow Mn^{2+} + HAsO_2$ $HAsO_4^{2-} \rightarrow HAsO_2$

Table 16.19. *Chemical reactions of As depending on the nature of the water environment*

The choice of the most appropriate oxidation technology should be based on several considerations, including cost, integration with the existing treatment, disinfection requirements and possible by-products.

16.2.2. *Arsenic treatments*

Arsenic removal treatments are numerous. They include adsorption processes onto iron and aluminum hydroxide flocs, onto activated aluminum, iron oxides, iron sulfides, manganese dioxide, by co-precipitation with lime, by separation using high-pressure membranes (reverse osmosis, nanofiltration) and by means of a biological treatment.

16.2.2.1. *Adsorption treatments onto hydroxide flocs*

The different arsenic removal processes, both physical and chemical, will be discussed in the following section.

16.2.2.1.1. Oxidation of As^{3+} into As^{5+}

Many chemical treatments require a prior oxidation of As^{3+} into As^{5+}. Actually, most treatment methods are efficient for removing arsenic (V) because it exists in oxy-anionic forms which are more highly charged than the species of arsenic (III), and which interact with and/or are better adsorbed onto different materials and supports such as membranes, metal hydroxides, resins or polymers. This is the reason why most arsenic treatment methods require a preliminary oxidation stage. The oxidation stage is as follows:

$$AsO_3^{3-} + H_2O \rightarrow AsO_4^{3-} + 2\ H^+ + 2\ e^-$$

$$As^{3+} \rightarrow As^{5+} + 2\ e^-$$

$$E = Eo + \frac{0.06}{2}\log\frac{[AsO4^{3-}][H^+]}{[AsO3^{3-}]}$$

$$E0 = 0.56\ V$$

$$E = 0.06\ V \text{ for } As^{3+} = 99\%$$

$$E = 0.23\ V \text{ for } As^{5+} = 99\% \text{ at pH 7}$$

Given the fact that the oxidation of As^{3+} by oxygen in the air is very slow, strong chemical oxidants such as chlorine, chlorine dioxide, $KMnO_4$ and ozone are used. UV radiation is ineffective.

The oxidant choice will depend on the quality of raw water, bearing in mind the specific demands for each of them:

– THM formation with chlorine in the presence of DOC, and chlorine demand in the presence of iron;

– chlorites formation with ClO_2;

– high demand for O_3 in the presence of iron, Mn, sulfides or DOC;

– need to control the demand for $KMnO_4$ depending on the variations of As, and demand for $KMnO_4$ in the presence of Fe, Mn and DOC.

With hydrogen peroxide, the conversion rate into arsenic (5+) reaches 100% when the As^{3+}/H_2O_2 ratio is 1–4. This reaction is pH dependent given that arsenous acid (H_3AsO_3) is not oxidized by H_2O_2.

Oxidants	Chemical reactions	Oxidant quantities mg per mg As
Chlorine (Cl_2)	$Cl_2 + H_3AsO_3 + H_2O \rightarrow 2Cl^- + H_2AsO_4^- + 3H^+$	0.95
Sodium hypochlorite (NaOCl)	$NaOCl + H_3AsO_3 \rightarrow NaCl + H_2AsO_4^- + H^+$	0.99
Chlorine dioxide (ClO_2)	$2ClO_2 + H_3AsO_3 + H_2O \rightarrow 2ClO_2^- + H_2AsO_4^- + 3H^+$	1.80
Ozone (O_3)	$O_3 + H_3AsO_3 \rightarrow O_2 + H_2AsO_4^- + H^+$ (pH 6.5) $O_3 + H_3AsO_3 \rightarrow O_2 + HAsO_4^{2-} + 2H^+$ (pH 8.5)	0.64
Potassium permanganate ($KMnO_4$)	$2MnO_4^- + 3H_3AsO_3 \rightarrow 2MnO_2 + 3H_2AsO_4^- + 5H^+ + H_2O$ (pH 6.3–8.3)	1.06

Table 16.20. *Oxidation reactions with various oxidants*

The quantity of hypochlorite required to oxidize 50 $\mu g \cdot L^{-1}$ As^{3+} in the presence of 1.2 $mg \cdot L^{-1}$ of Fe^{2+} is:

$$H_3AsO_3 + OCl^- \rightarrow H_2AsO_4^- + H^+ + Cl^-$$

Note that 1 mole of OCl^- reacts with 1 mole of arsenic.

The oxidation reaction of Fe^{2+} is:

$$2Fe^{2+} + OCl^- + H_2O \rightarrow 2Fe(OH)_3 + Cl^- + 4\,H^+$$

Note that 1 mole of OCl^- reacts with 2 moles of iron.

The molar concentration of As is:

$$\text{As} = \frac{50\ \mu gL^{-1}}{(74.92\ g{\cdot}mole^{-1})(10^6\ \mu g{\cdot}g^{-1})} = 6.67 \times 10^{-7} mole{\cdot}L^{-1}$$

$$\text{Fe} = \frac{1.2\ mgL^{-1}}{(55.85\ g{\cdot}mole^{-1})(10^3 mg{\cdot}g^{-1})} = 2.15 \times 10^{-5}\ mole{\cdot}L^{-1}$$

The quantity of OCl^- is then:

$$6.67 \times 10^{-7} + (2.15 \times 10^{-5})/2 = 1.141 \times 10^{-5}\ \text{mol}\cdot\text{L}^{-1} \text{ of } OCl^-$$

or:

$$1.141 \times 10^{-5} \times 51{,}450 = 0.59\ \text{mg}\cdot\text{L}^{-1}$$

This is a theoretical calculation, because in practice the dosages are higher due to the demand of the oxidant by other mineral or organic compounds.

16.2.2.1.2. Coagulation with iron or aluminum salts

Coagulation–flocculation–settling–filtration is a treatment technique used for removing turbidity, color, a part of organic matter as well as metals in the particulate form.

The coagulation process prompts the aggregation of suspended solids, forming flocs which can later be removed by sedimentation and/or filtration. Coagulation is generally described as a three-stage process: the formation of coagulants, particle destabilization and interparticle collisions.

The first two stages occur in rapid mixing, while the third stage takes place during flocculation, under slower stirring conditions.

Aluminum and iron (III) salts, such as ferric chloride and aluminum sulfate, are the most common coagulants used for drinking water treatment. Coagulation using aluminum and iron (III) salts can be used for removing dissolved mineral contaminants such as arsenic. Arsenic forms insoluble compounds with certain elements which are then co-precipitated. The principle of arsenic removal with iron or aluminum salts is based on adsorption/complexation mechanisms. For example, the arsenic removal mechanism by means of a ($FeCl_3$) coagulant shows that the floc surface is likened to an oxo-acid polymer. H^+ and OH^- ions can be assimilated to acid/base reactions, and adsorptions, to complexation reactions.

The adsorption of H^+ or OHs^- is interpreted in terms of acid/base reactions on the surface of the hydroxyl groups of the oxides formed, whereas complexation reactions are related to the adsorption of other cations or anions. Adsorption results in the formation of surface complexes between the soluble arsenic and the active sites of $Fe(OH)_3$-formed hydroxides:

$$Fe(OH)_3\ (s) + H_3AsO_4 \rightarrow FeAsO_4\cdot 2H_2O + H_2O$$

$$\equiv Fe\text{-}OH + H_2AsO_4^- + H^+ \rightarrow Fe\text{-}H_2AsO_4 + H_2O$$

$$\equiv \text{Fe-OH} + H_3AsO_3 + H^+ \rightarrow \equiv \text{Fe-}H_2AsO_3 + H_2O$$

where ≡ Fe-OH represents the surface's active site.

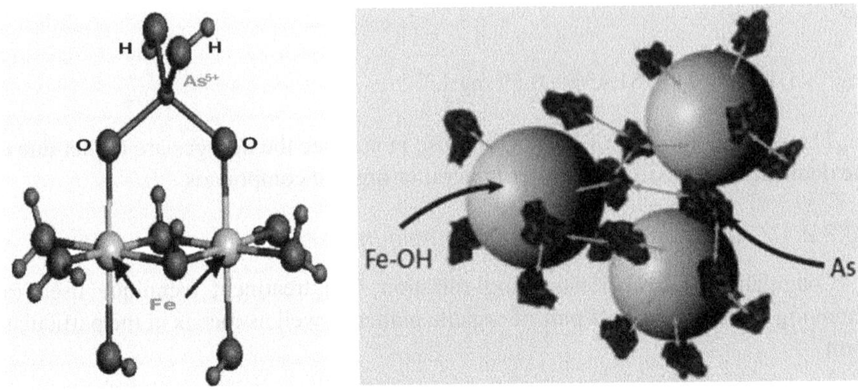

Figure 16.44. *Formation of the FeH_2AsO_4 complex. For a color version of this figure, see www.iste.co.uk/gaid/watertreatment3.zip*

The fixation of arsenic by iron hydroxide requires the prior oxidation of As^{3+} into As^{5+}, and the pH range is 5.5–8.5 for ferric chloride.

During the adsorption process, arsenic adheres to the surface of the hydroxide floc and complexation takes place as the flocs continue to agglomerate. Adsorption is the dominant mechanism for the removal of arsenic using coagulants. Surface complexation models assume that the number of binding sites capable of retaining arsenic is proportional to the coagulant dose. As a result, a saturation of the active surface is obtained for high initial arsenic concentrations.

Several factors affect the coagulation process such as coagulant dosage, pH, turbidity, natural organic matter (NOM), anions and cations in solution, zeta potential and temperature.

Using iron salts and in an optimal pH range, As^{5+} is removed over 95%, whereas As^{3+} is only removed by 50–60%. It is $FeCl_3$ which is the most efficient (>95%), whereas $Fe_2(SO_4)_3$ and $FeClSO_4$ lead to results between 80 and 85%. In contrast, $FeSO_4$ leads to a much lower efficiency, around 25%.

The pH is optimum until the value of 7.7. Efficiency rapidly decreases when pH becomes higher than 8.5. On the other hand, a good removal of organic matter by coagulation leads to a better removal of arsenic.

Coagulation treatments with aluminum salts can either involve monomeric forms: $AlCl_3$, $Al_2(SO_4)\cdot 18H_2O$, or pre-polymerized forms. The pre-polymerized forms are aluminum hydroxyl complexes with OH/Al ratios ranging from 1.5 to 2.7 for the coagulants prepared in situ, or around 2–2.1 for the marketed forms.

The chemical reactions with aluminum sulfate are:

– dissolution of the coagulant:

$$Al_2(SO_4)\cdot 18H_2O \rightarrow 2Al^{3+} + 3SO_4^{2-} + 18H_2O$$

– aluminum precipitation (acid medium):

$$2Al^{3+} + 6H_2O \rightarrow 2Al(OH)_3 + 6H^+$$

– co-precipitation:

$$H_2AsO_4^- + Al(OH)_3 \rightarrow \text{Al-As (complex)} + \text{other products}$$

Table 16.21 shows that these compounds lead to poorer results than unpolymerized forms. The greater the complexation with OH^- ions, the lower the removal.

Coagulant used	$Al_2(SO_4)_3\cdot 18H_2O$	$AlCl_3$	WAC	Aqualenc
As^{5+} removal %	75	75	35	30

Table 16.21. *Influence of aluminum salt used for the removal of As^{5+}*

With aluminum salts, the efficiencies obtained are significantly lower than those obtained with iron salts. As^{5+} is removed between 30 and 75%, whereas As^{3+} does not exceed a 10% removal.

With aluminum sulfate, the adsorption capacity is 0.1 M $As\cdot M^{-1}Al(OH)_3$ at pH 6.0–6.5, that is, 0.025 mg As per mg pure $Al_2(SO_4)_3$.

The mathematical equations (or models) for the removal of arsenates during coagulation with alum is to attribute all arsenic removal to soluble arsenate adsorbing onto or coprecipitating with fresh particulate a $Al(OH)_3$. Only fresh particulate $Al(OH)_3$ formed by precipitation of soluble aluminum added during the coagulation stage can be considered. Moreover, because soluble Al cannot adsorb any arsenic, it is necessary to quantify the concentration of added coagulant that remained soluble and did not form particulate $Al(OH)_3$.

As it happens with iron salts at the optimal coagulation dose, the quantity of arsenic removed is a function of the initial quantity. These results can be explained because iron hydroxide is less soluble than aluminum hydroxide and works across a wider range (5.5–8.5). The mechanism is larger with Fe because all the Fe^{3+} added will be converted to particulate $Fe(OH)_3$ and removed during treatment. Iron hydroxides influence transport of iron-oxide-sorbed arsenate and removal during treatment.

On the other hand, aluminum sulfate operates within a more restricted range (6.0–6.5).

The surface complexation model assumes that the number of binding sites capable of retaining arsenic is proportional to the coagulant dose. As a result, a saturation of the active surface is obtained for high initial arsenic concentrations. But considering that in water the initial concentrations of arsenic are relatively low compared to the surfaces developed by hydroxide sludge, saturation phenomena are rarely obtained. The removal percentage of arsenic becomes independent from the initial arsenic concentration, in particular, within the limits at which this element is observed in drinking water.

It is then possible to quantify the concentration of particulate iron oxide surface (Fe ads), soluble arsenate [As], and arsenate that is sorbed or coprecipitated with the iron oxide solid [As] removed. At the low ratios of As-metal hydroxide, a simple linear isotherm may be derived from a Langmuir isotherm as a model which explains the arsenate retained on sorption sites of the iron hydroxides surface.

Applying Langmuir's law for the adsorption of As:

$$-\ [\mathrm{As}]\%\mathrm{removed} = 100\% \frac{\mathrm{Kads}\,.[\mathrm{FeOH}]}{(1+\mathrm{Kads}[\mathrm{As}]\mathrm{in})}[\mathrm{As}]\mathrm{in}$$

$$-\ [\mathrm{As}]\%\ \mathrm{removed} = 100\% \frac{\mathrm{Kads}\,.[\mathrm{FeOH}]}{(1+\mathrm{Kads}[\mathrm{FeOH}])}$$

where:

– Kads: adsorption constant;

– [FeOH]: total concentration of active sites, which is proportional to the coagulant dose;

– [As]in: initial concentration of arsenic in water.

Adsorption of As is related to the number of active sites available which are represented by iron hydroxides. The adsorption kinetics will be effective as long as free sites are attractive for As.

This equation can be written as:

$$[As]\%\ removed = 100\% \frac{Kads\,.\,[Fe]}{(1 + Kads[Fe]}$$

[Fe] is the dosage of Fe coagulant in mM.

Kads is a constant in mM^{-1}. Its value ranges between 80 and 110 mM^{-1}.

$$[\mathrm{As}]\%\ \mathrm{removed} = 100\% \frac{\mathrm{Kads}\,.\,((\mathrm{FeCl3}\,, \mathrm{mg.L}-1).\frac{\mathrm{Fe\ mM}}{162.5\ \mathrm{FeCl3}})}{(1+\ \mathrm{Kads}\,.\,((\mathrm{FeCl3}\,, \mathrm{mg.L}-1).\frac{\mathrm{Fe\ mM}}{162.5\ \mathrm{FeCl3}})}$$

This equation is expected to be valid under conditions of a small coagulant dose range and at constant pH.

The hydroxide sludge saturation capacity is 0.14 mg $As^{5+} \cdot mg^{-1}$ ($Fe(OH)_3$) hydroxide.

Since 1 $mg \cdot L^{-1}$ of pure $FeCl_3$ produces 0.66 $mg \cdot L^{-1}$ of $Fe(OH)_3$ sludge, this means that the addition of 5 $mg \cdot L^{-1}$ of (pure) $FeCl_3$ leads to the formation of 3.30 mg of $Fe(OH)_3$ sludge, implying the theoretical removal of 0.461 mg of As^{5+} (i.e. 460 $\mu g \cdot L^{-1}$).

Experimental tests show that ferric chloride removes arsenate As^{5+} at a ratio of 92 µg $As^{5+} \cdot mg^{-1}$ of pure $FeCl_3$, or 37.8 µg $As^{5+} \cdot mg^{-1}$ of $FeCl_3$ (commercial use, 41%) at $pH < 8.0$; or 31.6 µg As^{5+} per mg Fe^{3+}.

Other lab tests show that the Fe^{3+} dose (mg $Fe \cdot L^{-1}$) can be deduced from:

$$\mathrm{Fe}(\mathrm{mg{\cdot}L^{-1}}) = 0.375 \frac{(\mathrm{As\ in} - \ \mathrm{As\ out})}{(\mathrm{As\ out})} = 0.375 \frac{(\mathrm{As\ removed})}{(\mathrm{As\ out})}$$

As in and As out are the inlet and outlet arsenic (As^{5+}) concentrations ($\mu g \cdot L^{-1}$):

1 mM Fe = 55.8 mg/L Fe = 162.3 mg/L $FeCl_3$ pure = 418 mg/L $FeCl_3$ (41%)

1 mg/L Fe = 2.9 mg/L $FeCl_3$ pure = 7.1 mg/L $FeCl_3$ (41%)

All the experimental models presented depend on the conditions of the tests: inlet As concentration, final target of As, raw water quality, interference of other ions or compounds, pH, etc.

In the presence of a mixture of Fe^{3+} and Al^{3+} coagulants, a rapid estimation of arsenic removal can be obtained using the ratio:

$$\% \text{ removal As} = 100\,\% \frac{\mathrm{K\,(Fe + Al)}}{(1 + \mathrm{K(Fe + Al)})}$$

where:

– Fe and Al: injected concentrations of iron and aluminum salts expressed in mM;

– K: reaction constant equal to 80–110 mM.

The oxidation–flocculation–filtration process is recommended for arsenic concentrations above 20 $\mu g \cdot L^{-1}$, with variability up to 150 $\mu g \cdot L^{-1}$.

The oxidation–coagulation–flocculation–filtration process is recommended for surface waters containing turbidity (<20 NTU) and variable arsenic concentrations.

The oxidation–coagulation–flocculation–settling–filtration process is recommended for surface water containing turbidity (>20 NTU) and variable arsenic concentrations.

In the presence of turbid surface water, coagulation–flocculation–settling is necessary to treat such turbidity. Arsenic is previously oxidized (As^{3+} to As^{5+}) and then treated simultaneously with suspended solids by means of a coagulant and a polymer. In this case, Veolia offers Multiflo® and Actiflo® as a standard solution (in concrete or in metal), or as a packaged solution. For example, with Actiflo®, the arsenic removal mechanism is based on the microsand/iron hydroxide-polymer agglomerate onto which arsenic is adsorbed. Concentrations higher than 1,000 $\mu g \cdot L^{-1}$ were treated in regions of Asia, with variable turbidities between 10 and 500 NTU.

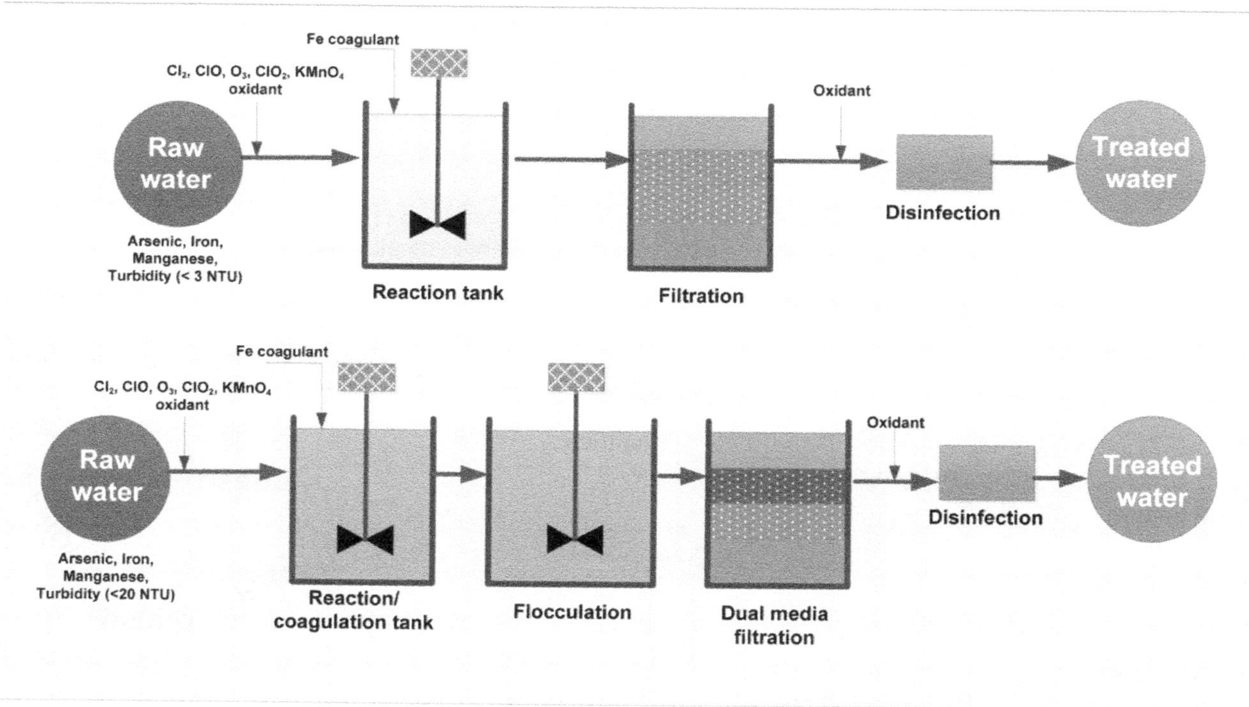

Figure 16.45. *Treatment with or without turbidity (<20 NTU). For a color version of this figure, see www.iste.co.uk/gaid/watertreatment3.zip*

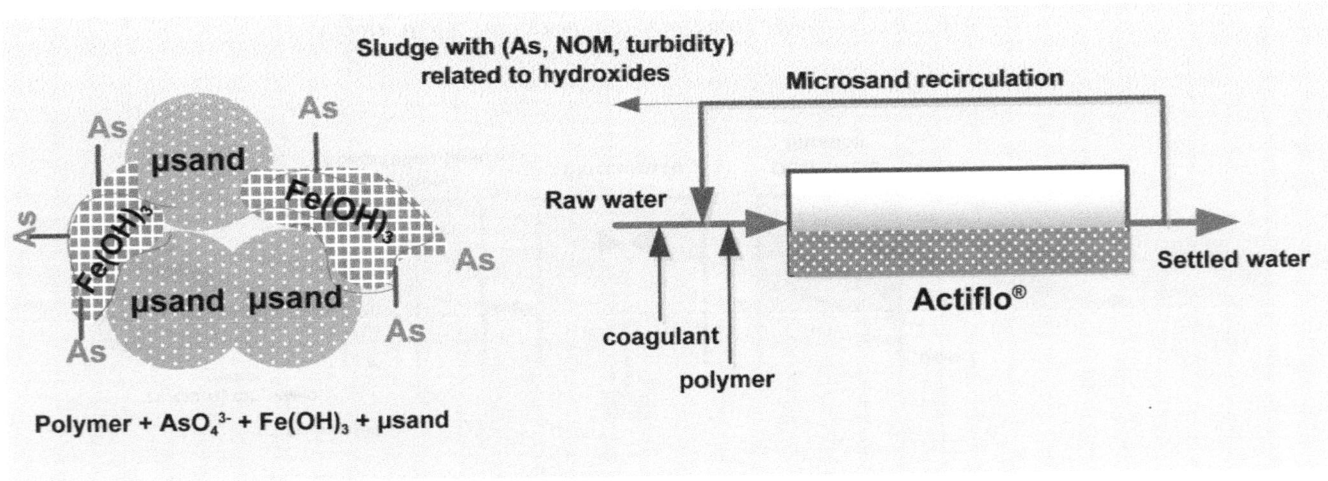

Figure 16.46. *Arsenic removal mechanism with Actiflo®. For a color version of this figure, see www.iste.co.uk/gaid/watertreatment3.zip*

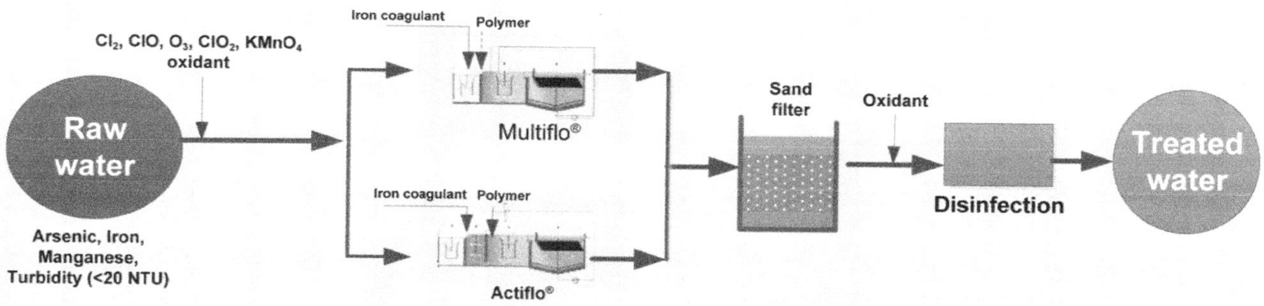

Figure 16.47. *Arsenic removal in turbid water. For a color version of this figure, see www.iste.co.uk/gaid/watertreatment3.zip*

16.2.2.2. *Lime precipitation*

A lime treatment is commonly used for reducing water hardness. Hardness is mainly due to the presence of calcium and magnesium ions. Lime supplies hydroxide ions which raise the pH, and then promote the precipitation of calcium and magnesium into $CaCO_3$ and $Mg(OH)_2$. Lime is also used for the precipitation of metals.

The removal of arsenic is strongly dependent on pH. The removal of As^{3+} and As^{5+} is low at pH < 9.5, but efficiency is greater than 98% for As^{5+} at pH 10.5, whereas As^{3+} is removed by over 75%. This is the reason why the conventional calcium softening at pH 8.5 is insufficient for the removal of arsenic. The co-precipitation of As^{5+} with $Mg(OH)_2$ is the primary mechanism for arsenic removal. As with iron and aluminum salts, As^{5+} is better removed than As^{3+}.

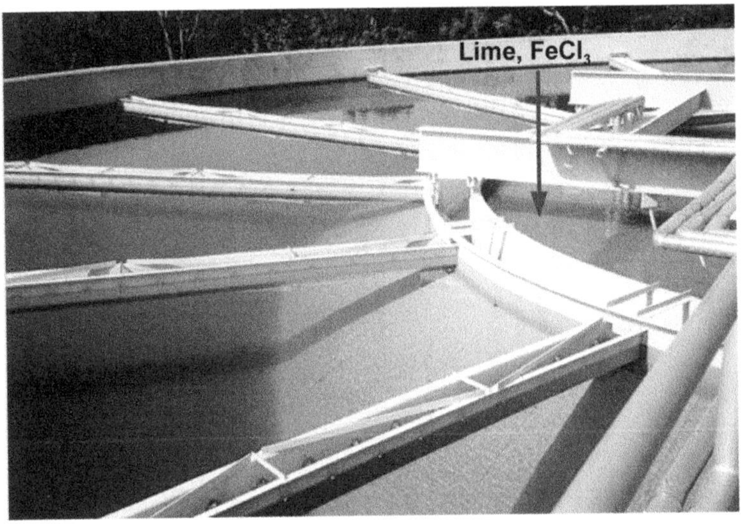

Figure 16.48. *Arsenic removal with lime and ferric chloride in a clariflocculator (Veolia site). For a color version of this figure, see www.iste.co.uk/gaid/watertreatment3.zip*

Moreover, low orthophosphate concentrations can limit arsenate removal with magnesium hydroxide.

In the presence of manganese in the raw water, a removal of As^{5+} can be obtained by sorption with $Mn(OH)_2$. Ferric chloride added before the softening stage with lime may increase removal efficiencies.

16.2.2.3. *Removal of arsenic by adsorption onto a solid material*

Various materials treat arsenic via adsorption mechanisms such as activated alumina (AA), iron oxyhydroxides, inert materials coated with iron, manganese dioxide. These materials are placed in reactors operating within a fixed bed.

16.2.2.3.1. Activated alumina (Al_2O_3)

Adsorption onto AA is a physicochemical process whereby the ions present in water are removed by adsorption onto the available sites on AA.

AA is mainly implemented in filters and removes compounds such as fluorides, arsenic, selenium, silica and NOM. Once saturated, the adsorption sites are regenerated with a strong base (NaOH).

The parameters affecting the efficiency of AA are arsenic's oxidation state, pH and contact time. AA comes in the form of a porous granular material, which has ion exchange properties associated with its large specific surface (300–500 $m^2 \cdot g^{-1}$).

As with many other processes, AA is more efficient against As^{5+} than As^{3+}. The sorption of As^{5+} occurs for pH values between pH 6.0 and 8.0, for which the AA surfaces are positively charged. Adsorption of As^{5+} is strongly pH dependent and exhibits strong affinity for AA at pH 7.6.

Anions (including arsenic) are best adsorbed at pH < 8.2 because the surface of AA has a net positive charge, which is balanced by the adsorption of anions such as hydroxides, fluorides and arsenates. The optimum pH for arsenic removal is between 5.5 and 6.0, and the AA adsorption capacity deteriorates as the pH increases from 6.0 to 8.2.

The reactors' design is based on a contact time of 3–6 min depending on the arsenic concentration to be removed. A pre-oxidation is required in the presence of As^{3+}.

The competition of the other ions present in raw water is observed according to the following scale:

$$OH^- > H_2AsO_4^- > Si(OH)_3O^- > F^- > HSeO_3^- > TOC > SO_4^{2-} > H_3AsO_3 > HCO_3^-$$

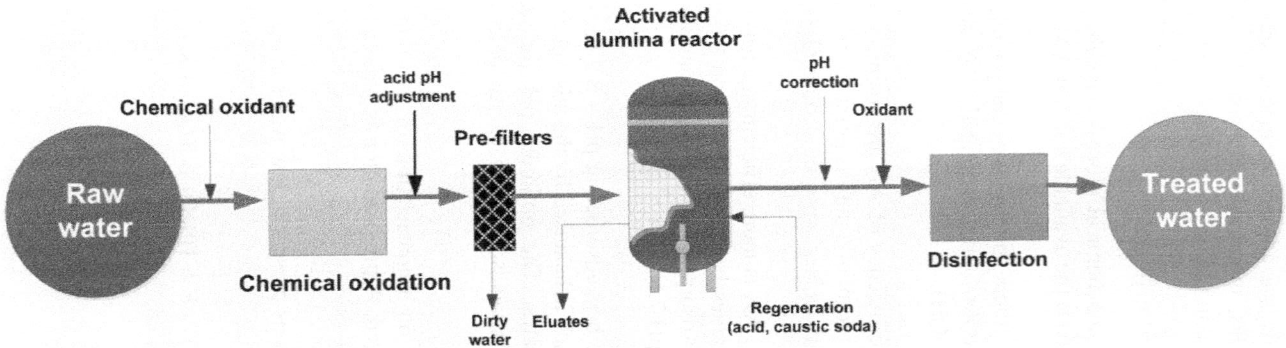

Figure 16.49. *Schematic diagram of As removal onto activated alumina. For a color version of this figure, see www.iste.co.uk/gaid/watertreatment3.zip*

The exchange capacity of AA in relation to As is more than 15,000 volumes/volumes of material for water with a 5.5–6 pH and contains the following concentrations: $Cl^- < 250\ mg{\cdot}L^{-1}$, $F^- < 2\ mg{\cdot}L^{-1}$, $SO_4^{2-} < 350\ mg{\cdot}L^{-1}$, dry residue $< 1{,}000\ mg{\cdot}L^{-1}$, $SiO_2 < 50\ mg{\cdot}L^{-1}$, $Fe < 0.5\ mg{\cdot}L^{-1}$, $Mn < 0.05\ mg{\cdot}L^{-1}$ and TOC $< 4\ mg{\cdot}L^{-1}$.

AA favors a good retention of arsenic at valence 3+ (50–70%) and 5+ (80–90%). The adsorption capacity is 30–112 mg of arsenic per gram of AA.

There is a competition effect with chloride and sulfate ions. The effect of sulfates is all the greater when pH is < 7. In the presence of humic acids, the adsorption capacities are reduced when the removal of arsenic is desired. This is valid for raw water with a turbidity < 0.5 NFU. When turbidity is higher than the standard, sand filtration should be provided, injecting the reagents upstream of the AA filter. Due to the operating conditions in an acid medium, it is important to bring the water to calco-carbonic equilibrium before distribution.

The AA is regenerated on site by adding 20 volumes of sodium (2N)/volume of material (pH 12–14). An acid wash with 15 volumes of H_2SO_4 (2N) for neutralization is often performed in order to dissolve the precipitates formed during the basic cleaning. A rinse with water is applied with the equivalent to seven volumes of material. However, these regeneration conditions (pH 12) tend to dissolve some of the AA (2%), which is why it is important to regularly supplement the necessary volumes of material.

After three or four regenerations, the AA must be completely renewed. The regeneration eluates containing As are sent to a specific treatment. AA is very effective and can be regenerated in situ. However, the adsorption efficiency is only high at acidic pH and arsenites must be pre-oxidized into arsenates.

<table>
<tr><th colspan="2">Physical characteristics of activated alumina</th><th colspan="2">Physicochemical composition (%)</th></tr>
<tr><td>Specific surface</td><td>$250\ m^2{\cdot}g^{-1}$</td><td>Al_2O_3</td><td>92.2</td></tr>
<tr><td>Total pore volume</td><td>$0.395\ cm^3{\cdot}g^{-1}$</td><td>Na_2O</td><td>0.90</td></tr>
<tr><td>Porosity</td><td>0.562</td><td>Fe_2O_3</td><td>0.08</td></tr>
<tr><td>Density</td><td>$620–830\ kg{\cdot}m^{-3}$</td><td rowspan="2">SiO_2</td><td rowspan="2">0.09</td></tr>
<tr><td>Abrasion</td><td>1.6% of weight</td></tr>
</table>

Table 16.22. *Physical characteristics and composition of activated alumina*

Figure 16.50. *Retention of arsenic onto activated alumina (Veolia site). For a color version of this figure, see www.iste.co.uk/gaid/watertreatment3.zip*

16.2.2.3.2. Oxyhydroxide materials and iron hydroxides

Adsorption onto adsorbent materials based on iron (iron oxides, oxyhydroxides and hydroxides) includes the amorphous forms of hydrous ferric oxide (FeO-OH), goethite (-FeOOH) and hematite (-Fe_2O_3), which are inert materials and adsorbents removing both As^{3+} and As^{5+} from water. The amorphous form of Fe(OOH) has the highest adsorption capacity because it has the highest developed surface. Amorphous iron oxides are known to have a large adsorption capacity for arsenic and to exert a strong influence on its mobility.

The removal mechanism is identical to the one obtained with iron salts, that is to say, adsorption/complexation.

	GEH (Germany)	**Kemira (Finland)**	**Evers (Germany)**	**Severn Trent (UK)**
Brand name	GFH	CFH 12	Everzit As	Bayoxide E 33
Active site	$Fe(OH)_3$ + β FeOOH	FeO(OH)	FeO(OH)	FeO(OH)
Shape	Granular	Granular	Granular	Granular
Grain size (mm)	0.32–2	1–2	0.5–2	0.5–2
Apparent density ($kg \cdot m^{-3}$)	0.9–1.2	1.15	0.625–0.675	0.53
Specific surface ($m^2 \cdot g^{-1}$)	190–250	52	300	120–200
Porosity (%)	75	72–80	75	85
Humidity (%)	10	<10	<1.5	<20
Optimum pH	5.5–6.5	6.5–7.5	6.5–7.5	6.0–8

Table 16.23. *Characteristics of some arsenic-adsorbing media*

Granular ferric oxyhydroxide (GFH) is a highly porous material (micropore volume 0.0394 $cm^3 \cdot g^{-1}$, mesoporous volume ~0.10 $cm^3 \cdot g^{-1}$) with a specific area of 190–250 $m^2 \cdot g^{-1}$.

Figure 16.51. *Some arsenic-adsorbing materials available on the market. For a color version of this figure, see www.iste.co.uk/gaid/watertreatment3.zip*

The treatment of arsenic with iron oxyhydroxide does not lead to the release of iron into the treated water or to the adsorption of Ca^{2+}, Mg^{2+}, Cl^- and SO_4^{2-}.

A very low retention of fluorides and nitrates is observed, whereas the presence of P has no influence on the adsorption mechanism. No variation in pH or dissolved oxygen is noticed.

A contact time of 5 min is often used, but this time can be longer, depending on the As concentration to be removed. A maximum velocity of 15 $m \cdot h^{-1}$ is recommended in order to avoid leaks of iron hydroxide fine particles into treated water. The contact time range is 3–12 min.

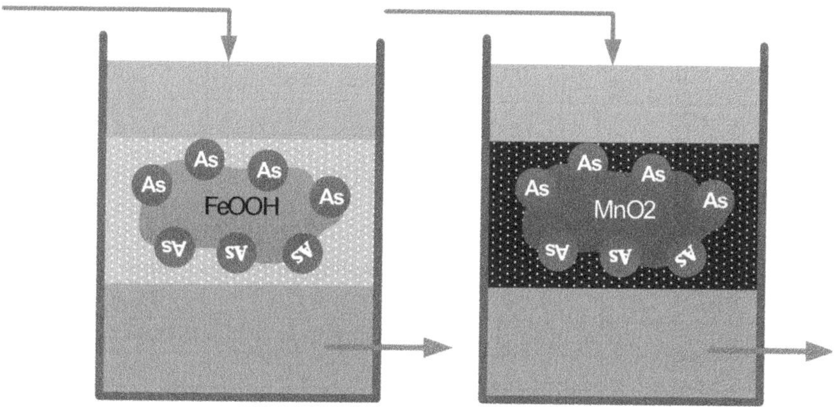

Figure 16.52. *Retention of arsenic on iron oxyhydroxide (GFH) and manganese dioxide (MnO_2). For a color version of this figure, see www.iste.co.uk/gaid/watertreatment3.zip*

Iron hydroxide removes As^{3+} as well as As^{5+}. The adsorption capacities strongly depend on pH. GFH has an adsorption capacity of 8 mg As/g FeOOH at pH 7. Table 16.24 summarizes the adsorption capacities of GFH for a contact time of 4.8 min at various pH and As^{3+} and As^{5+} concentrations. This capacity increases over 60% for the same pH from an initial As concentration of 20 $\mu g \cdot L^{-1}$ at 80 $\mu g \cdot L^{-1}$.

Table 16.25 shows the adsorption capacities of GFH as a function of pH and three contact times. The inlet arsenic concentration is 30 $\mu g \cdot L^{-1}$. The 20 min contact time increases the adsorption capacity by about 30%, compared to a 4.8 min contact time.

The GFH process is recommended when the arsenic concentration in raw water does not exceed 20 $\mu g \cdot L^{-1}$.

Inlet As ($\mu g \cdot L^{-1}$)	**20**	**30**	**60**	**80**	**130**
pH	**Adsorption capacity ($mg \cdot g^{-1}$)**				
6.5	3.4	5.3	6.3	8.9	8.2
7	2	3.3	5.4	4.7	4.3
7.5	1.2	1.9	3.2	3	2.8

Table 16.24. *Adsorption capacity of GFH in relation to arsenic*

Contact time (min)	**4.8**	**10**	**20**
pH	**Adsorption capacity ($mg \cdot g^{-1}$)**		
6.5	5.3	6.6	7.3
7	3.3	3.9	4.5
7.5	1.9	2.4	2.9

Table 16.25. *Adsorption capacity of GFH as a function of contact time*

Inlet As ($\mu g \cdot L^{-1}$)	**Treated water volume/bed volume**
20	$V/V = 292{,}429\ (10^{-0.248}.pH)\ 1{,}000$
30	$V/V = 230{,}331\ (10^{-1.0205}.pH)\ 1{,}000$
50	$V/V = 85{,}329\ (10^{-0.9461}.pH)\ 1{,}000$
80	$V/V = 39{,}732\ (10^{-0.9025}.pH)\ 1{,}000$
130	$V/V = 50{,}067\ (10^{-1.0078}.pH)\ 1{,}000$

Table 16.26. *BV on GFH media for different pH and arsenic concentrations*

Figure 16.53. *Pressure filters including adsorbent material (FeOOH). For a color version of this figure, see www.iste.co.uk/gaid/watertreatment3.zip*

16.2.2.3.3. Manganese dioxide (MnO_2)

The oxidation chemical mechanism of As^{3+} by manganese dioxide unfolds in two phases, involving the reduction of Mn^{4+} into Mn^{3+}:

$$2MnO_2 + H_3AsO_3 \rightarrow 2MnOOH\bullet + H_3AsO_4$$

$MnOOH\bullet$ is an intermediate reaction product (Mn^{3+}). This reaction is followed by the reaction of As^{3+} with $MnOOH\bullet$:

$$2MnOOH\bullet + H_3AsO_3 + 2H^+ \rightarrow 2Mn^{2+} + H_3AsO_4 + 3H_2O$$

The overall reaction is then:

$$2MnO_2 + 2H_3AsO_3 + 4H^+ \rightarrow 2Mn^{2+} + 2H_3AsO_4 + 2H_2O$$

The retention capacity of MnO_2 in relation to arsenic is 1.15 $mg{\cdot}g^{-1}$ MnO_2. An additional reaction could include the adsorption of As^{5+} by MnO_2:

$$MnO_2 + H_3AsO_3 \rightarrow MnO_2AsOOH + H_2O$$

MnO_2AsOOH represents the As^{5+} surface complex.

The reactions imply that As^{3+} is oxidized by MnO_2, followed by the adsorption of the reaction product of As^{5+} during the MnO_2 solid phase. The estimated As-Mn interatomic bond is 3.22Å for arsenic's two valences. It is possible for As^{3+}

oxidation to lead to a modification of the surface of MnO_2, allowing for the creation of new reaction sites for As^{5+}.

The removal of arsenite As^{3+} and arsenate As^{5+} by manganese oxide is an important reaction, both for the natural cycle of As and its applications and for the treatment of drinking water.

As^{5+} is better adsorbed at pH 7 than at pH values between 7.1 and 8 because under these pH conditions, the electrical charge on the material's surface is less negatively charged and the $HAsO_4^{2-}/H_2AsO_4^-$ ratio is weaker. Its value is 1.7 at pH 7, 2.14 at pH 7.1 and 13.5 at pH 8.

In the presence of iron and manganese, the interferences of these two metals modify the adsorption mechanisms, leading to the following reactions:

$$2Mn^{2+} + MnO_2\,(s) + 2H_2O \rightarrow Mn_3O_4(s) + 4H^+$$

$$2Fe^{2+} + 3MnO_2\,(s) + 4H_2O \rightarrow 2Fe_2O_3 + Mn_3O_4\,(s) + 8H^+$$

The application of this technique is valid when the initial concentration of As in raw water is lower than 20 $\mu g \cdot L^{-1}$. Otherwise, regeneration is too frequent.

MnO_2 affinity for the various elements present in water is evolves according to the following scale:

$$Mn^{2+}; Fe^{2+} > As^{5+} > PO_4^{3-} > SO_4^{2-}\,;\, CO_3^{2-}$$

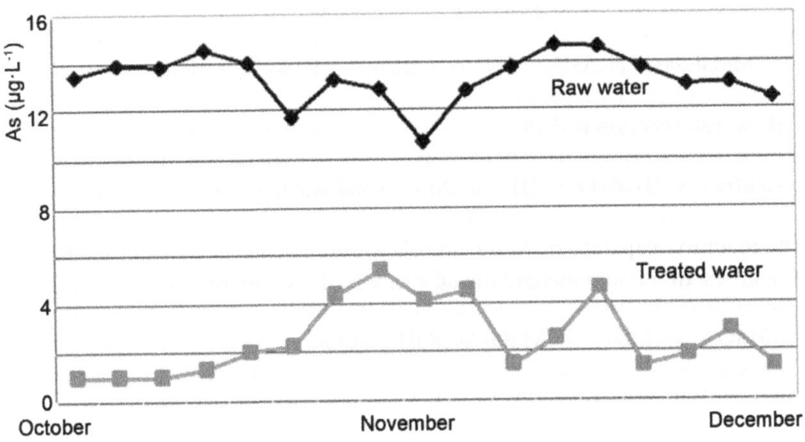

Figure 16.54. *Treatment of arsenic on an MnO_2 filter. For a color version of this figure, see www.iste.co.uk/gaid/watertreatment3.zip*

Figure 16.54 shows the filtration results for groundwater containing a total As concentration varying between 11 and 15 μg·L^{-1}. The contact time is 6 min and the pH is 7.2. The concentration of treated water varies between 1 and 5 μg·L^{-1}. There was no chemical regeneration during this period. The iron concentration in raw water is lower than 100 μg·L^{-1}.

16.2.2.3.4. Implementation

Manganese dioxide is a so-called "active" material in relation to arsenic in its dissolved form, which means that its surface is reactive: in this case, adsorption essentially involves a physical reaction bringing into play van der Waals forces. This is called "physiosorption".

Raw water crosses the material from top to bottom; treated water is collected at the bottom of the filter and then returns to the neutralization station. The filter is a closed reactor functioning under pressure.

The support of the filtering layer consists of a metal floor 12 mm thick, fitted with long-tailed OTV strainers. The density of these strainers is 40 per m^2 of draining surface.

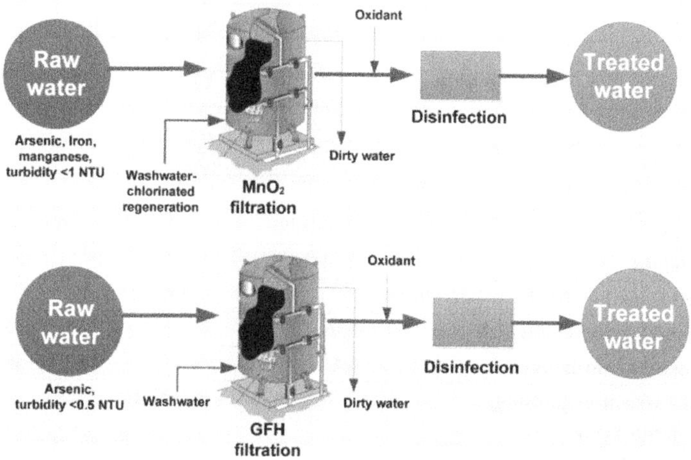

Figure 16.55. *Arsenic removal on MnO_2 filters and GFH. For a color version of this figure, see www.iste.co.uk/gaid/watertreatment3.zip*

They are fixed to the floor by flanged nuts screwed under the floor. The strainer heads are embedded in a 4/7 pumice layer 0.10 m thick, over which the adsorbent material filtering layer is set. The maximum head loss will be 0.5 bar.

The new material is loaded into the filter through the existing hydro-ejector, which is connected to the filter's loading pipe. The new material load is defined beforehand.

As the treatment progresses, the adsorption sites become saturated with arsenic. When the material reaches saturation, the arsenic-laden manganese dioxide is regenerated.

Regarding the "filtration" function, the material progressively becomes clogged as particles are retained. After a certain clogging threshold, the filter is washed with water.

16.2.2.3.5. Design parameters

Table 16.27 presents a comparative scale between adsorption onto MnO_2 and GFH. The difference in the adsorption capacities makes GFH advantageous in terms of material regeneration (MnO_2) or renewal (GFH).

Parameters	MnO_2	GFH
As^{3+} concentration	$30\ \mu g \cdot L^{-1}$	
Raw water pH	7	
Total inlet flow	100 m^3/h over 24 h	
Number of filters	1 u	
Flow rate	>15 m/h	
Unit area	6.7 m^2	
Filter diameter	2,920 mm	
Volume of material	10.05 m^3	
Contact time	6 min	6 min
Material height	1.5 m	1.5 m
Cylindrical filter height	3 m	
Material's adsorption capacity	0.36 g As/kg	4 g As/kg
Total capacity at saturation	4.4 kg of As	48.2 kg of As
Volume of raw water passing through the material	876,000 m^3	
Annual arsenic load	26.28 kg/year	
Regeneration frequency (MnO_2) or renewal (GFH)	Once a week	22 months

Table 16.27. *Comparison between adsorption onto MnO_2 and GFH*

As far as investment costs are concerned—the filters being identical—the difference originates in the cost of the GFH material, which is three to four times more expensive than MnO_2. In terms of operating costs, MnO_2 requires frequent regenerations, but is not renewed. On the other hand, GFH requires a total renewal of the material after about 2 years of operation.

Figure 16.56. *Arsenic removal on a manganese dioxide filtration unit. For a color version of this figure, see www.iste.co.uk/gaid/watertreatment3.zip*

16.2.2.3.6. Manganese dioxide and iron oxyhydroxide combination

The presence of arsenic, iron and manganese in the source water has led Veolia to patent a three-media filter. It contains a first layer of granular iron hydroxide or oxyhydroxide in order to retain arsenic and iron, an intermediate sand layer to prevent fine particles from the upper media from passing through the filter, and a manganese dioxide (MnO_2) lower layer. It is the Mangaflo+ process which has been implemented at a station in the east of France.

16.2.2.3.7. Modular arsenic treatment units (Opafer™)

OTV-Veolia has designed a range of modular units, inherited from the Filtraflo® As process, for removing the arsenic present in raw water. These compact, skid-mounted and fully automatable units are ready for use. They can be combined to treat different flow rates and arsenic concentrations, and are suitable for temporary or long-term use.

Arsenic is retained during the filtration of water on a bed of iron oxyhydroxide granules or MnO_2, or a combination of materials for removing iron, manganese and arsenic. This process does not require any cleaning or regeneration. Only backwashing should be regularly implemented.

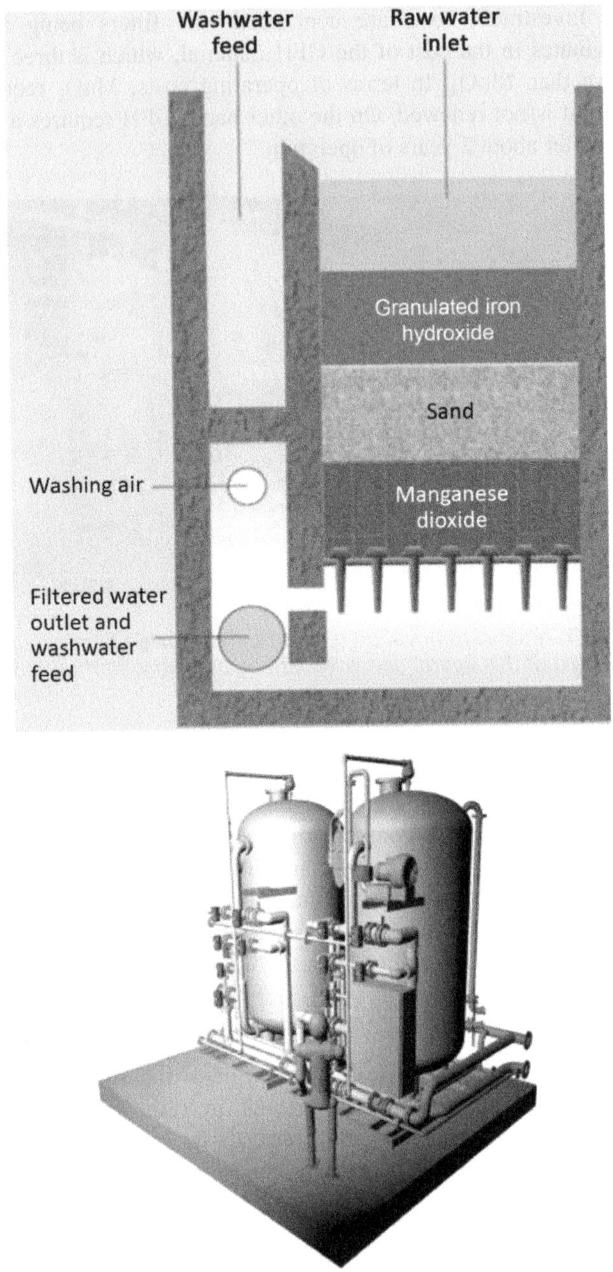

Figure 16.57. *Mangaflo+ process (Veolia patent) and Opaferunit™. For a color version of this figure, see www.iste.co.uk/gaid/watertreatment3.zip*

When saturated, iron oxyhydroxide is replaced so as to maintain treatment capacity. Its admission into a specialized discharge area is authorized as part of a commitment to sustainable development and responsible waste management.

The basic unit comprises:

– one or more filters connected to piping and valves;

– a washing pump with accessories;

– an electrical control cabinet with PLC automation or programmer.

As an option, a closed heat-insulated container, with heating for frost protection, can be supplied as well as additional equipment, such as a flow meter or remote monitoring equipment. The flow range is 25, 50 and 100 $m^3 \cdot h^{-1}$.

16.2.2.4. *Other treatments*

– Activated carbon can retain arsenic if it is specifically designed for this. Modified activated carbons with Fe bridging on their surface have been developed (Jacobi) in order to remove arsenic and micropollutants. Unmodified carbon has a low efficiency against arsenic.

The removal of As involves a chemical pre-oxidation because As^{5+} ions are more easily retained on a strongly anionic resin *ion exchanger* (in chloride or hydroxide form).

The exchange mechanism is as follows:

$$2RCl + HAsO_4^{2-} \rightarrow R2HAsO_4 + 2Cl^-$$

In regeneration:

$$R2HAsO_4 + 2NaCl \rightarrow 2RCl + Na_2HAsO_4$$

The disadvantage of anionic resins is that they also retain the other anions present in raw water, according to the following scale:

$$SO_4^{2-} > HAsO_4^{2-} > NO_3, CO_3^{2-} > NO_2^- > Cl^-$$

It is accepted that a sulfate concentration greater than 50 $mg \cdot L^{-1}$ or a dry residue greater than 500 $mg \cdot L^{-1}$ reduces the efficiency of ion exchangers by more than 50%.

As a result, the volume of water produced per volume of resin strongly depends on the initial concentration of sulfates and bicarbonates, as well as the arsenic concentration.

The application of ion exchangers is based on a contact time of approximately 3 min.

– *Anionic resin regeneration* conditions require the use of a normal sodium chloride solution with 2 eq of chlorides/equivalents of resin. Five bed volumes are used for each regeneration.

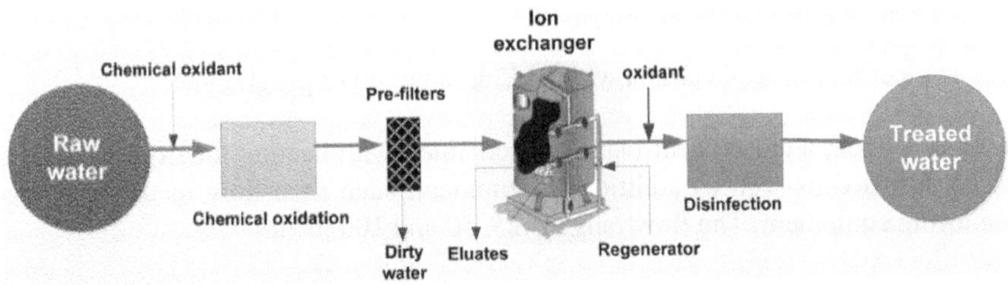

Figure 16.58. *Treatment of arsenic using ion exchange resins. For a color version of this figure, see www.iste.co.uk/gaid/watertreatment3.zip*

– *Microfiltration* and *ultrafiltration* have no direct effect on the removal of arsenic. A coagulation/flocculation stage with iron salts is necessary upstream of the membrane.

The design for microfiltration and ultrafiltration is identical to the one conventionally used when treating turbidity.

Backwashing (1–2 min duration every 30–45 min) recovers $Fe(OH)_3^-$ As flocs formed during the addition of $FeCl_3$.

– *Reverse osmosis* with a recovery rate greater than 85% allows for the retention of around 98% of arsenic V. If the arsenic is at valence III, removal will be around 60%.

– *Nanofiltration* achieves removals of over 95% of arsenic As^{5+}, but lower than 20% for arsenic As^{3+}. As^{5+} is present in monovalent ($H_2AsO_4^-$) and divalent ($HAsO_4^{2-}$) forms, whereas As^{3+} is present in the uncharged H_3AsO_3 form. The removal of arsenic is different for each of these species. These two species are correctly removed with tight NF membranes (NF 70 and NF 90, Dow/Filmtec) for a wide range of operating conditions with efficiencies greater than 95%. This suggests that it is the size exclusion which governs the separation mechanism, and not the charge interaction. Loose (porous) NF membranes show that the removal of As^{3+} (<20%) is much lower than that of As^{5+}.

The high efficiency of As^{5+} with the NF 200 membrane is attributed to the electrostatic repulsion of the co-ion $HAsO_4^{2-}$ (which is the predominant species at pH 8) due to the negatively charged membrane.

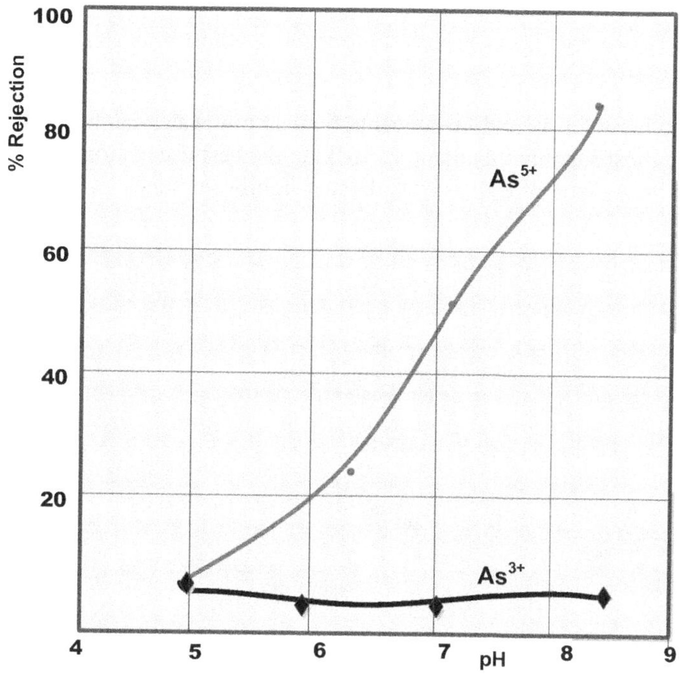

Figure 16.59. *Rejection of arsenic on NF 200 (T: 25°C, scanning speed: 0.2 m·s^{-1}, flow: 28.8 lmh)*

– The ability of some *bacteria* to integrate the oxidation–reduction reactions of metals and metalloids into their energy metabolism gives them an active role in the biogeochemical cycle of these elements. A wide range of bacteria isolated from different contaminated environments have been studied for their ability to synthesize arsenite and oxidize As^{3+} enzymatically. They include heterotrophic bacteria as well as chemotrophic bacteria, in which the As^{3+} ions serve as electron donors. In this case, the energy produced is used for fixing the CO_2 that bacteria use for their growth.

For the removal of arsenic from groundwater containing dissolved iron, biological iron removal (*Thiobacillus ferroxidans*) leads to a removal of As^{3+} by adsorption. As^{3+} is an energy substrate usable by aerobic bacteria in slightly acidic or neutral environments (pH 6–8). This makes preliminary chemical oxidation unnecessary, even if other tests have shown that bacterial oxidation makes it possible to oxidize As^{3+} into As^{5+}. On the other hand, raw water must be enriched with divalent iron (by injecting $FeSO_4$), if the Fe/As ratio is not sufficient. The recommended Fe/AS ratio is 40/1.

16.2.2.5. *The factor influencing the process choice: the Fe/As ratio in raw water*

The ratio between As and Fe concentrations in raw water can be used as a guide to select the most suitable treatment, by making the best use of iron's capacity to fix arsenic. An Fe/As ratio of 20/1 is considered a significant value.

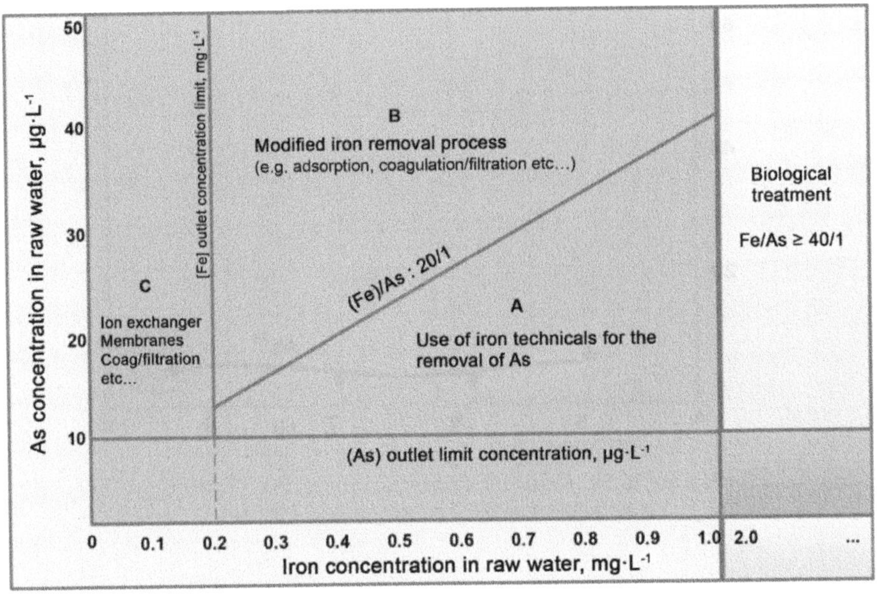

Figure 16.60. *Influence of the Fe/As ratio in raw water when choosing the most suitable process. For a color version of this figure, see www.iste.co.uk/gaid/watertreatment3.zip*

Due to the major role played by iron in arsenic removal, the iron concentration in the resource is an important parameter with regard to the technical choice to be adopted.

Figure 16.60 illustrates three domains:

– high iron concentrations (zone A): iron removal processes such as adsorption and co-precipitation can be used to facilitate arsenic removal. A water resource with an Fe/As ratio greater than or equal to 20/1 is suitable for the implementation of an iron removal technique to reduce arsenic. This ratio shows that 1 $mg \cdot L^{-1}$ of iron is expected to remove up to 50 $\mu g \cdot L^{-1}$ of arsenic. A 40:1 ratio favors biological treatment;

– moderate iron levels (zone B): for an iron/arsenic ratio lower than 20/1, a treatment process such as coagulation/filtration with the addition of an iron salt as a coagulant is one of the possible solutions;

– low iron levels (zone C): in this case, technologies which are not influenced by the presence of iron such as ion exchange, membranes and adsorption processes, etc., are considered. A resource with an iron concentration equal to or lower than the standard (0.2 $mg \cdot L^{-1}$) is less likely to be made up of iron particles capable of creating arsenite/Fe^{3+} complexes or precipitation.

16.2.2.6. *Advantages and disadvantages of the various technical solutions*

Table 16.28 summarizes the advantages and disadvantages of the various technical solutions discussed so far. It thus appears that:

– coagulant precipitation/co-precipitation (particularly iron salts) is used because it also removes other contaminants (turbidity, color, organic matter);

– while the use of ion exchange resins or AA is advantageous, it comes up against the disposal of eluates, the competition for resins on the part of other ions and the need to work in an acid medium for AA;

– membranes are not often or rarely used for the sole treatment of arsenic. These are justified when the salinity provided by the different ions is high (sulfates, nitrates, carbonates, etc.);

– MnO_2 as an adsorbent material is used less than iron oxides. The frequency of regenerations can be a limiting parameter. It does not lose its exchange capacity after regeneration or is it replaced;

– the iron oxyhydroxide adsorbent material is currently the most widely used because it is easy to implement and presents a high adsorption capacity. As it cannot be regenerated, its cost (investment and operation) is limited to low flow rates;

– the removal of arsenic can also be obtained by adsorption onto fluidized sand which is permanently coated with iron (II), itself oxidized by oxygen or hydrogen peroxide;

– carbonate removal with lime is justified when softening is desired. The sludge obtained cannot be recycled, which limits the use of this technology. A two-stage treatment will be preferred: removal of As and softening;

– online coagulation upstream of membrane filtration (MF/UF) gives very good results. The coagulant acts as an adsorbent and the membrane, as a physical separator.

<table>
<tr><th>Technical solutions</th><th>Removal % As^{5+}</th><th>Advantages</th><th>Disadvantages</th></tr>
<tr><td>Oxidation/ precipitation
Chemical oxidation</td><td>>95%</td><td>Simple to implement and fairly quick process</td><td>Chemical reagents
Competition between metals</td></tr>
<tr><td>Coagulation and co-precipitation:
- aluminum sulfate
- ferric chloride</td><td>70–80%
>95%</td><td rowspan="2">Simple to implement
Commonly used chemical reagents
Acceptable investment cost</td><td>Toxic sludge production
Low reduction of As^{3+}
Pre-oxidation required</td></tr>
<tr><td>Lime</td><td>>95%</td><td>Toxic sludge
pH > 10
Simultaneous water softening</td></tr>
<tr><td>Adsorption onto supports:
- Fe oxyhydroxide (GFH)
- manganese dioxide
- activated alumina (AA)</td><td>>90%
>80%
>90%</td><td>Well-known technical solutions (filtration) and easy to implement</td><td>AA and GFH applications only for low flows in closed filters
Filtration equipment
Washing equipment
Toxic wastewater
Regeneration (AA and MnO_2)
Replacement (GFH)
Operating cost depending on the support</td></tr>
<tr><td>High-pressure membranes
Nanofiltration
Reverse osmosis</td><td>>90%
>95%</td><td>High and stable efficiency
Simultaneously removes other constituents present in the water</td><td>Heavy equipment
Complex operation
High investment cost
High operating cost
High water losses
Remineralization required
Discharge of a toxic concentrate</td></tr>
</table>

Table 16.28. *Advantages and disadvantages of different arsenic removal processes*

In all cases, the treatment choice is based on the following parameters:

– the reliability of the treatment, guaranteeing distributed water with an arsenic concentration lower than 10 $\mu g \cdot L^{-1}$ (100% of the time);

– minimization of investment and operating costs;

– the generation (or not) of processing by-products and their disposal;

– simplicity of operation;

– minimization of chemical reagents and water losses;

– integration of the station into an existing system (borehole/reservoir), taking care to disrupt its operation as little as possible during the new works phase;

– the importance of treatment interruptions (clogging velocity, wash frequency and volume, material regeneration).

16.3. Removal of selenium (Se)

Selenium has a wide spectrum of uses, ranging from agriculture to industrial processes (electrical, metallurgical, glass and ceramic, chemical, food and pharmaceutical, agricultural, etc.). Selenium is a natural and essential nutrient for humans, animals and some plants.

However, selenium can also have adverse effects on health and reproduction when excessively consumed. In nature, selenium is mainly found in sedimentary rocks rich in organic matter (e.g. source rock from petroleum, coal, phosphorites and carbonaceous shales) as well as sulfide minerals.

Due to its natural occurrence in geological materials used as industrial raw materials, selenium is often present in the wastewater from oil and gas extraction, petroleum refining, power generation coal-fired electricity, metal mining and processing and from the production of phosphate fertilizers.

Selenium can also be present in irrigation water and runoff water from farms located in areas with seleniferous soils.

The concentration of selenium in groundwater is generally low and below the quality limit set at 10 $\mu g \cdot L^{-1}$. It tends to be higher in well-oxygenated groundwater due to the high mobility of Se^{6+}, which is less sensitive to adsorption by metal oxides (pH > 7) than Se^{4+}.

In surface waters, Se^{6+} is also weakly adsorbed onto oxides and clays at pH $\geq$ 7. Therefore, the oxidation of Se^{4+} into Se^{6+} leads to persistent mobility of selenium in water, especially in drainage water from arid agricultural areas. Se^{6+} concentrations observed in rivers vary between 0.005 and 0.3 $mg \cdot L^{-1}$.

16.3.1. *The chemistry of selenium*

Selenium belongs to the metalloid family. Its presence is closely related to that of sulfur. The molar mass of elemental selenium is 78.96 $g \cdot mol^{-1}$. At room temperature, it is available in a solid, stable form, insoluble in water and the usual organic solvents. It exists in different physical forms: a red or black amorphous form and a red or gray crystalline form. Selenium is present in the environment in both inorganic and organic forms.

The inorganic forms of selenium are:

– H_2Se selenide (oxidation state –2);

– elemental selenium, Se° (oxidation state 0);

– SeO_3^{2-} and $HSeO_3^-$ selenite (oxidation state +4);

– SeO_4^{2-}, $HSeO_4^-$ selenate (oxidation state +6).

However, in fresh waters, selenium is mainly present in the two anionic forms: selenite (SeO_3^{2-}) and selenate (SeO_4^{2-}), which correspond to selenium's oxidation states IV and VI.

In groundwater, Se^{6+} predominates in the [SeO_4^{2-}] form. Se^{4+} can also be found in the [SeO_3^{2-}] form, but in much lower concentration. The chemical speciation and behavior of selenium strongly depend on the medium's pH and oxidation–reduction potential.

In natural waters (pH between 6.0 and 9.0), under oxidizing conditions, Se^{6+} is predominant in the divalent ionic SeO_4^{2-} form. Se^{4+} essentially exists in the SeO_3^{2-} form.

16.3.1.1. *Selenium (6+)*

Selenic acid (H_2SeO_4) is a fairly strong acid, dissociated in an aqueous solution:

$$H_2SeO_4 + H_2O \rightarrow HSeO_4^- + H_3O^+$$

$$HSeO_4^- + H_2O \rightarrow SeO_4^{2-} + H_3O^+$$

Se^{6+} essentially exists in the SeO_4^{2-} form (pKa = 0 for $H_2SeO_4/HSeO_4^-$ and pKa = 1.7 for $HSeO_4^-/SeO_4^{2-}$).

16.3.1.2. *Selenium (4+)*

Selenous acid (H_2SeO_3) is a diacid:

$$H_2Se_2O_3 + H_2O \rightarrow HSeO_3^- + H_3O^+$$

$$HSeO_3 + H_2O \rightarrow SeO_3^{2-} + H_3O^+$$

In natural waterways, the main forms of Se^{4+} are the $HSeO_3^-$ and SeO_3^{2-} species. The preponderance of one over the other depends on the pH (pKa = 2.35 for $H_2SeO_3/HSeO_3^-$; pKa = 8 for $HSeO_3^-/SeO_3^{2-}$).

Se^{4+} is the most frequent form found in reductive conditions:

– at a pH below 8.15, the hydrogen-selenite ion $HSeO_3^-$ is the dominant form;

– at a pH greater than 8.15, the divalent anion SeO_3^{2-} will be the dominant form.

The distribution and diverse associations of selenium essentially depend on pH, the oxidation–reduction conditions and the reaction kinetics. The effects of pH and oxidation–reduction potential on the different oxidation states of selenium are shown in Figure 16.61.

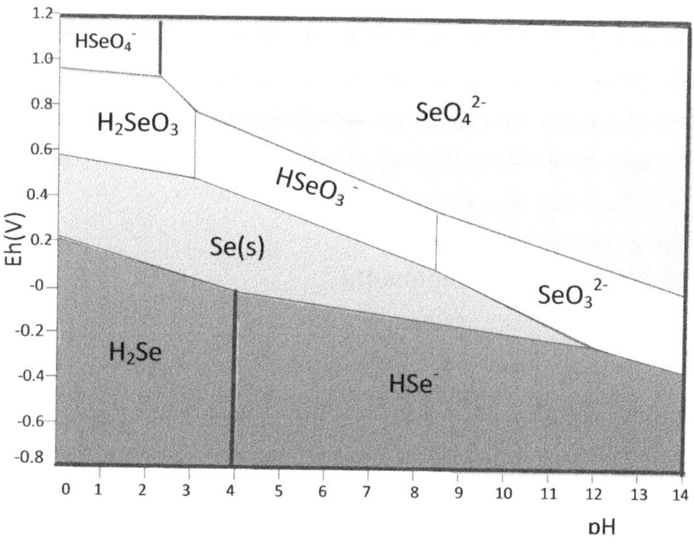

Figure 16.61. *Eh(V)-pH potential diagram for selenium. For a color version of this figure, see www.iste.co.uk/gaid/watertreatment3.zip*

Reduced species such as elemental selenium (Se°) and selenides (HSe^-) are insoluble, which is why they are usually discharged as a colloidal suspension into surface waters. The organic forms of selenium found in natural waterways are produced by microbiological assimilation and degradation.

This diagram in Figure 16.61 shows that:

– between 400 and 450 mv for a pH between 6 and 7.5, $HSeO_3^-$ (selenite) is the predominant species;

– at 400 mv and pH > 8, it is the SeO_4^{2-} (selenate) species which is dominant. The same applies to oxidation–reduction potentials > 600 mv and for pH values greater than 5. At –200 mV, elemental selenium and selenide are present.

In the aquatic ecosystem, selenium results from soil leaching. The contents vary depending on the soils' richness and water pH.

Selenium is mainly found in the inorganic form, namely in oxidation states IV and VI in aerated water. The distribution of these two species depends on the type of water, depth, aeration and pH.

Dissolved selenium can remain free in solution, be absorbed by an organism or be retained by organic and/or inorganic colloidal matter. This last possibility is the predominant one. As a matter of fact, 70–85% of organic and inorganic selenium can be bound to colloidal particles, and therefore, be removed in the form of sediments.

Surface waters (rivers and lakes) generally have concentrations of total dissolved selenium ranging from 0.1 to 0.3 $\mu g \cdot L^{-1}$, unless the water is alkaline and drains volcanic regions.

16.3.2. *Selenium removal treatments*

16.3.2.1. *Removal of Se by co-precipitation treatments (Fe, Al, $Ca(OH)_2$)*

Coagulation with iron or aluminum salts is not an efficient method for removing selenates (Se^{6+}). However, efficiencies are higher for selenites (Se^{4+}). The co-precipitation of selenium involves its adsorption and inclusion into the hydroxide conductor, and therefore, the formation of a mixed solid phase. Surface adsorption is one of the main co-precipitation mechanisms.

16.3.2.1.1. Iron salts

Removal of Se^{4+}

The efficiency of ferric chloride coagulation is between 85 and 95% for a pH range between 6 and 7. This efficiency is reduced by 15–20% when the pH rises to 8.3.

The factors influencing the Fe/Se precipitation are the type of coagulant used ($FeCl_3$, $FeSO_4$, $Fe_2(SO_4)_3$), pH and contact time.

With $FeCl_3$, a Se^{4+} removal efficiency of 84–94% is obtained with $FeCl_3$ dosages (commercial use, 41%) varying between 5 and 20 $mg \cdot L^{-1}$ and a pH between 6 and 8. The contact time is 10 min. Some studies have suggested that prechlorination may affect the removal of Se^{4+} because chlorine tends to oxidize Se^{4+} into Se^{6+} and thus reduce the performances of $FeCl_3$.

Removal of Se^{6+}

Conventional coagulation–flocculation techniques are ineffective for removing Se^{6+}. Despite 200 $mg \cdot L^{-1}$ dosages of $FeCl_3$ at pHs ranging from 6 to 8, the removal of Se^{6+} remains below 10%.

In the presence of ferrous (Fe^{2+}) ions, selenate ions are reduced to selenite ions which are then adsorbed onto the surface of the hydroxides ($Fe(OH)_3$ formed. The mechanism is as follows:

$$SeO_4^{2-} + 2\,Fe(OH)_2 + H_2O \rightarrow SeO_3^{2-} + 2\,Fe(OH)_3$$

16.3.2.1.2. Aluminum salts

Removal of Se^{4+}

The coagulation of aluminum salt is only 20% efficient (maximum level reached) for removing Se^{4+}.

Removal of Se^{6+}

Conventional coagulation–flocculation techniques are ineffective for removing Se^{6+}. Despite Al sulfate dosages higher than 180 $mg \cdot L^{-1}$ and a pH between 6 and 7, the removal of Se^{6+} remains below 10%.

16.3.2.1.3. Electrocoagulation

The iron hydroxides generated during electrocoagulation (iron electrodes) remove over 95% of Se^{4+} using an anode current density of 4.8 mA per cm^2. The removal mechanism is identical to that of iron salts.

16.3.2.1.4. Lime

Lime precipitation is not an efficient treatment method for removing selenate (Se^{6+}). However, efficiencies are higher for selenites (Se^{4+}).

Hence, over 50% removal of Se^{4+} is obtained by precipitation with lime at pH 11.5, and only 30% is obtained at pH 9.5. On the other hand, less than 10% is observed with Se^{6+}, even at a very high pH.

16.3.2.2. *Solid support adsorption treatment*

The most commonly used adsorbents are as follows: AA, iron oxyhydroxide, sand coated with ferric oxide, manganese dioxide and activated carbon. But their efficiency differs from adsorbent to adsorbent.

16.3.2.2.1. Activated alumina

The removal mechanism involves an exchange of the selenium present in the water for AA hydroxide ions. When the exchange sites are saturated, the AA is regenerated. This regeneration is done with soda (0.5%), rinsing and neutralization with acid (sulfuric acid for Se^{6+} and hydrochloric acid for Se^{4+}).

Since AA is amphoteric, adsorption is pH dependent. At a pH below 8 (AA's charge neutralization), the surface of AA has a positive charge and will adsorb the anions present in the water.

AA exhibits an optimal adsorption capacity for Se^{4+} when the pH is between 5 and 6. The interference effects of the bicarbonate ions present in water show that an increase from 50 to 200 $mg \cdot L^{-1}$ of the bicarbonate concentration reduces the removal efficiency of Se^{4+} about 20% at a pH between 6.0 and 6.5.

Selenate (Se^{6+}) is not removed for pH $\geq$ 6. A reduction over 50% is obtained at pH 5. Contact times in the reactor are generally calculated between 5 and 8 min. The adsorption of Se^{6+} is disadvantaged-by the interference of other ions, such as sulfate ions, for example. An increase in the concentration of sulfates from 5 to 500 $mg \cdot L^{-1}$ decreases the number of treated water bed volumes from 450 to 15 for a pH value of 6. When the sulfate/Se^{6+} ratio is low, the adsorption capacity of AA is higher.

16.3.2.2.2. Iron oxyhydroxide

The adsorption by ferric oxyhydroxide is particularly efficient for the removal of Se^{4+} when the pH is between 6.0 and 7.0. A removal of between 85 and 95% is observed. Thus, for a pH greater than 7.5, efficiency decreases as a function of the increase in pH. An adsorption capacity of 0.5–1.2 mg of Se^{4+}/g is obtained for pH values of 8.3 and 7.2, respectively.

The adsorption of Se^{4+} depends on the type of oxyhydroxide: amorphous ferrihydrites are more efficient than crystalline ferri-oxyhydroxides.

Selenate (Se^{6+}) is not removed by the action of iron oxyhydroxide for pH ≥ 6. A reduction over 50% is obtained at pH 5. Contact times in the reactor are generally calculated between 5 and 8 min. As selenite ions are adsorbed far more efficiently than selenate ions, this implies that Se^{6+} ions must be reduced to Se^{4+} before precipitation/adsorption onto these materials. The adsorption of Se^{6+} can be strongly affected by the presence of sulfates, bicarbonates and silica.

16.3.2.2.3. Adsorption onto iron-based media

Veolia offers a Zero-Selenium® treatment process which removes this metal from an iron-based material (Hydrex 9670) whose effectiveness has been proven on different types of water containing varying concentrations of Se. The reactor consists of a rising circular filter with a material height of 1.5–2.5 m with a contact time of 10 min. Flow rates are between 9 and 15 $m \cdot h^{-1}$.

Figure 16.62. *Hydrex 9670. For a color version of this figure, see www.iste.co.uk/gaid/watertreatment3.zip*

Figure 16.63. *Water reoxygenation tank. For a color version of this figure, see www.iste.co.uk/gaid/watertreatment3.zip*

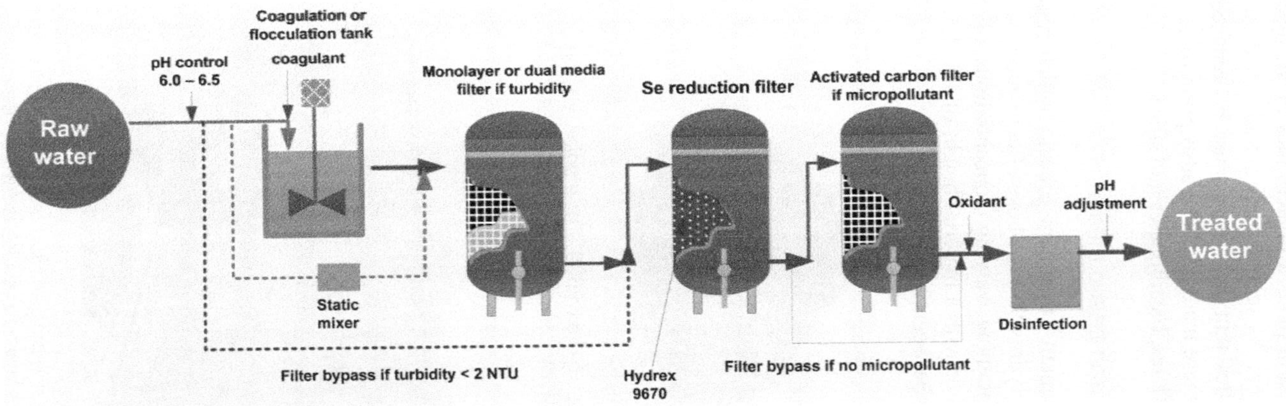

Figure 16.64. *Se treatment station depending on different raw water quality options. For a color version of this figure, see www.iste.co.uk/gaid/watertreatment3.zip*

The filter incorporates a feed water distribution ring, which provides an even flow distribution. When the head loss signals it, the filter is backwashed. This volume is estimated at 5% of the processed flow. The pH which was adjusted to 6.0–6.5 upstream of the Hydrex material is then readjusted in the treated water. Sand filtration (optional) or activated carbon filtration in case of micropollutants' presence is included in the treatment.

The water produced can be distributed after disinfection. In the case of turbid raw water, adhesion in a sand filter or in a dual media filter makes it possible to remove turbidity so that it is < 2 NTU at the inlet of the Se reduction filter.

The adsorption capacity of Hydrex 9670 is 1 mg $Se \cdot g^{-1}$ of material. With Hydrex 9670, the adsorption of the selenite ion Se^{4+} leads to a removal over 98% after only 10 min of contact time for an initial concentration of Se^{4+} of 0.5 $mg \cdot L^{-1}$ and a pH < 7.5. With a concentration of Se^{6+} of 23 $\mu g \cdot L^{-1}$, the 5 $\mu g \cdot L^{-1}$ threshold is only exceeded after 75,000 BV.

16.3.2.3. *Other treatments*

– Powdered activated carbon is not effective to treat this element.

– Resins exchangers of strongly basic anions make it possible to remove Se^{6+} more efficiently than Se^{4+}. The number of bed volumes to be treated before Se^{6+} leakage is higher when the sulfate concentration decreases. An increase of more than 700 bed volumes is expected when a strongly basic anion exchange resin treats source water containing sulfate at a concentration of 50 $mg \cdot L^{-1}$ compared to a sulfate concentration of 192 $mg \cdot L^{-1}$, because the affinity for sulfates is almost as great as for Se^{6+}. A molar ratio of 0.3 between sulfate ions/selenate ions showed a decrease in efficiency of around 10%. A 50% decrease in efficiency is obtained when this molar ratio rises to 2.

Strongly basic anion exchange resins are used for the removal of selenium, which is present in its anionic form. The regeneration salt for resins is sodium chloride.

Anion exchange resins show better affinity for Se^{6+} than for Se^{4+}. The removal of selenium is between 90 and 97%, either in selenate or selenite form (Se^{6+} or Se^{4+}). On the other hand, the exchange capacities are different and more favorable for Se^{6+}.

The resin's behavior toward the hydrogen-selenite ion ($HSeO_3^-$) and in relation to the selenite ion is identical. Column tests on groundwater containing a dissolved salt concentration of 712 $mg \cdot L^{-1}$, a sulfate concentration of 192 $mg \cdot L^{-1}$ and a

Se^{4+} concentration of 0.1 $mg{\cdot}L^{-1}$ have made it possible to obtain treated water with a residual concentration of Se^{4+} of 10 $\mu g{\cdot}L^{-1}$, for a contact time of 5 min.

The exchange capacity of strongly anionic resins is 31.9 mg Se^{4+} per liter of resin. An exchange capacity over 60 mg of Se^{6+} per liter of resins can be reached. Chlorides are less competitive than sulfates but should also be considered. Polyamine-type basic resins show low selenate ion removal efficiencies for pH values between 3 and 12.

16.3.2.3.1. Reverse osmosis/nanofiltration

High-pressure membranes force water through a membrane, in which case ionic species such as Se^{4+} or Se^{6+} are discharged into the concentrate. The efficiency of a high-pressure membrane treatment system depends on several factors, including raw water quality, membrane type, threshold molecular weight, and system recovery capacity.

A selenium removal efficiency (Se^{4+} and Se^{6+}) in the range between 80 and 99% is obtained with reverse osmosis and nanofiltration membranes. The hydraulic efficiencies vary between 50 and 75% and the supply pressures are around 15–20 bar. The pH plays an important role and should be kept below 7.5.

16.3.2.4. *Biological treatment*

Selenate and selenite can be biologically reduced to their selenium's elemental form, after which it becomes insoluble. The biological process leads to a reduction of Se^{6+} and Se^{4+} ions to insoluble elemental selenium (Se°).

$$Se^{6+} + \text{bacteria} + DOC \rightarrow Se^{4+} + Se^{\circ}$$

Selenate Selenite Particle elemental Se

The microbial reduction, which is carried out under anoxic or anaerobic conditions, depends on the medium's oxidation–reduction potential and uses a source of organic carbon as an electron donor. Se^{6+} and Se^{4+} ions act as electron acceptors. The general formula for this process is given as:

$$4CH_3COO^- + 3SeO_4^{2-} \rightarrow 3Se^{\circ} + 8CO_2 + 4H_2O + 4H^+$$

The presence of nitrates (anoxic conditions) interferes with the Se biological removal process. Selenates can be biologically reduced to selenites, then elemental selenium, under anaerobic conditions and in the presence of a source of organic carbon. Parameters such as temperature, pH, contact time, concentration, speciation of selenium and the addition of a carbon source are involved in anaerobic biological reactions. As with biological removal of nitrates or sulfates, interference from these ions can be observed. However, these interferences can be reduced when specific growth conditions are implemented in the biological reactor or when bacteria are introduced into the reactor so that mixed populations can develop therein. The results obtained show Se^{4+} residuals of around 50 $\mu g \cdot L^{-1}$ for water initially containing 1,000 $\mu g \cdot L^{-1}$ of Se^{6+}.

Applied to industrial water, the AnoxKaldnes™ MBBR (Moving Bed Biofilm Reactor) biological process ensures a compact and reliable biodegradation of selenates. Veolia offers a biological treatment comprising pH adjustment followed by an anoxic treatment stage with two moving bed biological reactors (Figures 16.65 and 16.66). These tanks are fitted with AnoxKaldness™ supports with a specific surface area equal to 800 m². Upstream of this stage, a reductive agent (sodium bisulfite) is added to remove any oxygen present in the feed water.

$$SeO_4^{2-} \rightarrow SeO_3^{2-} \rightarrow Se$$

$$\downarrow$$

heterotrophic bacteria

The dosage of sodium bisulfite is automatically controlled depending on the ORP. The primary purpose of the MBBR is to reduce nitrates, and then to reduce selenates and selenites to elemental selenium. Two contact tanks are deployed: one for the removal of nitrates and the other for the reduction of selenium.

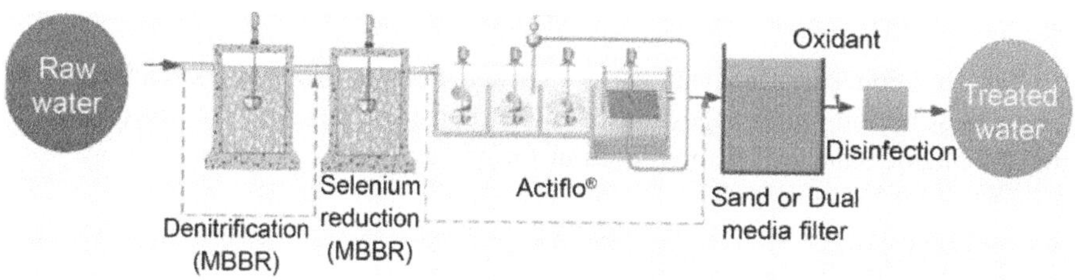

Figure 16.65. *Se biological treatment process with MBBR (Moving Bed Biofilm Reactor, Veolia). For a color version of this figure, see www.iste.co.uk/gaid/watertreatment3.zip*

The chemical reagents introduced into MBBR tanks include the addition of a readily biodegradable carbon source and phosphoric acid to act as a phosphorus source.

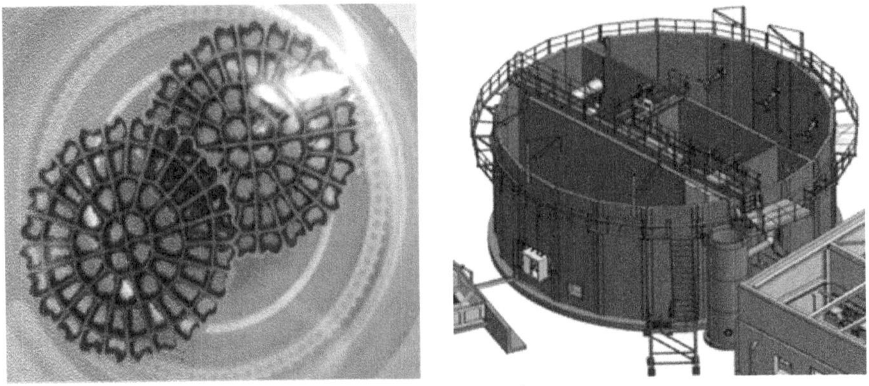

Figure 16.66. *Anoxkaldness™ K5 supports used in MBBRs (Veolia). For a color version of this figure, see www.iste.co.uk/gaid/watertreatment3.zip*

16.4. Nickel removal

The presence of nickel in water has a geological origin (underground water). It can be released into surface waters by the deterioration and erosion of geological materials. It enters the aquatic environment through industrial discharges (the manufacture of batteries, electroplating, jewelry, cast iron parts, etc.) in the form of leachates but also through pipes. Because of its resistance to oxidation and corrosion, it is used in coins, for plating iron, copper, brass. It is particularly appreciated for the high-performance alloys it forms.

High Ni^{2+} concentrations can cause diarrhea, anemia, hepatitis, kidney damage. In humans, the long-term ingestion of nickel can lead to kidney or liver damage, lung and bladder cancer. The quality limit for Ni^{2+} is 20 $\mu g \cdot L^{-1}$.

16.4.1. *The chemistry of nickel*

The most common oxidation state is +2, but compounds of Ni(0), Ni(I), Ni(III) and even Ni(IV) are known. It can be found in oxidation state –1 in the $[Ni(CO)_6]^{2-}$ ion, +1 in the $[NiBr(PPh_3)_3]$ compound , +3 in the $[NiF_6]^{3-}$ ion and +4 in the $[NiF_6]^{2-}$ ion. It is a transition element that forms many complexes, including hexa-aquanickel $[Ni(H_2O)_6]^{2+}$.

16.4.2. *Nickel (Ni) removal treatment*

The most commonly used techniques for nickel removal are physicochemical processes, such as coagulation/flocculation, adsorption onto activated carbon, ion exchange resins and high-pressure membrane technologies.

16.4.2.1. *Coagulation/flocculation and precipitation: electrocoagulation*

The electrocoagulation carried out with aluminum electrodes reaches 25% removal efficiency at pH 6.5. This efficiency increases to 65% for an 8.5 pH. However, the particulate Al residual at pH 8.5 is high and requires filtration. After filtration, this residual decreases to values < 0.1 $mg \cdot L^{-1}$. With the same electrodes, a removal efficiency of 99.5% is obtained with a Ni^{2+} concentration of 25 $mg \cdot L^{-1}$ in raw water, a 35 min contact time and a pH of 7.8. This efficiency decreases to 80% when the contact time is 10 min.

Electrocoagulation performed with iron electrodes enables a removal efficiency of 25% for a pH of 6.5. This efficiency increases to 80% for a pH value of 8.5 (Figure 16.67).

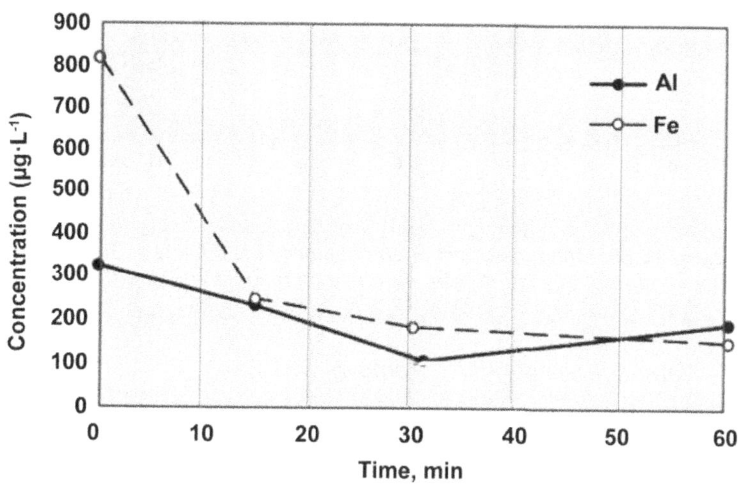

Figure 16.67. *Electrocoagulation with Al and Fe electrodes (pH 6.5)*

The precipitation pH for Ni^{2+} can be obtained using the ratio:

$$\text{pH} = \frac{(logKsp - log[Ni^{2+}] + 14n)}{n}$$

where:

– Ksp: solubility product of $Ni(OH)_2$ equal to 6×10^{-16};

– $[Ni^{2+}]$: nickel concentration in $mol \cdot L^{-1}$;

– n: index related to the number of hydroxides (n = 2 for $Ni(OH)_2$).

For $[Ni^{2+}] = 10^{-7.2}$ $mol \cdot L^{-1}$, the precipitation pH in $Ni(OH)_2$ is equal to 10. This confirms that metals with high Ksp values precipitate at high pH.

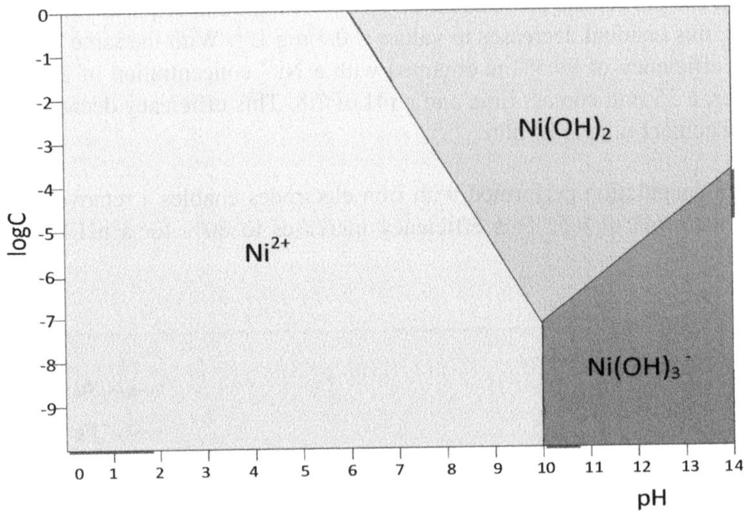

Figure 16.68. *Influence of pH on the solubility of $Ni(OH)_2$ at 25°C. For a color version of this figure, see www.iste.co.uk/gaid/watertreatment3.zip*

16.4.2.2. *Water softening and Ni^{2+} removal*

Transition metals easily precipitate as hydroxides $(Me(OH)_n)$. The equilibria between precipitates and soluble forms are governed by the thermodynamic constants of the species and the conditions in solution.

The precipitation of nickel is favored by the increase in pH (pH 10):

$$Ni^{2+} + 2OH^- \rightarrow Ni\,(OH)_2 \quad pK = 14.7$$

16.4.2.2.1. With caustic soda

Precipitation with caustic soda requires high dosages. To pass from an initial Ni^{2+} concentration of 38–15 $\mu g \cdot L^{-1}$, it is necessary to add no less than 270 $mg \cdot L^{-1}$ of caustic soda (30%) and to reach a pH of 9.5. A caustic soda supplement with a total dosage of 450 mg·L^{-1} (a pH value of 10.6) decreases the Ni^{2+} concentration to less than 5 µg·L^{-1}. But this excess in sodium hydroxide leads to an excess in the maximum acceptable concentration (200 mg·L^{-1}) of Na^{+} acceptable in drinking water standards.

16.4.2.2.2. With lime

Lime precipitation leads to the following reaction:

$$NiSO_4 + Ca(OH)_2 \rightarrow Ni(OH)_2 + CaSO_4$$

A Ni^{2+} concentration of 15 µg·L^{-1} is obtained by catalytic water softening with lime for approximately 450 mg·L^{-1} of $Ca(OH)_2$ per liter. These injection rates correspond to pH values greater than 8.5, for a raw water quality with a Ca hardness equal to 380 mg·L^{-1} $CaCO_3$, an alkalinity equal to 310 mg·L^{-1} $CaCO_3$ and a Ni^{2+} concentration of 38 µg·L^{-1}.

Beyond this treatment rate, nickel reduction continues and the tests indicate that it decreases below 5 µg·L^{-1} for an injection rate of 500 mg·L^{-1} (corresponding to a pH equal to 10.6).

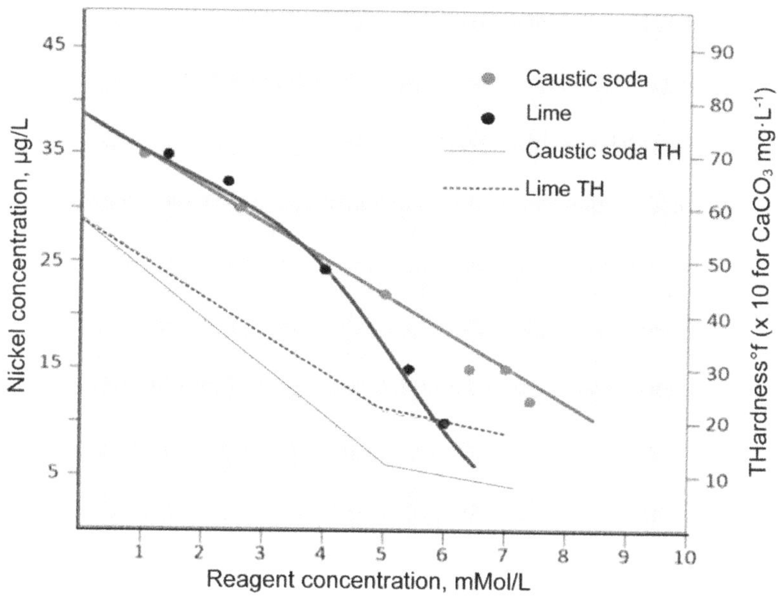

Figure 16.69. *Nickel removal during catalytic water softening with caustic soda and lime. For a color version of this figure, see www.iste.co.uk/gaid/watertreatment3.zip*

16.4.2.3. *Adsorption onto different materials*

16.4.2.3.1. Activated carbon

The effectiveness of powdered activated carbon is represented by the Freundlich equation:

$$X/M = 1.02\ Cout^{1.59}$$

which shows that the passage from an initial concentration of 80 $\mu g \cdot L^{-1}$ to a final concentration (Cout) of 20 $\mu g \cdot L^{-1}$ involves a quantity of powdered activated carbon of 29.7 $mg \cdot L^{-1}$. Granular activated carbon (Chemviron F100 and F200 activated carbons) results in adsorption capacities of 4.5 and 5.5 $mg \cdot g^{-1}$, respectively.

16.4.2.3.2. Manganese dioxide

Manganese oxide has the property of being charged in an acidic or basic medium:

$$MnO_2 + H_2O \rightarrow MnO_3H_2$$

In the presence of H^+ ions, this becomes:

$$MnO_3H_2 + H^+ \rightarrow Mn(OH)_3^+$$

In the presence of OH ions$^-$, the reaction is the following:

$$MnO_3H_2 + OH^- \rightarrow MnO(OH)_3^-$$

These charged groups can then exchange cations and anions, depending on the environmental conditions. So with Ni^{2+}, the reaction becomes:

$$2MnO(OH)_3^- + Ni^{2+} \rightarrow \{MnO(OH)_3\}_2\ Ni$$

The ions are trapped by the manganese oxide that can then be regenerated, either in an acid medium (in the case of cation retention) or in a basic medium (in the case of anion retention).

16.4.2.3.3. Sand coated with iron or Mn

The Metclean® process involves injecting an oxidizing agent (O_3, H_2O_2, ClO_2, $KMnO_4$) into a fluidized sand bed reactor, as well as the continuous addition of Fe^{2+} or Mn^{2+} ions. Particles resulting from the oxidation of iron or manganese coat the sand (0.3–0.6 mm), which then becomes a catalytic reaction support. The surface of sand coated with iron or manganese has a high adsorption capacity and the advantage of being continuously renewed. The contact time is of about 10 min. An efficiency over 85% is obtained with Mn-coated sand at pH 8.

16.4.2.4. *Membrane techniques*

Microfiltration/ultrafiltration membranes have no effect on nickel ions dissolved in water.

The NF membrane resulted in a Ni^{2+} retention rate of around 80% with an NF200-type membrane. With an NF 90-type membrane, a Ni^{2+} removal over 96% is obtained. A Ni^{2+} removal over 95% is obtained with RO membranes, regardless of the initial nickel concentration in the water to be treated. Thus, a final Ni^{2+} concentration <10 $\mu g \cdot L^{-1}$ is obtained for raw water whose Ni^{2+} concentration is 1,000 $\mu g \cdot L^{-1}$.

16.4.2.5. *The Metclean® process (Veolia)*

Adsorption processes have long been recognized for their efficiency in the treatment of heavy metals present in drinking water, wastewater and industrial water. The MetClean® technology – patented by Veolia Water Systems – is based on a fluidized bed reactor, which treats many metals by adsorption, including not only arsenic, but also cadmium, chromium, mercury, molybdenum, nickel, selenium, zinc, copper, vanadium, barium and strontium.

The MetClean® process can remove up to 99% of heavy metals in a single step. The by-product obtained with the MetClean® process is a granular compound with a dryness rate of 80–90%, that is to say, a volume 20 times lower than competing processes. This results in an economic gain on the management of by-products, which compensates for the investment and operating costs. The compactness of MetClean® facilities makes them an economical solution, suitable for many applications.

In this process, the water to be treated is injected into the reactor at a high speed and fluidifies a bed of granules made up of sand. Adding iron, manganese and an oxidizing agent prompts the adsorption of metals onto the granules' surface and their oxidation. As this process takes place continuously, the granular size gradually increases, until their replacement (Figure 16.70).

The process generates a by-product, corresponding to the fraction of granules which must be renewed; made up of about 10% water, it can be easily evacuated and recovered.

The operating principle is based on the following considerations. The adsorption and oxidation mechanisms of Fe^{2+} and/or Mn^{2+} take place on the surface of an inert fluidized material. Dissolved metals and metalloids are adsorbed/co-precipitated into the material. The number of sites/volume ratio must be sufficient for the reaction to take place on the material, and not in a liquid medium. The MetClean® process leads to removal efficiencies between 90 and 98% obtained for the following elements: As, Cd, Cr, Hg, Mo, Ni, Se, Zn, Cu, V, Ba, Sr. A concentration lower than 10 $\mu g \cdot L^{-1}$ in As is obtained at the reactor's outlet for an inlet concentration of 1,000 $\mu g \cdot L^{-1}$ (pH < 7).

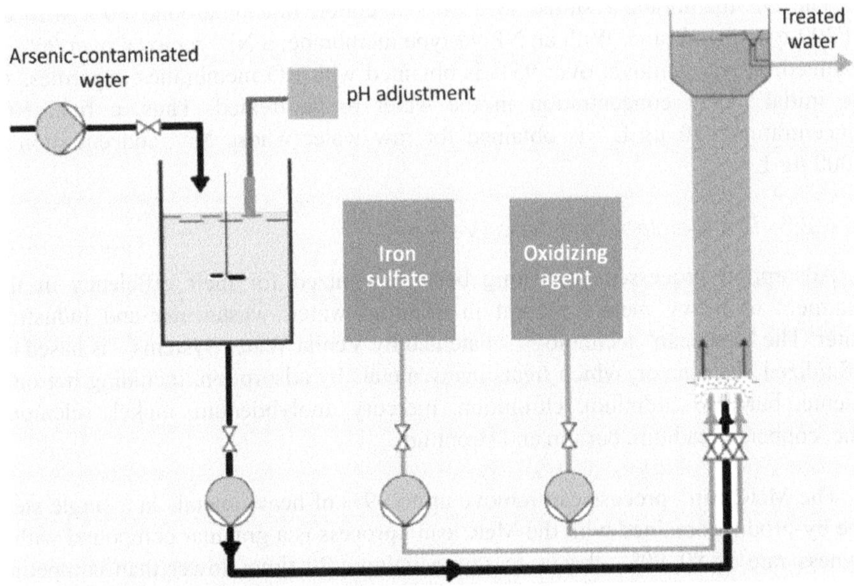

Figure 16.70. *Diagram of the MetClean® process (Veolia). For a color version of this figure, see www.iste.co.uk/gaid/watertreatment3.zip*

The adsorption onto Fe^{2+}/Mn^{2+} and the oxidation into Fe^{3+}/Mn^{4+} permanently regenerate the material's active surface and, in virtue of the adsorption/co-precipitation mechanisms, heavy metals are fixed onto the active surface surrounding the grain of sand (Figure 16.71).

The parameters governing the mechanisms include:

– the reaction time in the fluidized bed;

– the rate of chemical reactions on the material's surface and in water;

– the number of reactive sites on the material's surface per reactor volume.

The choice of iron or manganese as a complexing agent depends on the type of metal.

The flow rate is between 30 and 35 $m^3 \cdot m^{-2} \cdot h^{-1}$. The total reactor height is 5–6 m, with a static material height of 1.5–5 m. Contact times vary between 3 and 10 min, depending on the metal to be removed and its concentration.

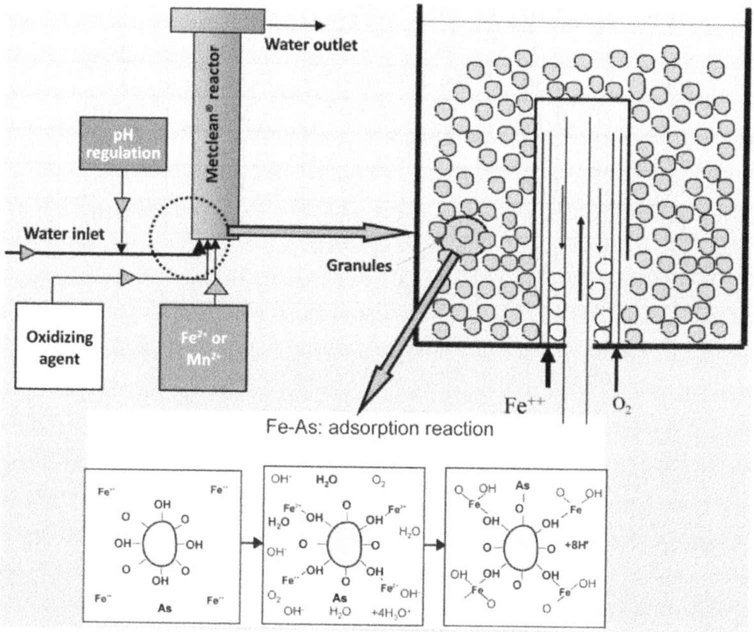

Figure 16.71. *Metal reactions and adsorption during the MetClean® process. For a color version of this figure, see www.iste.co.uk/gaid/watertreatment3.zip*

The operating pH is related to the type of metal to be removed. It is between pH 5 and 8.5. New sand is reinjected into the reactor when a granule load is removed. The removal frequency depends on the head loss created in the reactor as the granules grow thicker. The grain size of fresh sand is 0.4–0.6 mm.

The MetClean® process (Figure 16.72) can treat most metals in solution.

These applications are useful in the following domains:

– drinking water: water contamination by arsenic or nickel, released by pyrite oxidation, is a fairly widespread process at a global scale which can be treated by a MetClean® unit;

– groundwater: it is contaminated by metal leaching (e.g. the case of chromium stemming from the surface treatment of metals, landfill sites) and can be treated with a containerized MetClean® solution.

Metal	Choice of complexing agent	Molar ratio	pH	Metal	Choice of complexing agent	Molar ratio	pH
As^{3+}	Fe	1:10	6–7	Ni^{2+}	Mn	1:15	7.4–8
As^{5+}	Fe	1:5	6–8	Pb^{2+}	Mn	1:5	7–8
Ba^{2+}	Mn	1:30	7.9–8.5	PO_4^{3-}	Fe	1:5	<8
Cd^{2+}	Mn	1:10	7.5–8	Se^{6+}	Mn	1:30	7.5–8
Co^{2+}	Mn	1:10	7.5–8	U	Fe	1:20	6.3–6.6
Cr^{3+}	Fe	1:10	6–7	V	Fe	1:5	5.5–8
Cr^{6+}	Fe	1:4	7.3–7.5	Zn^{2+}	Mn	1:10	7.5–8
Cu^{2+}	Mn	1:10	7.5–8.5				
Mo^{2+}	Fe	1:20	5.4–5.6				

Table 16.29. *Choice of complexing agent, operating pH and molar ratios for various heavy metals*

Figure 16.72A. *MetClean® process. For a color version of this figure, see www.iste.co.uk/gaid/watertreatment3.zip*

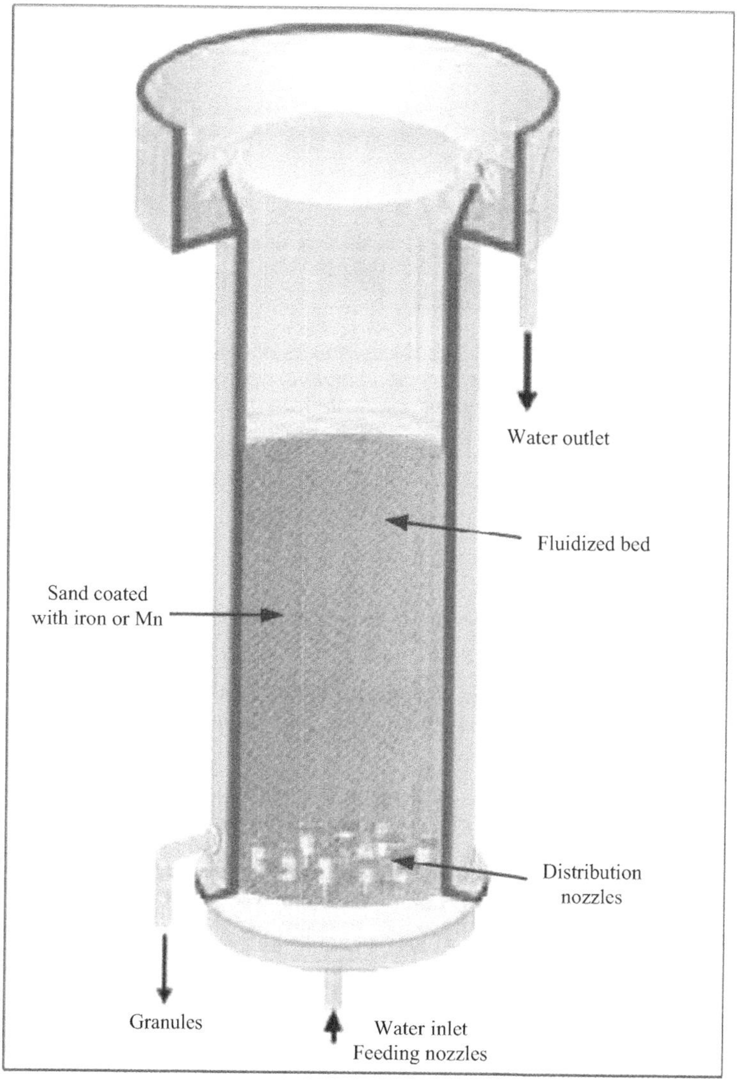

Figure 16.72B. *MetClean® process. For a color version of this figure, see www.iste.co.uk/gaid/watertreatment3.zip*

This technology can be implemented at a new station or at an existing station before sand filtration. The resulting sludge is not liquid, but there is the production of granules made of microsand and removed metals (Figure 16.73), which are fixed onto the support metal (iron or manganese). Granule density is about 2.5 $g \cdot cm^{-3}$.

Arsenic	Chromium	Nickel
Adsorbed onto iron granules	Adsorbed onto chromate granules	Adsorbed onto manganese granules

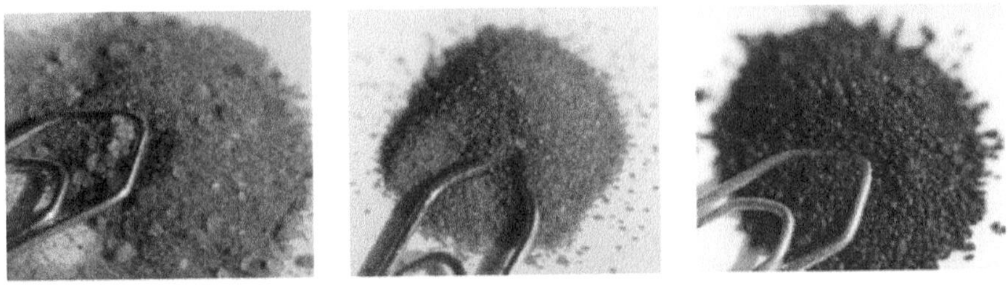

Figure 16.73. *MetClean® process granules containing various metals removed from raw water. For a color version of this figure, see www.iste.co.uk/gaid/watertreatment3.zip*

16.5. References

Iron and manganese

Appelo, C.A.J., Drijver, B., Hekkenberg, R., Jonge, M. (1999). Modeling in situ iron removal from ground water. *Ground Water*, 37(6), 811–817.

Aydin, S., Tufekci, N., Arayici, S., Ozturk, I. (2000). Catalytic effects of high Mn (IV) concentrations on Mn (II) oxidation. *Water Sci. Technol.: Water Supply,* 42(1–2), 387–392.

Barnes, A., Sapsford, A.J., Dey, M., Williams K.P. (2009). Heterogeneous Fe (II) oxidation and zeta potential. *J. Geochem. Exploration*, 100, 192–198.

Baylar, A., Bagatur, T., Emiroglu, E. (2007). Aeration efficiency with nappe flow over stepped cascades. *Proc. Instit. Civil Eng.: Water Manag.*, 160(1), 43–50.

Bergeon, T., Oger, M.L., Cleret, D. (1995). Présence de métaux non ferreux dans les eaux potables. Report.

Bergeon, T., Faivre, M., Santurette, N. (2002). Traitement de deferrisation et de démanganisation. *Générale des Eaux*, 3(14).

Burger, M.S., Krentz, C.A., Mercer, S.S., Gagnon, G.A. (2008). Manganese removal and occurrence of manganese oxidizing bacteria in full-scale biofilters. *J. Water Supply Res. T.*, 57(5), 351–359.

Carlson, K.H. and Knocke, W.R. (1999). Modeling manganese oxidation with $KMnO_4$ for drinking water treatment. *J. Amer. Water Works Assoc.*, 125(10), 892–896.

Coughlin, W.R. and Matsui, I. (1976). Catalytic oxidation of aqueous Mn (II). *J. Catalysis*, 41, 108–123.

Czekalla, C., Mevius, W., Hanert, H. (1985). Quantitative removal of iron and manganese by micro-organisms in rapid sand filters (in situ investigations). *Water Supply*, 3(1), 111–123.

Davison, W. and Seed, G. (1983). The kinetics of the oxidation of ferrous iron in synthetic and natural waters. *Geochim. Cosmochim. Acta*, 47(1), 67–79.

De Vet, W.W.J.M. (2011). Biological drinking water treatment of anaerobic groundwater in trickling filters. PhD Thesis, Delft University of Technology.

De Vet, W.W.J.M., Rietveld, L.C., Van Loosdrecht, M.C.M. (2009). Influence of iron on nitrification in full-scale drinking water trickling filters. *J. Water Supply: Res. Technol. – Aqua*, 58(4), 247–256.

Emerson, D. and Moyer, C. (1997). Isolation and characterization of novel ironoxidizing bacteria that grow at circumneutral pH. *Appl. Environ. Microb.*, 63(12), 4784–4792.

Emmenegger, L., King, D.W., Sigg, L., Sulzberger, B. (1998). Oxidation kinetics of Fe (II) in a eutrophic Swiss lake. *Environ. Sci. Technol.*, 32(19), 2990–2996.

Faivre, M., Papet, J.F., Lebeau, T. (2001). Guide filtration : objectifs, conception, exploitation. *Générale des Eaux,* 10.

Gaid, K., Ghouilem, D., Ait Amar, A. (1989). La démanganisation des eaux sur sable traité. *Journal of Technology*, 7, 98–105.

Kirby, C.S., Dennis, A., Kahler, A. (2009). Aeration to degas CO_2, increase pH, and increase iron oxidation rates for efficient treatment of net alkaline mine drainage. *Appl. Geochem.*, 24(7), 1175–1184.

Knocke, W.R., Hamon, J., Thompson, C. (1988). Soluble manganese removal on oxida-coated filter media. *J. Am. Water Works Assoc.*, 80, 65–69.

Knocke, W.R., Van Venschoten, J.E., Kearney, M.J., Soborski, A.W., Reckhow, D.A. (1991). Kinetics of manganese and iron oxidation by potassium permanganate and chlorine dioxide. *J. Am. Water Works Assoc.*, 83(6), 80–87.

Liang, L., McNabb, A.J., Paulk, J.M., Gu, B., McCarthy, J.F. (1993). Kinetics of Fe (II) oxygenation at low partial pressure of oxygen in the presence of natural organic matter. *Environ. Sci. Technol.*, 27, 1864–1870.

McAllan, J., Banks, D., Beyer, N., Watson, I. (2009). Geochemistry: Exploration. Alkalinity, temporary (CO_2) and permanent acidity: An empirical assessment of the significance of field and laboratory determinations on mine waters. *Environ. Anal.*, 9, 299–312.

Merkle, P.B., Knocke, W.R., Gallagher, D.L., Little, J.C. (1997). Dynamic model for soluble Mn^{2+} removal by oxide-coated filter media. *J. Environ. Eng.*, 123, 650–658.

Olancyzuk-Neyman, K., Czescik, P., Lasinska, E., Bray, R. (2001). Evaluation of the effectivity of selected filter beds for iron and manganese removal. *Water Sci. Technol.: Water Supply*, 1(2), 159–165.

Pumiceand, A. and Heertjes, P.M. (1975). Removal of manganese from ground water by heterogeneous autocatalytic oxidation. *Trans. Inst. Chem. Eng.*, 53, 154–164.

Roccaro, P., Barone, C., Mancini, G., Vagliasindi, F.G.A. (2007). Removal of manganese from water supplies intended for human consumption: A case study. *Desalination*, 210, 205–214.

Santana-Casiano, J.M., González-Dávila, M., Millero, F.J. (2010). Comment on "Oxygenation of Fe (II) in natural waters revisited: Kinetic modelling approaches, rate constant estimation and the importance of various reaction pathways" by Pham and Waite (2008). *Geochim. Cosmochim. Acta*, 74(17), 5150–5153.

Sharma, S.K., Kappelhof, J., Groenendijk, M., Schippers, J.C. (2001). Comparison of physicochemical iron removal mechanisms in filters. *J. Water Supply Res. T.*, 50(4), 187–198.

Sharma, S.K., Petrusevski, B., Schippers, J.C. (2005). Biological iron removal from groundwater: A review. *J. Water Supply Res. T.*, 54(4), 239–247.

Stumm, W. and Morgan, J.J. (1996). *Aquatic Chemistry, Chemical Equilibria and Rates in Natural Waters*, 3rd ed. Wiley Interscience, New York.

Sung, W. and Morgan., J.J. (1980). Kinetics and product of ferrous iron oxygenation in aqueous systems. *Environ. Sci. Technol.*, 14(5), 561–568.

Teunissen, K. (2007). Iron removal at groundwater pumping station Harderbroek. Master's Thesis, Delft University of Technology.

Teunissen, T., Abrahamse, A., Leijssen, H., Rietveld, L., Van Dijk, H. (2008). Removal of both dissolved and particulate iron from groundwater. *Drink. Water Eng. Sci. Discuss.*, 1, 87–115.

Viraraghavan, T., Subramanian, K.S., Tanjove, S. (1996). Removal of arsenic in drinking water by manganese green sand filtration. *Adv. Filtr. Sep.Technol.*, 10, 502–507.

Wong, J.M. (1984). Chlorination-filtration for iron and manganese removal. *J. Am. Water Works Assoc.*, 76, 76–79.

Arsenic

Badruzzaman, M., Westerhoff, P., Knappe, D.R.U. (2004). Intraparticle diffusion and adsorption of arsenate onto granular ferric hydroxide (GFH). *Water Res.*, 38(18), 4002–4012.

Bajpai, S. and Chaudhuri, M. (1999). Removal of arsenic from ground water by manganese dioxide-coated sand. *J. Environ. Eng.*, 125, 782–787.

Bakshi, S., He, Z.L., Harris, W.G. (2014). Biochar amendment affects leaching potential of copper and nutrient release behavior in contaminated sandy soils. *J. Environ. Qual.*, 43(6), 1894–1902.

Bang, S., Korfiatis, G.P., Meng, X. (2005a). Removal of arsenic from water by zerovalent iron. *J. Hazard Mater.*, 121, 61–67.

Campos, V. (2002). The effect of carbon steel wool in removal of arsenic from drinking water. *Environ. Geol.*, 42, 81–82.

Caniyilmaz, S. (2005). Arsenic removal from groundwater by Fe-Mn oxidation and microfiltration. PhD Thesis, University of Pittsburgh.

Chen, H.W., Frey, M.M., Clifford, D., McNeill, L.S., Edwards, M. (1999). Arsenic treatment considerations. *J. Am. Water Works Assoc.*, 91, 74–85.

Cheng, R.C., Liang, S., Wang, H.C., Beuhler, M.D. (1994). Enhanced coagulation for arsenic removal. *J. American Water Works Assoc.*, 86(9), 79–90.

Clifford, D.A., Ghurye, G., Tripp, A.R. (2003). Arsenic removal from drinking water using ion-exchange with spent brine recycling. *J. Am. Water Works Assoc.*, 35, 119–130.

Dixit, S. and Hering, J.G. (2003). Comparison of arsenic (V) and arsenic (III) sorption onto iron oxide minerals: Implications for arsenic mobility. *Environ. Sci. Technol.*, 37, 4182–4189.

Driehaus, W., Jekel, M., Hildebrandt, U. (1998). Granular ferric hydroxide – A new adsorbent for the removal of arsenic from natural water. *Water SRT – Aqua*, 47(1), 30–35.

Edwards, M. (1994). Chemistry of arsenic removal during coagulation and Fe–Mn oxidation. *J. Am. Water Works Assoc.*, 86(9), 64–78.

Gaid, K. (2002). L'élimination de l'Arsenic dans les eaux à potabiliser. In *Conférence sur la pollution des eaux souterraines*, 30 May, Thiers.

Gaid, K. and Christensen, T.C. (2000). Elimination de l'arsenic dans les eaux à potabiliser : cas d'applications. In *Journées Informations Eaux*, 13–15 September, University of Poitiers.

Gaid, K. and Cleret, D. (2001). Élimination de l'arsenic dans les eaux à potabiliser. In *Congrès AGTHM*, 23–26 April, Paris.

Gaid, K., Ravarini, P., Raguenes, I. (1998). L'élimination de l'arsenic sur l'usine d'eau potable de Baudricourt (Vosges). *L'Eau, L'Industrie, Les Nuisances*, 205, 54–58.

Gaid, K., Pichon, JF., Thomas, P. (2004). L'hydroxyde de fer : un matériau innovant et efficace pour l'élimination de l'arsenic dans les eaux souterraines. *L'Eau, L'Industrie, Les Nuisances*, 273, 35–42.

Gholami, M.M., Mokhtari, M.A., Aameri, A., Alizadeh Fard, M.L.R (2006). Application of reverse osmosis technology for arsenic removal from drinking water. *Desalination*, 200(1–3), 725–727.

Gude, J.C.J., Rietveld, L.C., Van Halem, D. (2017). As (III) oxidation on MnO_2 during groundwater treatment. *Water Res.*, 111, 41–51.

Gulledge, J.H. and O'Connor, J.T. (1973). Removal of As (V) from water by adsorption on aluminium and ferric hydroxides. *J. Am. Water Works Assoc.*, 65(8), 548–552.

Hanson, A., Bates, J., Heil, D. (2000). Removal of arsenic from ground water by manganese dioxide-coated sand. *J. Environ. Eng.*, 126, 1160–1161.

Hathaway, S.W. and Rubel Jr., F. (1987). Removing arsenic from drinking water. *J. Am. Water Works Assoc.*, 79(8), 61–65.

Hering, J.G., Chen, P.Y., Wilkie, J.A., Elimelech, M. (1996a). Arsenic removal by ferric chloride. *J. Am. Water Works Assoc.*, 88, 155–167.

Hering, J.-G., Chen, P.Y., Wilkie, J.A., Elimeleh, M., Liang, S. (1996b). Arsenic removal by ferric chloride. *J. American Water Works Assoc.*, 88(4), 155–167.

Hering, J.G., Chen, P.Y., Wilkie, J.A., Menachem, E. (1997). Arsenic removal from drinking water during coagulation. *J. Environ. Eng.*, 123(8), 800–807.

Highfield, D.E. (2002). Arsenic occurrence in metro phoenix groundwater and treatment by granular ferric hydroxide. PhD Thesis, University of Arizona, Tucson.

Jang, M., Min, S.H., Kim, T.H., Park, J.K. (2006). Removal of arsenite and arsenate using hydrous ferric oxide incorporated into naturally occurring porous diatomite. *Environ. Sci. Technol.*, 40(5), 1636–1643.

Jubinka, L. and Rajakovic, V. (1992). The sorption of arsenic onto activated carbon impregnated with metallic silver and copper. *Sep. Sci. Technol.*, 27(11), 1423–1433.

Kang, M., Kawasaki, M., Tamada, S., Kamei, T., Magara, Y. (2000). Effect of pH on the removal of arsenic and antimony using reverse osmosis membranes. *Desalination*, 131, 293–298.

Kartinen Jr., E.O. and Martin, C.J. (1995). An overview of arsenic removal processes. *Desalination*, 103(1–2), 79–88.

Kim, D.H., Kim, K.W., Cho, J. (2006). Removal and transport mechanisms of arsenics in UF and NF membrane processes. *J. Water Health*, 4(2), 215–223.

Kuhlmeier, P.D. and Sherwood, S.P. (1996). Treatability of inorganic arsenic and organoarsenicals in groundwater. *Water Environ. Res.*, 68, 946–951.

Lin, T.F. and Wu, J.K. (2001). Adsorption of arsenite and arsenate within activated alumina grains: Equilibrium and kinetics. *Water Res.*, 35, 2049–2057.

Lorenzen, L., Van Deventer, J.S.J., Landi, W.M. (1995). Factors affecting the mechanism of the adsorption of arsenic species on activated carbon. *Miner. Eng.*, 8(45), 557–569.

Masue, Y., Loeppert, R.H., Kramer, T.A. (2007). Arsenate and arsenite adsorption and desorption behavior on coprecipitated aluminum: Iron hydroxides. *Environ. Sci. Technol.*, 41, 837–842.

McNeil, L.S. and Edwards, M. (1997). Predicting as removal during metal hydroxide precipitation. *J. Am. Water Works Assoc.*, 89(3), 75–86.

Meng, X.G., Korfiatis, G.P., Bang, S., Bang, K.W. (2002). Combined effects of anions on arsenic removal by iron hydroxide. *Toxicol. Lett.*, 133(1), 103–111.

Mohan, D. and Pittman Jr., C.U. (2007). Arsenic removal from water/wastewater using adsorbents – A critical review. *J. Hazard. Mater.*, 142, 1–53.

Ngo, H.H., Vigneswaran, S., Hu, J.Y., Thirunavukkarasu, O., Viraraghava, T. (2002). A comparison of conventional and non-conventional treatment technologies on arsenic removal from water. *Water Sci. Technol.: Water Supply*, 2(5–6), 119–125.

Ning, R.Y. (2002). Arsenic removal by reverse osmosis. *Desalination*, 143(3), 237–241.

Oh, J.I., Urase, T., Kitawaki, H., Rahaman, M.M., Rahaman, M.H., Yamamoto, K. (n.d.). Modelling of arsenic rejection considering affinity and steric hindrance effect in nonofiltration membranes. *Water Sci. Technol.*, 42(3–4), 173–180.

Oh, J.I., Yamamoto, K., Kitawaki, H., Nakao, S., Sugawara, T., Rahaman, M.M., Rahaman, M.H. (2000). Application of low-pressure nanofiltration coupled with a bicycle pump for the treatment of arsenic-contaminated groundwater. *Desalination*, 132, 307–314.

Pattanayak, J., Mondal, K., Mathew, S., Lalvani, S.B. (2000). A parametric evaluation of the removal of As (V) and As (III) by carbon based adsorbents. *Carbon*, 38, 589–596.

Peter, M. (2005). Ion exchange. An overview of technologies useful for arsenic removal. *Ultrapure Water*, 22(5), 42–43.

Seidel, A., Waypa, J.J., Elimelech, M. (2001). Role of charge (Donnan) exclusion in removal of arsenic from water by a negatively charged porous nanofiltration membrane. *Environmental Engineering Science*, 18(2), 105–113.

Senn, A.C., Hug, S.J., Kaegi, R., Hering, J.G., Voegelin, A. (2018). Arsenate co-precipitation with Fe(II) oxidation products and retention or release during precipitate aging. *Water Res.*, 131, 334–345.

Singh, T.S. and Pant, K.K. (2004). Equilibrium, kinetics and thermodynamic studies for adsorption of As(III) on activated alumina. *Sep. Purif. Technol.*, 36, 139–147.

Smith, D.S. and Edwards, M. (2002). Bench-scale evaluation of innovative of arsenic removal process. *JAWWWA*, 94(9), 78–88.

Sun, H., Wang, L., Zhang, R., Sui, J., Xu, G. (2006). Treatment of groundwater polluted by arsenic compounds by zero valent iron. *J. Hazard Mater.*, 129, 297–303.

Thirunavukkarasu, O.S., Viraraghavan, T., Subramanian, K.S. (2003). Arsenic removal from drinking water using iron oxide-coated sand. *Water Air Soil Pollut.*, 142, 95–111.

Tran, L. (2017). Arsenic removal using manganese dioxide for arsenic oxidation process, case study – Skovby waterworks. Thesis, Université des sciences appliquées du Sud-Est de la Finlande, Mikkeli.

Wang, L.W. and Sorg. T.J. (2002). Technology selection and system design U.S. EPA arsenic removal technology demonstration program round 1, Water Supply and Water Resources. Report, Division National Risk Management Research Laboratory Cincinnati.

Wilkin, R.T. (2005). High-level arsenite removal from groundwater by zero-valent iron. *Chemosphere*, 59, 307–386.

Yuan, T., Luo, Q.F., Hu, J.Y., Ong, S.L., Ng,W.J. (2003). A study on arsenic removal from household drinking water. *J. Environ. Sci. Health*, 38A(9), 1731–1744.

Selenium

Barron, E. (2007). Mesure de l'exposition au sélénium : évaluations chimiques et épidémiologiques auprès de sujets alimentés par une eau potable à teneur élevée, dans la Vienne. PhD Thesis, Université de Poitiers.

Das, S., Lindsay, M.B.J., Essilfie-Dughan, J., Hendry, M.J. (2010). Dissolved selenium (VI) removal by zero-valent iron under oxic conditions: Influence of sulfate and nitrate. Report, Department of Geological Sciences University of Saskatchewan, Saskatoon.

Devoy, J. (2001). Mécanismes de sorption des espèces du sélénium (sélénite et séléniate) sur des minéraux à base de cuivre. PhD Thesis, Université Henri Poincaré, Nancy.

Gibson, B.D., Blowes, D.W., Lindsay, M.B.J., Ptacek, C.J. (2017). Mechanistic investigations of Se (VI) treatment in anoxic groundwater using granular iron and organic carbon: An EXAFS study. *J. Hazard. Mater.*, 241–242, 92–100.

Liang, L., Sun, W., Guan, X., Huang, Y., Choi, W., Bao, H., Li, J., Jiang, Z. (2014). Weak magnetic field significantly enhances selenite removal kinetics by zero valent iron. *Water Res.*, 49, 371–378.

Liang, L., Guan, X., Huang, Y., Ma, J., Sun, X., Qiao, J., Zhou, G. (2015). Efficient selenate removal by zero-valent iron in the presence of weak magnetic field. *Sep. Purif. Technol.*, 156, 1064–1074.

Meher, A.K., Jadhav, A., Labhsetwar, N., Bansiwal, A. (2020). Simultaneous removal of selenite and selenate from drinking water using mesoporous activated alumina. *Appl. Water Sci.*, 10(10), 2–12.

Onoguchi, A., Granata, G., Haraguchi, D., Hayashi, H., Tokoro, C. (2019). Kinetics and mechanism of selenate and selenite removal in solution by green rust-sulfate. *R. Soc. Open Sci.*, 6, 182147.

Peak, D. and Sparks, D.L. (2002). Mechanism of selenite adsorption on iron oxides and hydroxides. *Environ. Sci. Technol.*, 36, 1460–1466.

Sandy, T. and Disante, C. (2010). Review of available technologies for the removal of selenium from water. Report, North American Metals Council, Washington.

Sharma, V.K., Sohn, M., McDonald, T.J. (2019). *Remediation of Selenium in Water: A Review.* In *Advances in Water Purification Techniques*, Ahuja, S. (ed.). Elsevier, Amsterdam.

Smith, K., Lau, A.O., Vance, F.W. (2009). Evaluation of treatment techniques for selenium removal. In *70th Annual International Water Conference*, ESW, Orlando.

Su, T., Guan, X., Gu, G., Wang, J. (2008). Adsorption characteristics of As (V), Se (IV), and V (V) onto activated alumina: Effects of pH, surface loading, and ionic strength. *J. Coll. Interface Sci.*, 326, 347–353.

Yoon, I.H., Bang, S., Kim, K.-W., Kim, M.G., Park, S.Y., Choi, W.-K. (2016). Selenate removal by zero-valent iron in oxic condition: The role of Fe (II) and selenate removal mechanism. *Sci. Pollut. Res.*, 23, 1081–1090.

Zhang, P. and Sparks, D.L. (1990). Kinetics of selenate and selenite adsorption/desorption at the goethite/water interface. *Environ. Sci. Technol.*, 24, 1848–1856.

Zhang, Y., Wang, J., Amrhein, C., Frankenberger, W.T. (2005). Removal of selenate from water by zerovalent iron. *J. Environ. Qual.*, 34, 485–495.

Nickel

Alyüz, B. and Veli, S. (2009). Kinetics and equilibrium studies for the removal of nickel and zinc from aqueous solutions by ion exchange resins. *J. Hazard. Mater.*, 167, 482–488.

Choksi, P.M. and Joshi, V.Y. (2007). Adsorption kinetic study for the removal of nickel (II) and aluminum (III) from an aqueous solution by natural adsorbents. *Desalination*, 208, 216–231.

Coman, V., Robotin, B., Ilea, P. (2013). Nickel recovery/removal from industrial wastes: A review. *Resour. Conserv. Recycl.*, 73, 229–238.

Dermentzis, K., Christoforidis, A., Valsamidou, E. (2011). Removal of nickel, copper, zinc and chromium from synthetic and industrial wastewater by electrocoagulation. *Int. J. Environ. Sci.*, 1(5), 697–710.

Dhokpande, S.R., Kaware, J.P., Kulkarni, S.J. (2013). Research for removal of nickel from waste water – A review. *Int. J. Sci. Eng. Technol. Res.*, 2, 2162–2166.

Escudero, G., Espinoza, E., Rao, F. (2017). Chemical precipitation of nickel species from waste water. *Int. Res. J. Pure Appl. Chem.*, 15, 1–7.

Fu, F. and Wang, Q. (2011). Removal of heavy metal ions from wastewaters: A review. *J. Environ. Manag.*, 92, 407–418.

Gaid, K., Belabed, W., Kestali, N., Semsari, S. (1994). Evaluation de la toxicité de quelques métaux lourds l'aide du test Daphnie. *Techniques & Sciences et Méthodes*, 6, June, 331–335.

Hashim, M.A., Mukhopadhyay, S., Sahu, J.N., Sengupta, B. (2011). Remediation technologies for heavy metal contaminated groundwater. *J. Environ. Manag.*, 92, 2355–2388.

Heidmann, I. and Calmano, W. (2008). Removal of Zn (II), Cu (II), Ni (II), Ag (I) and Cr (VI) present in aqueous solutions by aluminium electrocoagulation. *J. Hazard. Mater.*, 152, 934–941.

Hoseinian, F.S., Rezai, B., Kowsari, E., Safari, M. (2018). Kinetic study of Ni (II) removal using ion flotation: Effect of chemical interactions. *Miner. Eng.*, 119, 212–221.

Ipek, U. (2005). Removal of Ni (II) and Zn (II) from an aqueous solution by reverse osmosis. *Desalination*, 174, 161–169.

Johnson, P.D., Girinathannair, P., Ohlinger, K.N., Ritchie, S., Teuber, L., Kirby, J. (2008). Enhanced removal of heavy metals in primary treatment using coagulation and flocculation. *Water Environ. Res.*, 80(5), 472–478.

Kadirvelu, K., Thamaraiselvi, K., Namasivayam, C. (2001). Adsorption of nickel (II) from aqueous solution onto activated carbon prepared from coirpith. *Sep. Purif. Technol.*, 24, 497–505.

Kumara, A., Baloucha, A., Pathanb, A.A., Abdullaha, Jagirania, M.S., Mahara, A.M., Zubaira, M., Lagharia, B. (2019). Remediation of nickel ion from wastewater by applying various techniques: A review. *Acta Chem. Malaysia*, 3(11), 1–15.

de Mello Ferreira, A., Marchesiello, M., Thivel, P.X. (2013). Removal of copper, zinc and nickel present in natural water containing Ca^{2+} and ions by electrocoagulation. *Sep. Purif. Technol.*, 107, 109–117.

Mohsen-Nia, M., Montazeri, P., Modarress, H. (2007). Removal of Cu^{2+} and Ni^{2+} from wastewater with a chelating agent and reverse osmosis processes. *Desalination*, 217, 276–281.

Murthy, Z. and Chaudhari, L.B. (2008). Application of nanofiltration for the rejection of nickel ions from aqueous solutions and estimation of membrane transport parameters. *J. Hazard. Mater.*, 160, 70–77.

Semsari, S. and Gaid, K. (1993). Evaluation de la toxicité de quelques métaux lourds à l'aide du test Daphnie. In *Tribune de l'eau*, September–October, 55–63.

Semsari, S. and Gaid, K. (1994). Etude de l'incidence de la salinité et de la dureté sur la toxicité des métaux : Cd, Cu, Ni, Al et Cr au moyen du test Daphnie. In *Prés. Séminaire Maghrébin sur l'environnement*, 4–6 May, Marrakech.

Tzanetakis, N., Taama, W., Scott, K., Jachuck, R., Slade, R., Varcoe, J. (2003). Comparative performance of ion exchange membranes for electrodialysis of nickel and cobalt. *Sep. Purif. Technol.*, 30, 113–127.

Wang, G., Li, A., Li, M. (2010). Sorption of nickel ions from aqueous solutions using activated carbon derived from walnut shell waste. *Desalin. Water Treat.*, 16, 282–289.

Yang, S., Li, J., Lu, J., Chen, Y., Wang, X. (2009). Sorption of Ni (II) on GMZ bentonite: Effects of pH, ionic strength, foreign ions, humic acid and temperature. *Appl. Radiat. Isot.*, 67, 1600–1608.

Zamboulis, D., Peleka, E.N., Lazaridis, N.K., Matis, K.A. (2011). Metal ion separation and recovery from environmental sources using various flotation and sorption techniques. *J. Chem. Technol. Biotechnol.*, 86, 335–344.

Index

A, B, C

D, E

F, H, I

L, M

N, O, P

R, S,

T, V

Summary of Volume 1

Chapter 5. Settling

Chapter 6. Flotation

Summary of Volume 2

Chapter 7. Removal of Natural Organic Matter

Chapter 8. Filtration

Chapter 9. Adsorption on Activated Carbon

Index

Summary of Volume 4

Chapter 17. Microfiltration and Ultrafiltration

Chapter 18. Nanofiltration and Reverse Osmosis

Chapter 19. Desalination by Reverse Osmosis

Index

Summary of Volume 5

Chapter 20. Calco-carbonic Equilibrium, Correction of Aggressivity and Remineralization

Chapter 22. Disinfection By-products

Chapter 23. Sludge Treatment